STUDIES OF HIGH TEMPERATURE SUPERCONDUCTORS
(Advances in Research and Applications)

STUDIES OF HIGH TEMPERATURE SUPERCONDUCTORS
(Advances in Research and Applications)

Volume 43
BSCCO Tapes and More on Josephson Structures and Superconducting Electronics

ANANT NARLIKAR
EDITOR
National Physical Laboratory, New Delhi, India and
Visiting Fellow, Peterhouse, Cambridge, U.K.

Nova Science Publishers, Inc.
New York

Senior Editors: Susan Boriotti and Donna Dennis
Coordinating Editor: Tatiana Shohov
Office Manager: Annette Hellinger
Graphics: Wanda Serrano
Editorial Production: Jennifer Vogt, Matthew Kozlowski, Jonathan Rose
and Maya Columbus
Circulation: Ave Maria Gonzalez, Indah Becker and Vladimir Klestov
Communications and Acquisitions: Serge P. Shohov
Marketing: Cathy DeGregory

Library of Congress Cataloging-in-Publication Data
Available upon request.

ISBN 1-59033-343-8

400 Oser Ave, Suite 1600
Hauppauge, New York 11788-3619
Tele. 631-231-7269 Fax 631-231-8175
e-mail: Novascience@earthlink.net
Web Site: http://www.novapubishers.com

Printed in the United States of America

CONTRIBUTORS TO VOLUME 43

S. Adachi, Superconductivity Research Laboratory, International Superconductivity Technology Centre, 1-10-13 Shinonome, Koto-ku, Tokyo, 135-0062 JAPAN

F.M. Araujo-Moreira, Department of Physics, Multidisciplinary Centre for Development of Ceramic Materials – MCDCM, Universidade Federal de Sao Carlos, Caixa Postal 676, Sao Carlos, SP, 13565-905 BRAZIL

Md. Ashrafuzzaman, Institute de Physique, Universite de Neuchatel, SWITZERLAND

P. Barbara, Physics Department, Georgetown University, Box 571228, Washington DC, 0057-1228, USA

Hans Beck, Institute de Physique, Universite de Neuchatel, SWITZERLAND

N.D. Browning, Department of Physics, University of Illinois at Chicago, 845 West Taylor Street, Chicago, IL 60607-7059, USA

A.B. Cawthorne, Neocera Inc., 10000 Virginia Manor Road, Suite 300, Beltsville, MD 20705-4215, USA

Wei Gao, School of Engineering, The University of Auckland, New Zealand

David B. Haviland, Nanostructure Physics, The Royal Institute of Technology (KTH), SCFAB, Roslagstullsbacken 21, 106 91 Stockholm, SWEDEN

M. Horibe, Superconductivity Research Laboratory, International Superconductivity Technology Centre, 1-10-13 Shinonome, Koto-ku, Tokyo, 135-0062 JAPAN

Y. Ishimaru,Superconductivity Research Laboratory, International Superconductivity Technology Centre, 1-10-13 Shinonome, Koto-ku, Tokyo, 135-0062 JAPAN

K. Kishida, Mechanical Properties Research Group, National Institute for Materials Science, 1-2-1 Sengen, Tsukuba, Ibaraki 305-0047, JAPAN

Sean Li, School of Materials Engineering, Nanyang Technological University, SINGAPORE

C.J. Lobb, Centre for Superconductivity Research, Department of Physics, University of Maryland at College Park, College Park, MD 20742-4111, USA

E.T. Serra, CEPEL-Centro de Pesquisas de Energia Eletrica, P.O.Box 68007, CEP 21944-970, Rio de Janeiro, RJ, BRAZIL

Y.P. Sun, Key Laboratory of Internal Friction and Defects in Solids, Institute of Solid State Physics, Chinese Academy of Sciences, Hefei 230031, PEOPLES REPUBLIC OF CHINA

K. Tanabe, Superconductivity Research Laboratory, International Superconductivity Technology Centre, 1-10-13 Shinonome, Koto-ku, Tokyo, 135-0062 JAPAN

Y. Tarutani, Superconductivity Research Laboratory, International Superconductivity Technology Centre, 1-10-13 Shinonome, Koto-ku, Tokyo, 135-0062 JAPAN

Yoshinori Uzawa, Kansai Advanced Research Centre, CRL, 588-2 Iwaoka, Iwaoka-cho, Nishi-ku, Kobe 651-2492, JAPAN

H. Wakana, Superconductivity Research Laboratory, International Superconductivity Technology Centre, 1-10-13 Shinonome, Koto-ku, Tokyo, 135-0062 JAPAN

Zhen Wang, Kansai Advanced Research Centre, CRL, 588-2 Iwaoka, Iwaoka-cho, Nishi-ku, Kobe 651-2492, JAPAN

Michio Watanabe, Semiconductors Laboratory, RIKEN (The Institute of Physical and Chemical Research), 2-1 Hirosawa, Wako-shi, Saitama 351-0198, JAPAN

Y. Wu, Superconductivity Research Laboratory, International Superconductivity Technology Centre, 1-10-13 Shinonome, Koto-ku, Tokyo, 135-0062 JAPAN

S.K. Xia, CEPEL-Centro de Pesquisas de Energia Eletrica, P.O.Box 68007, CEP 21944-970, Rio de Janeiro, RJ, BRAZIL

FUTURE VOLUMES PLANNED

Graphite-based Superconductors

Further Studies on Magnesium Diboride

Commercial Status and Future Prospects of HTS

CONTENTS OF EARLIER VOLUMES

Consolidated contents of Vol.1 to 22 are provided in Vol.23

PREFACE

More than fifteen years have elapsed since the high Tc phenomenon was discovered. While there is no dearth of theories to explain the occurrence of high temperature superconductivity, its origin and manifestations today are as contested or uncertain as before. Not only do many of the initial puzzles remain unresolved but the subject has continually generated a host of new discoveries and novel material systems that throw up even greater challenges and questions for theorists and experimentalists alike. Nevertheless, in the recent years, commendable progress has been made in exploitation of the phenomenon in various electrical and electronic applications. It is hardly surprising that superconductivity remains a very rich and competitive field of academic and applied research. This book series consolidates and builds the continually evolving theoretical and practical knowledge in the subject. To facilitate a holistic understanding of the phenomenon and promote its technological application, the series provides an international forum for the publication of recent experimental findings and new theoretical insights. Given its interdisciplinary premise, the "Studies of High Temperature Superconductors" is aimed at professional scientists and engineers, as well as graduate students of physics, chemistry, materials science, solid state electronics and engineering.

This 43rd volume of *Studies* covers two important areas. The first four chapters focus on the BSCCO tapes and describe their fabrication, microstructure and properties, while the subsequent five chapters describe the interesting features exhibited by Josephson structures, their arrays and unusual properties and interesting technological applications of low Tc NbN based JJs. These nine chapters of the volume are contributed by renowned authors working in Superconductivity in eight different countries and thus keeping up with the international character of *Studies*.

The present volume was prepared during my stay at the Inter-University Consortium at Indore and I am grateful to the Director, Dr B.A. Dasannacharya for providing various facilities and maintenance support.

Finally, a splendid co-operation of Mr Frank Columbus, President, Nova Science Publishers, Inc., N.Y., and his colleagues is acknowledged.

April, 2002

Anant Narlikar

CONTENTS

Chapter-2

ATOMIC SCALE INVESTIGATION OF GRAIN BOUNDARIES AND DEFECTS IN Bi-2223/Ag TAPES BY SCANNING TRANSMISSION ELECTRON MICROSCOPY

K. Kishida, Mechanical Properties Research Group, National Institute for Materials Science, 1-2-1 Sengen, Tsukuba, Ibaraki 305-0047, JAPAN
and
N.D. Browning, Department of Physics, University of Illinois at Chicago, 845 West Taylor Street, Chicago, IL 60607-7059, USA

Chapter-3

FABRICATION OF Bi-2223/Ag TAPES BY CONTROLLING PARTIAL MELTING TEMPERATURE AND COOLING RATE

S.K. Xia and E.T. Serra, CEPEL-Centro de Pesquisas de Energia Eletrica, P.O.Box 68007, CEP 21944-970, Rio de Janeiro, RJ, BRAZIL

Chapter-4
STUDIES ON THE ANOMALOUS MAGNETIZATION PEAK OF $Bi_2Sr_2CaCu_2O_{8+\delta}$ AND $(Bi,Pb)_2Sr_2CaCu_2O_{8+\delta}$ SINGLE CRYSTALS

Y.P. Sun, Key Laboratory of Internal Friction and Defects in Solids, Institute of Solid State Physics, Chinese Academy of Sciences, Hefei 230031, PEOPLES REPUBLIC OF CHINA

Chapter-5
DYNAMICS OF TWO-DIMENSIONAL JOSEPHSON JUNCTION ARRAYS

Md. Ashrafuzzaman and Hans Beck, Institut de Physique, Universite de Neuchatel, SWITZERLAND

Chapter-8

PARAMAGNETIC MEISSNER EFFECT IN GRANULAR SUPERCONDUCTORS STUDIED THROUGH THE MAGNETIC PROPERTIES OF 2D-JOSEPHSON JUNCTION ARRAYS

F.M. Araujo-Moreira, Department of Physics, Multidisciplinary Centre for Development of Ceramic Materials – MCDCM, Universidade Federal de Sao Carlos, Caixa Postal 676, Sao Carlos, SP, 13565-905 BRAZIL

P. Barbara, Physics Department, Georgetown University, Box 571228, Washington DC, 0057-1228, USA

A.B. Cawthorne, Neocera Inc., 10000 Virginia Manor Road, Suite 300, Beltsville, MD 20705-4215, USA
and
C.J. Lobb, Centre for Superconductivity Research, Department of Physics, University of Maryland at College Park, College Park, MD 20742-4111, USA

Chapter-9

LOW-NOISE QUASI-OPTICAL SUPERCONDUCTING MIXER BASED ON NIOBIUM NITRIDE FOR SUBMILLIMETER WAVELENGTHS

Yoshinori Uzawa and Zhen Wang, Kansai Advanced Research Centre, CRL, 588-2 Iwaoka, Iwaoka-cho, Nishi-ku, Kobe 651-2492, JAPAN

EFFECTS OF MECHANICAL DEFORMATION AND THERMAL TREATMENT ON MICROSTRUCTURE OF BSCCO SUPERCONDUCTORS

Sean Li
School of Materials Engineering, Nanyang Technological University, Singapore

Wei Gao
School of Engineering, The University of Auckland, New Zealand

1. INTRDOUCTION

It has been demonstrated that the critical current density (J_c) across a grain boundary decreases drastically with increasing misorientation angle, and the boundary behaves as a weak-link for misorientations greater than 10° in high-temperature (T_c) superconductors [1-3]. This results that polycrystalline materials have lower J_c than single crystal and thin films by a few orders [4]. However, the exact nature of the grain boundary weak links is still not clearly understood. It was suggested that the nature is most likely related to some of the following possibilities; (i) disturbed crystal structure or altered interatomic distances near the grain boundaries, (ii) deviation in chemistry at the grain boundaries (cations or oxygen stoichiometry), (iii) presence of impurity layers such as carbonaceous compounds or amorphous materials, (iv) microcracks caused by thermal contraction anisotropy or by inhomogeneous oxygenation, or (v) anisotropic superconductivity and resultant weak connections at large-angle boundaries. These subtle structural and compositional inhomogeneity and the presence of a large number of dislocations at the grain boundaries would certainly influence the electronic structure, especially the Cu-O bond distance/angle, a small change of which is associated with suppression or loss of superconductivity in high-T_c superconductors.

For many major applications, use of the bulk $(Bi,Pb)_2Sr_2Ca_2Cu_3O_{10+x}$ Bi2223 high-T_c superconductors has been hindered by their relatively lower J_c than of thin films. It is believed that this is caused by a number of materials problems including the grain boundary weak links and flux line movement [5]. A key problem in the applications of Bi2223 superconductors is the sharp drop of the J_c values in external magnetic fields [6]. Possible reasons for this phenomenon are the crystal anisotropy, granular grain boundaries and the low density of intrinsic defects acting as flux pinning sites. In addition, the critical "transport" current density (J_{ct}) for transmitting current over useful lengths is often very much smaller than the local J_c in the strong superconducting region with good flux pinning [7]. Thus, almost all research regarding the weak-link of superconductors involves in the transport properties of grain boundaries in superconducting oxides, because the properties of grain boundaries control the macroscopic electromagnetic character of the polycrystalline superconducting oxides.

It has been evidenced that the critical current variation with the position in a tape has shown substantial macroscopic heterogeneity in BSCCO tapes. Sectioned Bi2223 tapes showed a factor of three differences in the critical currents measured at the tape edges and at the tape centre [8, 9]. This trend in J_c presents a parabolic relationship with respect to the transverse cross section of the tapes. It has proposed that this relation strongly depends on the local crystallographic orientation distribution of high-T_c superconducting oxides. Furthermore, the critical current measured near the centre of the tape agrees well with the macroscopic critical current density. This has been evidenced by the determinations of the local J_c via analysis of magneto-optic image of flux penetration into the tape [10-12]. The analysis shows that the Bi2223 tapes have regions of very high J_c near the edges of the tape, but these regions may be disconnected from one another. These observations draw attention to the promise of BSCCO tapes if the microstructure and consequent J_c can be optimised throughout the tape cross-section and length by microstructural control. Moreover, the local J_c measurements were made on the top and bottom sides of the tape. The results also showed that the 10 μm thick layer of a superconductor next to the silver sheath was the actual current carrying region. Its thickness was approximately 10% of the nominal thickness of a superconducting core in the silver sheathed tape [13, 14]. The remaining current carrying region shows granular behaviour and much lower current-carrying capability. The wide distribution of the $J_c(x,y)$ also revealed by magneto-optical imaging, indicating that the performance of Bi2223 tapes can be significantly improved if a larger fraction of well aligned grains can be produced more uniformly throughout the tape cross-section.

The processing parameters affecting J_c are chemical composition, phase content, and particle size of the starting powder; while for the oxide core in the tape these are density, orientation of the superconducting crystals, and grain size, volume fraction and distribution of the various impurity phases. Thermo-mechanical processing includes temperature, time, and atmosphere for heat treatment; thickness reduction schedule, and number of iterative steps [15-17]. It has been proven that the mechanical deformation is an important process step for reducing weak links through increasing texture and uniformity of microstructure, leading to increase critical current density [18-21]. Powder-in-tube (PIT) technique, involving mechanical deformation and thermal treatment processes, is widely used to fabricate Bi2223 superconductor tapes in large-scale with high current carry capability for practical applications [22-25].

In this study, the effects of thermo-mechanical processing on the microstructure of BSCCO superconductors have been investigated. It includes the effects of mechanical deformation and thermal treatment on the texture extent, texturing mechanism, texturing kinetics, grain growth behaviors and transverse micro- and mesotexture distribution characteristics in BSCCO superconductor tapes etc. The crystal fracture and lattice distortion behaviors of the BSCCO oxides, and the effects of energy storage on the recrystallisation processes have also been studied to understand the mechanisms of the microstructural changes.

2.1. EFFECTS OF MECHANICAL DEFORMATION ON THE PHASE TRANSFORMATION OF Bi2223

Mechanical deformation is widely used to produce textured microstructures to improve the J_c in Bi2223 superconducting tapes [26-28]. Subsequent thermal treatment is employed to convert the Bi2212 phase to the high-T_c Bi2223 phase [29-32]. In order to have a high degree of grain alignment in the Bi2223 tapes, heavy mechanical deformation is used to the tapes containing mainly Bi2212, Ca_2PbO_4 and $(Bi,Pb)_2Sr_2CuO_x$ phases [33]. However, the effects of mechanical deformation on the phase transformation from Bi2212 to Bi2223 have not been clearly understood. Studies in the transformation and distribution of the Bi2223 phase in the tapes subjected to different extents of mechanical deformation could provide the information in the relations between mechanical deformation and phase transformation of BSCCO compounds, thus optimising the processes to produce high quality Bi2223 tapes.

In the studies, PIT technique was employed to prepare the BSCCO tapes. The powder material, which contained mainly Bi2212, Ca_2PbO_4 and $(Bi,Pb)_2Sr_2CuO_x$ phases, has a nominal cation ratio of Bi : Pb : Sr : Ca : Cu = 1.84 : 0.34 : 1.91 : 2.04 : 3.06. The BSCCO compounds in silver tubes were swaged and drawn into ϕ1mm wires by multi-step drawing. In order to release the stress introduced by swaging and drawing, the wires were pre-annealed at 250°C for 1 hr. Mechanical deformation was performed by rolling the wires into tapes with different thickness reduction. Each rolling step produced approximate 20% thickness reduction. The deformation ratio R is defined as the thickness reduction of the tape:

$$R = (t_o - t_f)/t_o \tag{1}$$

where t_o and t_f are the original and final thickness of the superconductor cores, respectively. The deformed tapes were annealed at 833°C for 15, 25 and 45 hrs to transform the Bi2212 to Bi2223 phase. After annealing, the silver sheath on the BSCCO superconducting oxide core was etched away with an aqueous solution of 50%NH_4 + 50% H_2O_2 at room temperature. X-ray diffraction of the specimen surface after removing the Ag sheets indicated that no reaction products due to the etching treatment could be detected.

To determine the effects of mechanical deformation on the phase transformation, X-ray diffraction was carried to measure the abundance of Bi2223 phase in the BSCCO superconducting tapes. The ratio of Bi2223 phase in the phase mixture, X_{Bi2223}, can be defined as:

$$X_{Bi2223} = \frac{I^{(0010)}_{Bi2223}}{0.88 I^{(008)}_{Bi2212} + I^{(0010)}_{Bi2223}} \tag{2}$$

where $I^{(0010)}_{Bi2223}$ and $I^{(008)}_{Bi2212}$ are the integrated intensities of the X-ray diffraction characteristic peaks for the (*0010*) of Bi2223 phase and the (*008*) of Bi2212 phase. To investigate the distribution of the Bi2223 phase abundance across the thickness of the tape, the surface of the BSCCO superconducting oxide core was removed by grinding with an extra fine SiC paper.

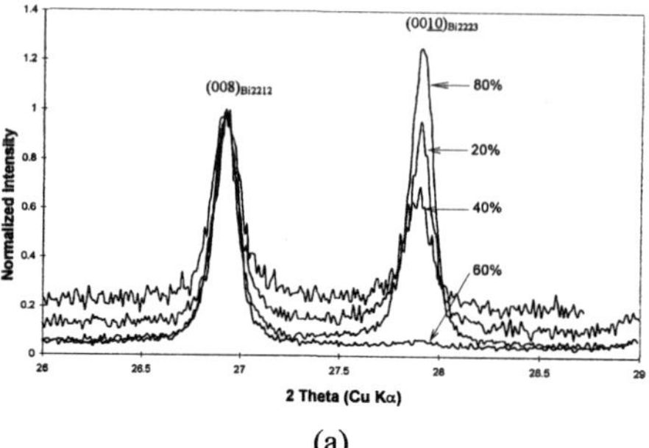

(a)

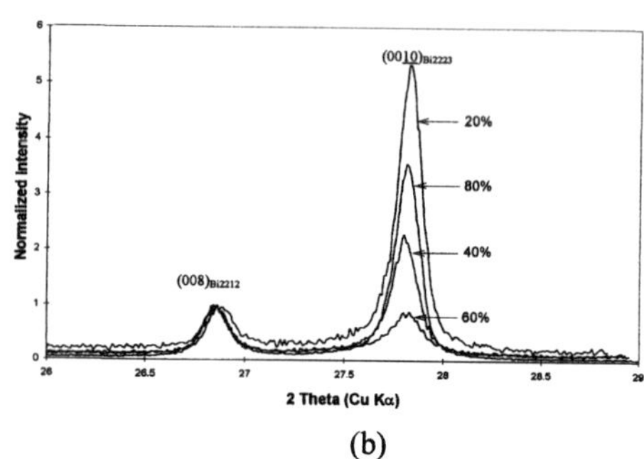

(b)

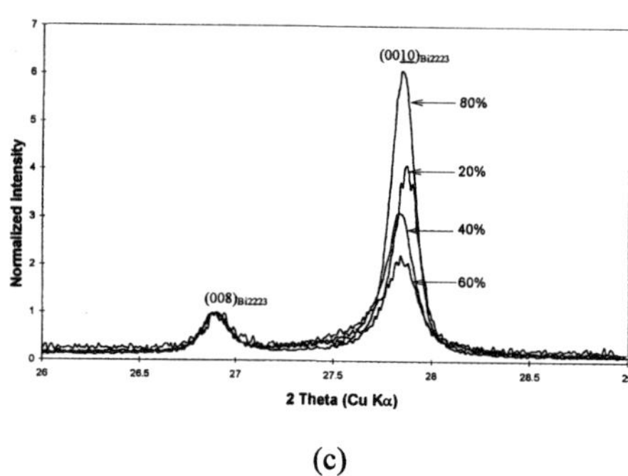

(c)

Fig. 1 X-ray spectra of the tapes with different deformation ratios after an annealing for (a) 15 hrs, (b) 25 hrs, (c) 45 hrs.

Figs. 1(a) to 1(c) are XRD spectra of the tapes with different mechanical deformation after annealing at 833°C for 15, 25 and 45 hrs, respectively. The (*008*) peak of Bi2212 phase and the (*0010*) peak of Bi2223 phase are located at ~26.9° and ~27.8° in these spectra, respectively. The intensities of the Bi2212 (*008*) peaks in the tapes with different mechanical deformations are normalised to a unit scale in these figures. Fig. 1(a) indicates that the intensity of the Bi2223 (*0010*) peak decreased as the mechanical deformation increased from 20% to 60%, but then increased dramatically as the mechanical deformation increased from 60% to 80%. A similar tendency was observed [Figs. 1(b) and 1(c)] for the tapes that had undergone 25 and 45 hrs of annealing, although the Bi2223 phase had a much higher intensity after a longer annealing duration. These results represent the effects of mechanical deformation on Bi2223 phase transformation.

The abundance of the Bi2223 phase formed after annealing for 15, 25 and 45 hrs was plotted versus the mechanical deformation (Fig. 2). The following features can be seen from these plots:

(1) With a low mechanical deformation (20-40%), the Bi2223 formed relatively slowly during the first 15 hrs of annealing. However, the abundance of Bi2223 phase reached ~0.8 after 25 hrs. Further annealing did not produce more Bi2223 phase.

(2) With a high deformation (80%), the Bi2223 phase formed relatively fast. 70% Bi2223 formed after 15 hrs of annealing. Further annealing only slightly increased the Bi2223 phase.

(3) With a deformation between these two regimes (60%), the formation of the Bi2223 phase was much slower. There was very little Bi2223 formed after 15 hrs of annealing, and it took a much longer annealing time (45 hrs) to produce a similar amount of Bi2223 as the tapes with lower or higher mechanical deformations.

The results above indicate that the mechanical deformation strongly affects the phase transformation of Bi2212 to Bi2223 in the early stages of annealing, implying that different mechanism may dominate the transformation kinetics above and below the 60% deformation ratio.

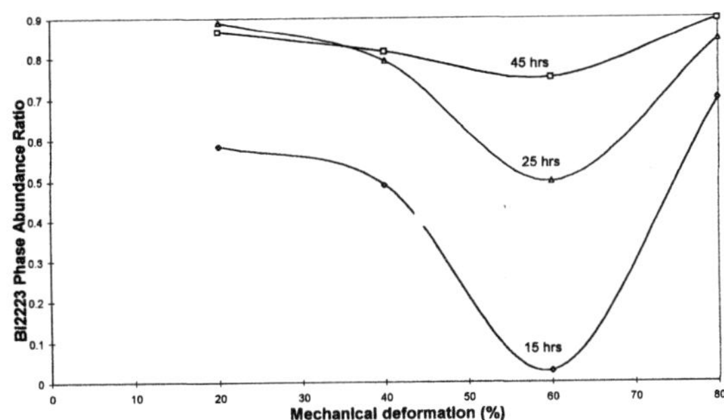

Fig. 2 Bi2223 Phase abundance versus the mechanical deformation ratios.

2.2. PHASE DISTRIBUTION ACROSS THE DEPTH OF OXIDE CORES

Fig. 3 shows the distributions of the Bi2223 phase across the depth of the oxide core from the silver sheet-core interface. The tapes had undergone 20% and 40% mechanical deformation followed by 25 hrs of annealing. It can be seen that a relatively high Bi2223 abundance was produced at the interface, and the Bi2223 phase abundance decreased across the depth of the core. In the 20% deformation tape, the abundance of Bi2223 dropped sharply to ~0.6 at 9 μm from the interface, while the abundance of Bi2223 remained ~0.8 at 13 μm from the interface in the 40% deformation tape. This result indicates that the tapes with greater deformation have better phase transformation towards the center of the tape, than the tapes with lower deformation.

Fig. 4 shows the Bi2223 phase abundance as a function of depth in the 40% deformation tapes after different annealing times. This measurement was carried out with another sample cut from the same wire used in the previous experiment. It can be seen that a short annealing time (15 hrs) produced a relatively low abundance of Bi2223 (~0.58) at the interface and the abundance of Bi2223 dropped sharply to 0.33 and 0.17 at 4 μm and 12 μm from the interface. The abundance of Bi2223 at the interface in this experiment is consistent with the result (~0.51) shown in Fig. 2, and it is believed that the small difference is due to the measurement error for the different samples. The tapes annealed for 25 and 45 hrs, however, have a much higher interface concentration and deeper distribution profiles for the Bi2223 phase. The different phase distribution profiles in the tapes after short and long annealing times showed that the Bi2223 phase transformation takes place from the interface towards the centre of the tape.

It was also noted that X-ray has a certain penetration depth and the XRD measurement provides an average information from the penetration layer (~2 μm in the present case). The present results provide semi-quantitative information for comparison purposes.

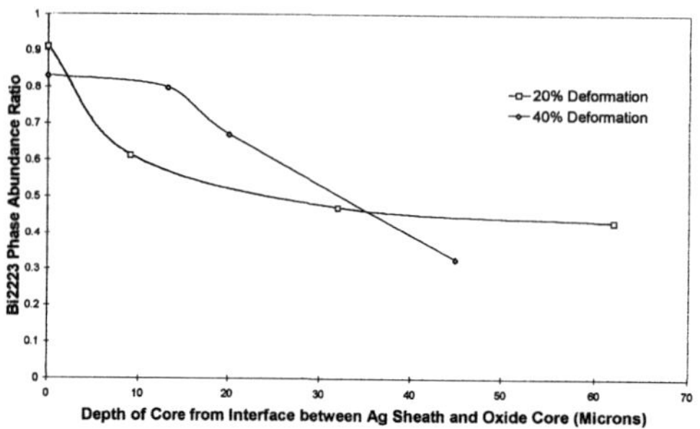

Fig. 3 Fraction distribution of Bi2223 phase across the oxide core depth for the samples with 20% and 40% deformation ratios annealed for 25 hrs.

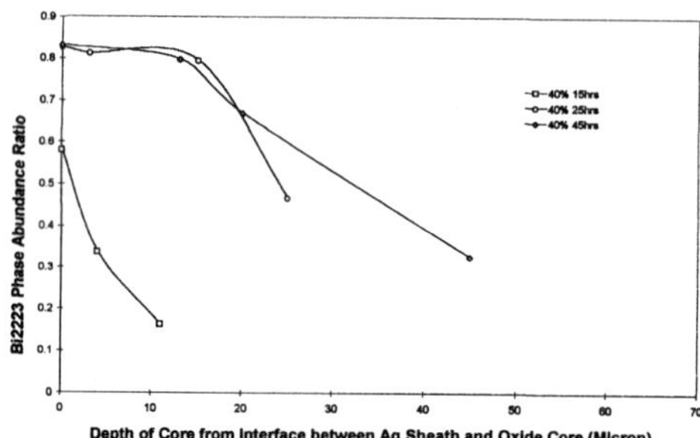

Fig. 4 Fraction distribution of Bi2223 phase across the oxide core depth for sample with a deformation ratio of 40% after annealing for different times.

It was proposed that the phase transformation in the BSCCO compounds is controlled by diffusion processes. This is consistent with the present observation that the concentration of Bi2223 is higher at the interface than inside the core. Mechanical deformation increases the material density in the tapes, and a greater deformation results in a higher density. It is also understandable that the density is higher near the interface than the inside of the core as the deformation starts at the surface. Also the phase abundance reaches a saturation limit from the interface to a certain region, implying a diffusion controlled phase transformation process. The thickness of the layer containing the saturated abundance of Bi2223 was not significantly influenced by the annealing time after ~25 hrs, and the thickness of this layer is similar to the thickness of the layer with good alignment of the Bi2223 grains [34]. Therefore, it is reasonable to assume that the Bi2223 grain alignment process is associated to the Bi2223 phase transformation in the tapes.

2.3. EFFECTS OF MECHANICAL DEFORMATION ON GRAIN SIZE WITH PHASE TRANSFORMATION

Fig. 5 plots the grain size after different annealing times as a function of mechanical deformation. At the interface area, the plate-like Bi2223 grains were aligned parallel to the interface (the normal of the grains are perpendicular to the interface). In this case, the plate-like Bi2223 grain size was defined as the size of the "plate" at the interface.

Fig. 5 shows that after 15 hrs annealing, the difference in grain size over the different deformation ratios is very small. The measured grain size was approximately 3 μm, and seems independent of the mechanical deformation. In fact, the grain size after a short annealing time represents the grain sizes for a mixture of the Bi2223 and Bi2212 phases. Particularly, little Bi2223 was formed in the tape with 60% deformation. Therefore, the grain size of the

60% deformed tape represents the recrystallised grain size of the Bi2212 phase. For the other deformation ratios, the grain size represents a mixture of the recrystallised Bi2212 grains and the newly formed Bi2223 grains.

After annealing for 25 and 45 hrs, the Bi2223 abundance in the 60% deformation tapes sharply increased from 0.02 to 0.5 and 0.75, Fig. 2. The grain size also increased dramatically from ~3 μm to ~8.3 μm, more than twice of the size of the grains in the tapes with other deformation ratios. The grain sizes in the BSCCO superconducting tapes with 20%, 40% and 80% deformation, however, did not change significantly with annealing time. This result will be explained below with the different nucleation and growth mechanisms associated with the different mechanical deformations.

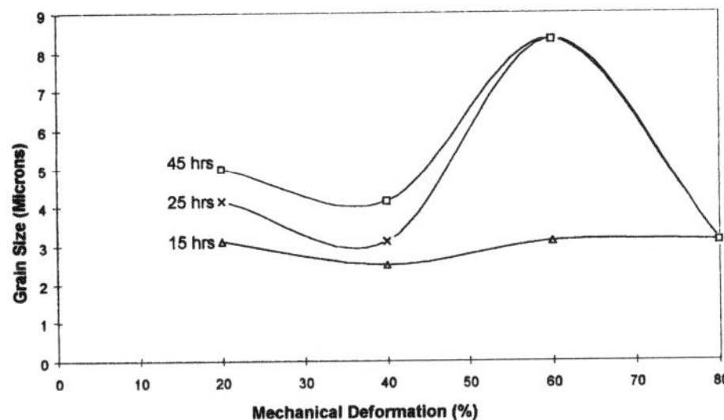

Fig. 5 Grain size of BSCCO with different annealing times versus mechanical deformation ratio.

2.4. MECHANISMS OF MECHANICAL DEFORMATION EFFECTS ON PHASE TRANSFORMATION

Fig. 2 displays the unusual effects of mechanical deformation on the Bi2223 phase transformation, as 60% deformation resulted in the slowest transformation kinetics. The phase transformation processes in the BSCCO compounds are complicated, but a tentative explanation can be suggested below.

There are two major effects to the BSCCO tapes during the mechanical processing. (1) A part of deformation energy is absorbed by the oxide core as fracture energy, splitting the Bi2212 crystals and producing more interface area. (2) A part of the energy is stored in the crystals as the result of lattice distortion of the Bi2212 crystals. In this case, the first effect took place when the mechanical deformation ratio was relative low while the second effect occurred when higher mechanical deformation was applied to the BSCCO tapes. However, these effects have different contributions to the phase transformation process of the BSCCO superconducting materials. Obviously, the area of the fractured surfaces increased as the mechanical deformation ratio increased, thus increasing the surface energy. This surface energy drove the fractured Bi2212 crystals to recrystallize, hence reducing the energy of the system during annealing at 833°C. However, the lattice distortion increased the internal energy of the Bi2212 and the balance phases; thus facilitating phase transformation to form Bi2223 instead of the recrystallized Bi2212 crystals. There were two possible types of annealing behaviour for the Bi2212 phase, either the Bi2212 recrystallized to minimise the surface area or reacted with the balance phases to form the Bi2223 phase. Therefore, the annealing behaviour of the deformed Bi2212 crystals was determined either by the surface energy or by the chemical potential and internal energy of the Bi2212 crystals. When the surface energy was higher than the chemical potential and the internal energy, recrystallization of the Bi2212 crystals took place, while the chemical reaction occurred when the chemical potential and internal energy dominated the systems. Therefore, it could be inferred that recrystallization was probably a favourable process when the deformation ratio was less than 60% and the trend increased with the deformation ratio. This is because that the surface area rapidly increased as the deformation ratio increased, until the surface area reached a "saturation" value. This represents the trend that Bi2223 formation decreased with increasing mechanical deformation. However, as the deformation ratio exceeded 60%, the lattice distortion of the Bi2212 crystals increased rapidly, and therefore the internal energy would increase with increasing mechanical deformation.

Fig. 6 is a schematic illustration, showing that the driving force for the Bi2223 phase formation is enhanced by the internal energy. T_e is the equilibrium temperature of Bi2212 and Bi2223 phases. Annealing the sample at temperatures above T_e ($T > T_e$), allows the Bi2223 phase to form because of the lower free energy, while the formation of the Bi2212 phase takes place at lower annealing temperatures ($T < T_e$). As the internal energy of the Bi2212 phase is increased by the mechanical deformation, the free energy of Bi2212 phase becomes $G'_{Bi2212} = G_{Bi2212} + E_{internal}$. It is similar to the original free energy curve of the Bi2212 phase (G_{Bi2212}) without the mechanical deformation, but is shifted up to G'_{Bi2212} and forms the new free energy curve – G'_{Bi2212}. The G'_{Bi2212} curve is parallel to the G_{Bi2212} curve, as shown in Fig. 6. Obviously it results in a decrease in the equilibrium temperature of the Bi2212 and Bi2223 phases from T_e to T'_e. As a result of the reduced equilibrium temperature, the driving force for the Bi2223 phase formation from the overheating (833°C - T'_e) increases from ΔG to $\Delta G'$, as shown in Fig. 6. In this case, the chemical potential and the internal energy replace the surface energy as the dominant energies of the system. Therefore, after annealing the formation of the Bi2223 phase increased as the deformation ratio increased, when the ratio was higher than 60%.

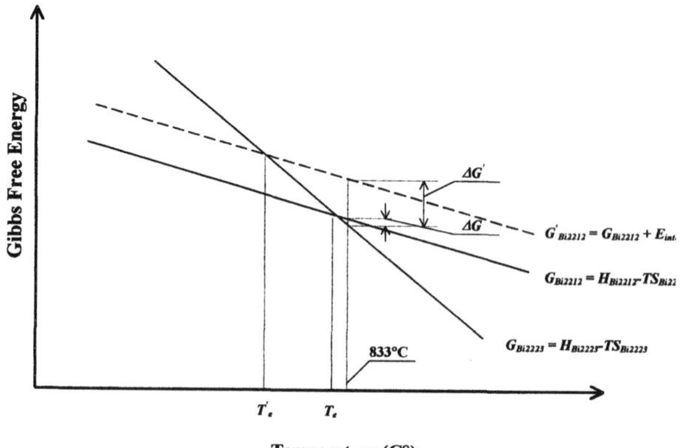

Fig. 6 A schematic illustration shows the mechanism that the driving force of the Bi2223 phase formation was enhanced by the internal energy.

As discussed above, the effects of mechanical deformation on the phase transformation are complex. Apart from the change of interface area and internal energy, the effects on the phase transformation may also be related to a number of other factors, such as the effective diffusivity, concentration gradient and grain size. For example, the fine grain size lead to an increase in the grain boundary (interface) area. Therefore, the effect of the surface diffusion would be stronger than that of the lattice diffusion during the phase transformation and grain growth processes. This implies that the formation of Bi2223 should increase with increasing deformation ratio. In addition, the mechanical deformation might also influence the number of nuclei and the nucleation rate. Generally speaking, the nucleation rate is independent of the number of the nucleation sites. The relations between the number of nuclei (N), the nucleation rate (R), the area of the nucleation sites (A) and the annealing time (t) can be expressed as:

$$N = RAt \qquad (3)$$

Obviously, for a certain nucleation rate, higher mechanical deformation exposes more surfaces and therefore provided more nucleation sites, thus producing a large amount of Bi2223 nuclei that facilitate the phase transformation from Bi2212 to Bi2223. However, the different mechanical deformation processes may result in different grain boundary area, thus affecting the deformation ratio corresponding to the minimum formation of Bi2223. It is obvious that mechanical deformation influences both the thermodynamics and kinetics of the BSCCO phase formation.

3. EFFECTS OF MECHANICAL DEFORMATION AND PHASE TRANSFORMATION ON GRAIN ALIGNMENT IN Bi2223 TAPES

It is clear that the mechanical deformation strongly affects the kinetics of phase transformation from Bi2212 to Bi2223. Mechanical deformation also affects the grain alignment of the Bi2223 phase. Grain alignment of superconducting oxides has a strong effect on J_c of the polycrystalline BSCCO superconductors. PIT technique is widely used to induce the highly textural microstructure in the polycrystalline BSCCO superconductors. However, the lack of understanding of the grain alignment mechanisms induced by mechanical deformation and thermal treatment in the BSCCO system is one of the major hurdles standing in the way of improving microstructure and J_c in BSCCO superconducting materials.

To investigate the mechanisms of grain alignment induced by the mechanical deformation and phase transformation, the specimens were prepared by rolling with various deformation ratios and then annealing at 833°C for 60 hrs. The Lotgering method was employed to quantify the degree of c-axis orientation alignment in the polycrystalline Bi2223 superconductors. X-ray diffraction was carried out to determine the Lotgering factor. The Lotgering factor, F, is defined in the following equations [35]:

$$F = (P-P_o)/(1-P_o) \qquad (4)$$

$$P = \Sigma I_{(00l)} / \Sigma I_{(hkl)} \qquad (5)$$

where P is the sum of the integrated intensities for all $(00l)$ diffractions divided by the sum of all intensities of (hkl) diffractions in the textured specimen. P_o is an equivalent parameter for a randomly oriented specimen. The factor F varies from 0 for an entirely non-oriented specimen to 1 for a completely oriented specimen. Grain alignment extents were measured by using Lotgering Factors before and after annealing.

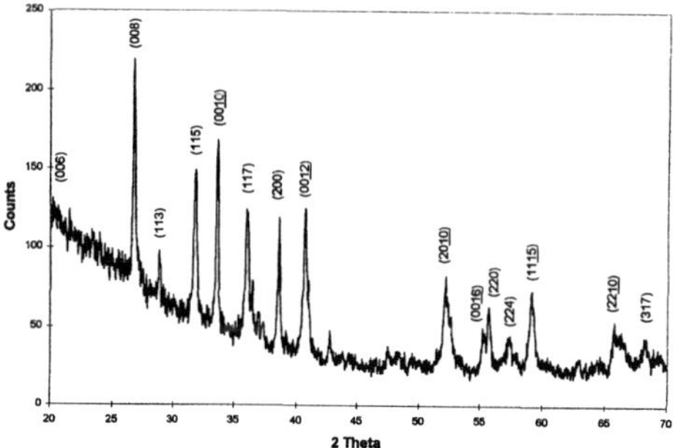

Fig. 7 X-ray spectra of Bi2212 tape with 24% mechanical deformation before annealing.

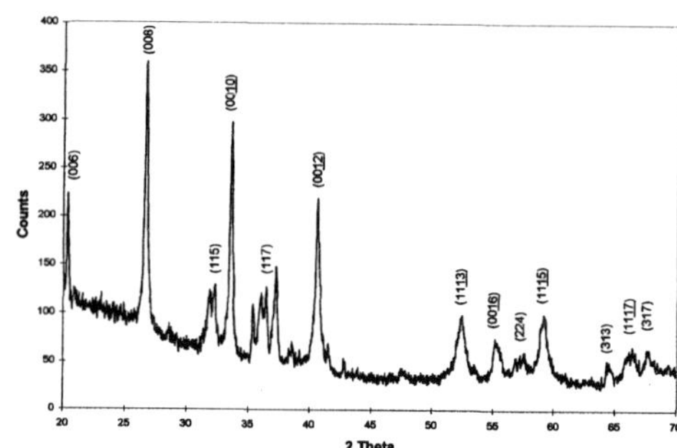

Fig. 8 X-ray spectra of Bi2212 tape with 80% mechanical deformation before annealing.

The X-ray spectra of the Bi2212 tapes, after mechanical deformation with 24% and 80% thickness reduction and before annealing, are shown in Figs. 7 and 8. The Lotgering Factors of the Bi2212 specimens after 24% and 80% deformations were 0.24 and 0.53, respectively. It can be clearly seen that in the Bi2212 tape, the relative intensities of *(00l)* peaks in the specimen with 80% deformation are much stronger, and the intensities of *(hkl)* peaks including *(113)*, *(115)*, *(117)*, *(200)*, $(20\underline{10})$ and $(11\underline{15})$ are much weaker compared with the specimen of 24% reduction. Figs. 9 and 10 show the X-ray spectra after annealing and Bi2223 phase formation. The *(hkl)* peaks including *(015)*, *(019)*, *(110)*, $(01\underline{11})$, $(11\underline{12})$, $(01\underline{19})$, $(11\underline{18})$ and $(01\underline{21})$ from Bi2223 which existed in the specimen with 24% deformation disappeared in the specimen with 80% deformation. It is obvious that the high deformation ratio promoted the *(00l)* grain alignment in the Bi2212 specimens.

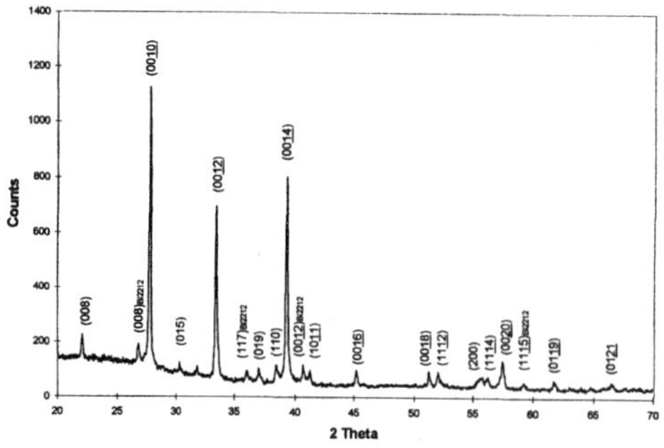

Fig. 9 X-ray spectra of Bi2212 tape with 24% mechanical deformation after phase transformation

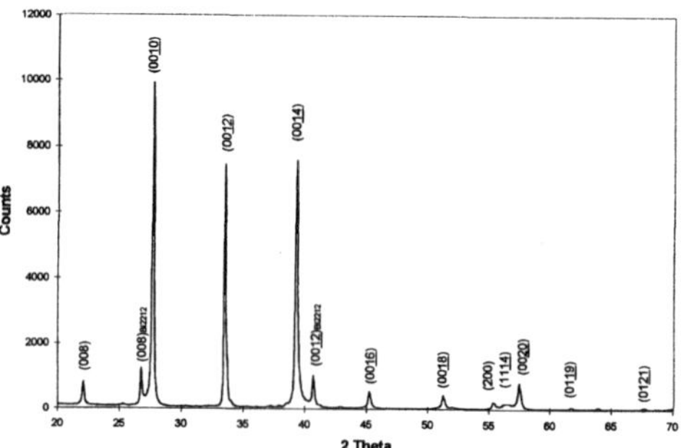

Fig. 10 X-ray spectra of Bi2212 tape with 80% mechanical deformation after phase transformation.

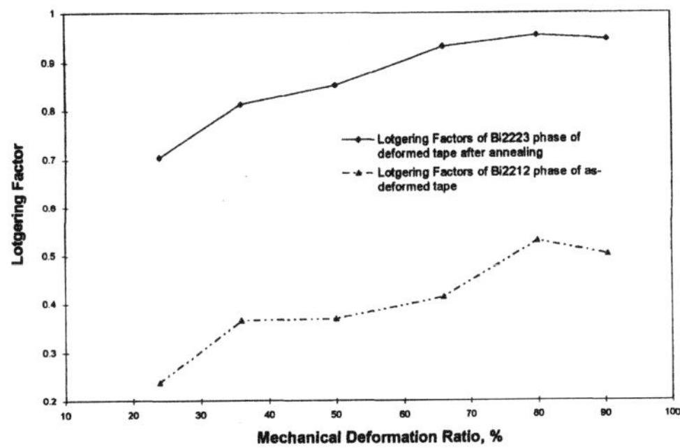

Fig. 11 Lotgering Factors versus mechanical deformation for the specimens after phase transformation.

Fig. 11 plots the Lotgering Factors as a function of mechanical deformation for the specimen subjected to mechanical deformation and phase transformation. The results indicate that: (1) The grain alignment extent on the surface of the Bi2212 tapes (before annealing) increases with increasing reduction ratio. (2) The grain alignment on the surface of Bi2223 (after annealing) also increases with increasing reduction, although the increase is saturated when the reduction ratio becomes high (70%-90%). (3) The phase transformation (from Bi2212 to Bi2223) improves the grain alignment significantly. Lotgering Factor of Bi2223 reached ~0.95 with 80% mechanical deformation after phase transformation sintering. Further increase of deformation to 90% did not produce any better alignment, probably because the grain alignment may be damaged by the formation of "sausage" with the heavy mechanical deformation. (4) The Bi2212 phase with better grain alignment leads to a higher degree of grain alignment in Bi2223 after annealing, revealing a relationship between the microstructures before and after phase transformation.

4.1. EFFECTS OF MECHANICAL DEFORMATION AND RECRYSTALLIZATION ON GRAIN ALIGNMENT IN Bi2223 TAPES

After BSSCO tapes were shaped and sintered, the additional mechanical deformations and thermal treatments were still effective in enhancing the J_c in Bi2223 superconductor tapes. The first mechanical deformation performed on the raw superconducting materials was aimed to induce a textured microstructure in the dominant Bi2212 phase, and the subsequent thermal treatment was used to obtain the desired Bi2223 phase. However, the additional mechanical deformation performed on the dominant Bi2223 phase is considered to improve the extent and distribution of the textured microstructures of the Bi2223 phase, and the subsequent re-annealing treatments mainly provide reconnection of the broken grains. It is believed that the annealing treatments have always been associated with deformation-texturing processes. This is a complicated thermo-mechanical process. The annealing processes generally involve recovery, recrystallisation and grain growth, being different with the phase transformation annealing that converts Bi2212 phase. Although suggestion was made to a possible "annealing texture" formation and "reaction induced texture" mechanisms [36], very little detailed studies on the mechanism have been done.

To have a better understanding on the effects of mechanical deformation and recrystallization on grain alignment, the following questions should be addressed:

(1) Does the recrystallization annealing improve the grain alignment?
(2) What is the relation of deformation and grain alignment?
(3) What is the relation of deformation and recrystallization grain size in the Bi2223 phase?

In order to obtain a pure Bi2223 phase, the wires were sintered at 832°C for 60 hrs in air and 850°C for 60 hrs again in an atmosphere consisting of 10%O_2 + 90%Ar. Mechanical deformation was performed by rolling these sintered wires into tapes with different thickness. The tapes with different deformation ratios were then annealed at 832°C for 60 hrs in air; and the recrystallization of Bi2223 oxide took place in this process and grain alignment extents were also measured using Lotgering Factors before and after annealing.

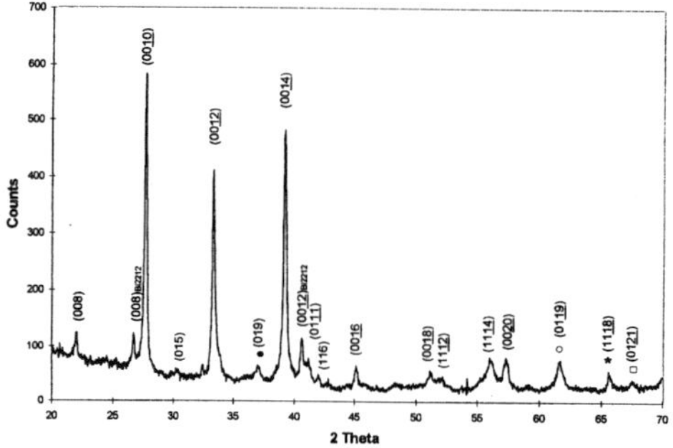

Fig. 12 X-ray spectra of a Bi2223 specimen with 81% deformation ratio before the recrystallization annealing.

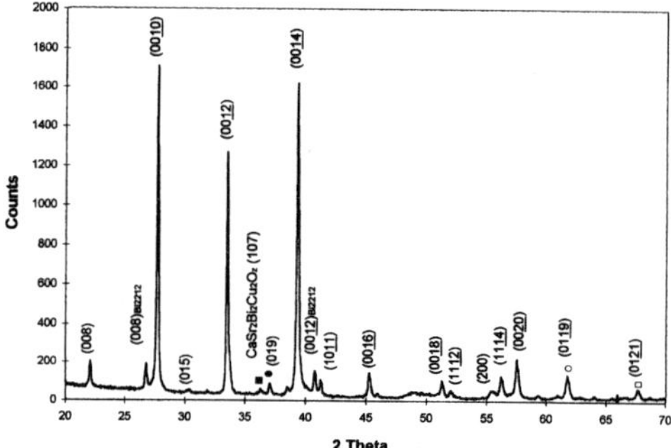

Fig. 13 X-ray spectra of a Bi2223 specimen with 81% deformation ratio after the recrystallization annealing.

Typical X-ray diffraction spectra of Bi2223 tapes with 81% accumulated mechanical deformation before and after annealing are shown in Figs. 12 and 13. Comparing the spectra of the rolled specimen with the non-oriented Bi2223 powder specimen [37], the intensities of (*00l*) diffractions including (008), (0010), (0012) and (0014) increased significantly, and the intensities of other (*hkl*) diffractions such as (105), (107), (109) and (110) decreased considerably. This result indicates that mechanical deformation produced a high degree of (*00l*) grain alignment. Comparing the specimens before and after annealing (Figs. 12 and 13), the (*00l*) peaks from the specimens after annealing are stronger and sharper, indicating a higher degree of (*00l*) orientation alignment.

Fig. 14 and Fig. 15 show the X-ray spectra of the textured Bi2223 tapes with different deformation ratios before and after annealing, respectively. Their Lotgering factors are plotted versus the deformation ratio in Fig. 16. Two clear tendencies can be seen from Fig. 16:

(1) The Lotgering factors after annealing are considerably higher than before annealing,
(2) The Lotgering factor decreases with increasing mechanical deformation ratio.

The effect of annealing on the grain alignment is influenced by the deformation ratio. With ~50% thickness reduction, annealing produced the strongest effect on the grain alignment in Bi2223. The Lotgering factor, *F*, increased from 0.80 to 0.92 after annealing. A smaller or greater deformation ratio results in weaker effect of annealing on the grain alignment. This evidence is clearly in favor of the mechanism of annealing-induced texture.

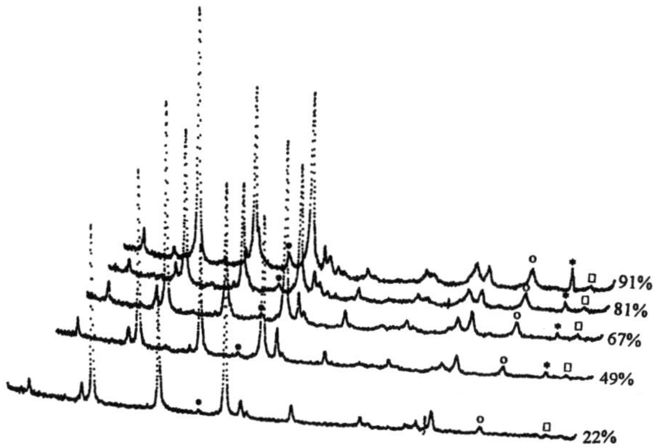

Fig. 14 X-ray spectra of Bi2223 tapes with different deformation ratios before annealing. $(01\underline{19})_{Bi2223}$ is marked with **O**, $(11\underline{18})_{Bi2223}$ is marked with ∗, $(01\underline{21})_{Bi2223}$ is marked with □, and $(019)_{Bi2223}$ is marked with •.

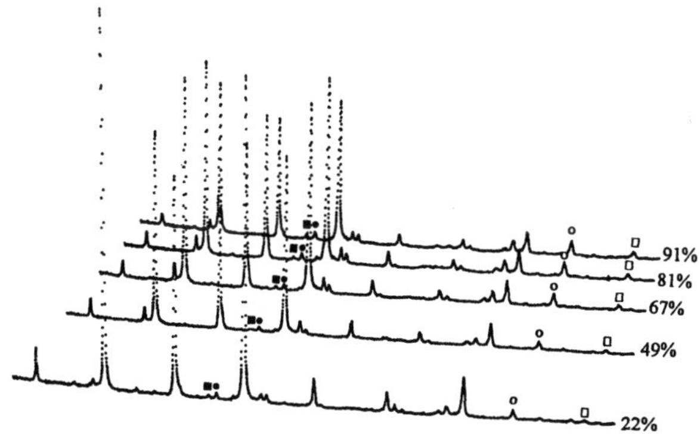

Fig. 15 X-ray spectra of Bi2223 tapes with different deformation ratios after annealing. $(01\underline{19})_{Bi2223}$ is marked with **O**, $(11\underline{18})_{Bi2223}$ is marked with ∗, $(01\underline{21})_{Bi2223}$ is marked with □, $(019)_{Bi2223}$ is marked with •, and *(017)* of $CaSr_2Bi_2Cu_2O_z$ is marked with ■.

Fig. 16 also shows that the Lotgering factors of grain alignment on the surface of the Bi2223 tapes decreases with increasing mechanical deformation ratio. This observation seems to go against general thought that higher deformation should produce better grain alignment. However, this can be explained with the microstructural observations shown in Figs. 17 and 18. Before deformation, the surface of the Bi2223 tapes already had a grain alignment resulted from the previous mechanical deformation and phase transformation. A heavy mechanical deformation damages the surface grain alignment. With a relatively low deformation (thickness reduction < 50%), the plate-like grains on the surface were broken into pieces of 10-20 μm, and slightly elongated along the rolling direction [Fig. 18(a)]. When a much higher force was applied to roll the wire, the Bi2223 grains were heavily smashed into small pieces as shown in Fig. 18(b), resulting in severe damage to the grain alignment on the surface of the tapes. This effect can be clearly seen from the X-ray diffraction spectra in Fig. 14. With 22% deformation, for example, the diffractions from (*019*) (marked with •) and (*11$\underline{18}$*) (marked with *) did not appear in the spectra. As the deformation ratio increased to 49% or greater, (*019*) and (*11$\underline{18}$*) diffractions appeared, indicating that some of the Bi2223 grains rotated from (*00l*) orientation to (*019*) and (*11$\underline{18}$*) orientations due to the deformation damage. During the subsequent annealing, however, those grains either rotated back to (*00l*), and/or coalesced into the grains with (*00l*) orientation through recrystallisation and growth. Therefore, the (*019*) diffraction has been reduced and the (*11$\underline{18}$*) peak disappeared after annealing, as shown in Fig. 15. This can also explain why the (*00l*) grain alignment was significantly improved by annealing.

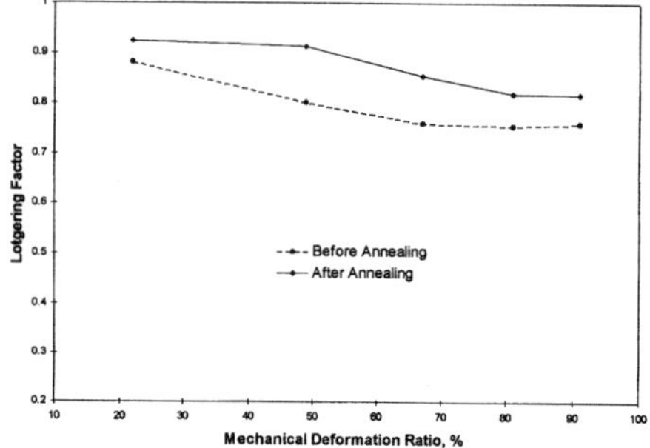

Fig. 16 Lotgering factors versus deformation ratio before and after the recrystallization annealing.

Fig. 17 SEM micrograph showing the surface morphology of Bi2223 wire before mechanical deformation. The Ag sheet was removed by etching.

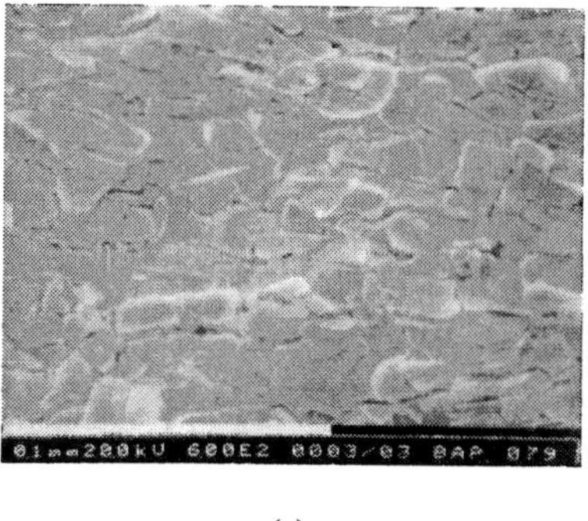

(a)

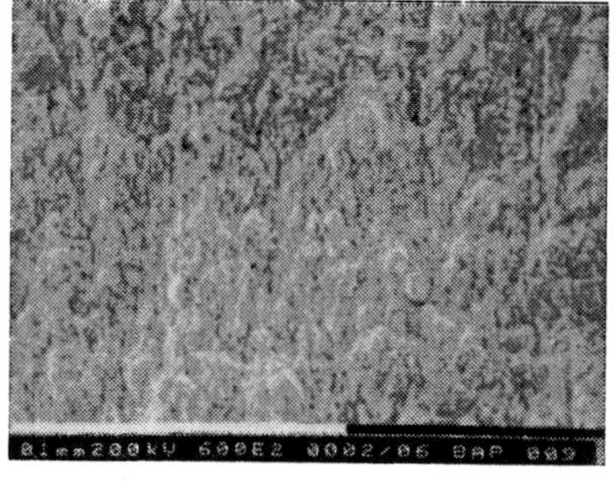

(b)

Fig. 18 SEM micrographs showing the surface morphology of Bi2223 tapes with (a) 22% and (b) 91% deformation ratio before annealing.

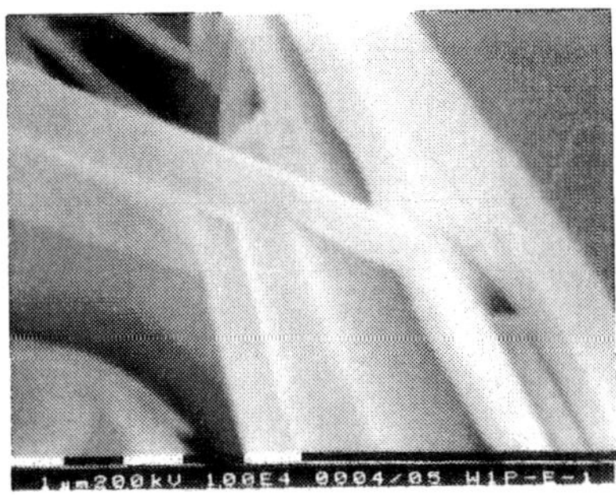

Fig. 19 Mechanical twin crystals in a Bi2223 tape with heavy deformation.

In addition, the high mechanical deformation ratio may introduce some types of structural defects such as stacking faults and twins, and possibly allotropic $CaSr_2Bi_2Cu_2O_z$ phase in residual $Bi_2Sr_2CaCu_2O_y$. As shown in Fig. 15, the (*107*) peak of $CaSr_2Bi_2Cu_2O_z$ appeared after deformation and the relative intensity increased with increasing deformation ratio. Fig. 19 shows a mechanical twin structure in a heavily deformed Bi2223 ribbon. The contributions of these types of structural defects can also degrade the (*00l*) orientation alignment in Bi2223, as shown in Fig. 16.

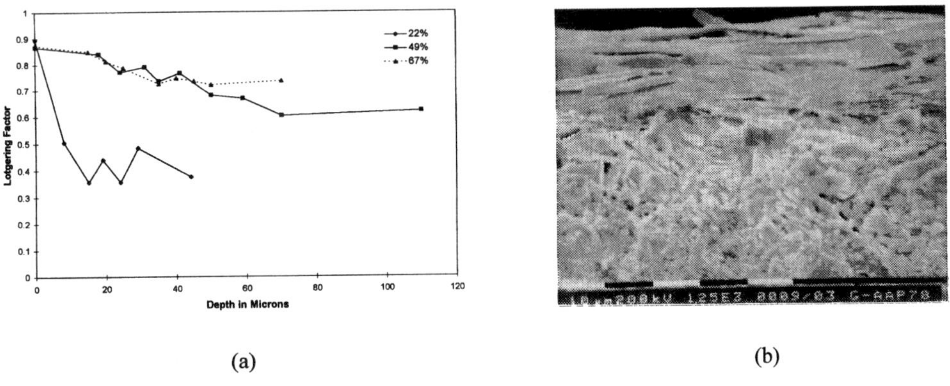

(a) (b)

Fig. 20 (a) Lotgering factor versus depth for specimens with thickness reduction of 22%, 49% and 67%, (b) SEM morphology shows that the high degree of texture microstructure was only formed on the surface of the Bi2223 tapes with 22% mechanical deformation.

Although heavy mechanical deformation does not improve the grain alignment on the top surface, it produces grain alignment in a greater depth of the specimens. Fig. 20 shows the measured Lotgering factors across the thickness of Bi2223 tapes produced by different deformation ratios, indicating that high deformation ratios produced grain alignment with a higher thickness and a different profile. Comparing with the effects of mechanical deformation and phase transformation on grain alignment, the following two tendencies are clear in the effect of mechanical deformation and recrystallization on grain alignment of Bi2223 in the tapes:

(1) Unlike Bi2212 shown in Fig. 6, the Lotgering Factors decrease with the increasing mechanical deformation for Bi2223 phase.

(2) Similar to Bi2212 phase, the Lotgering factors of Bi2223 phase after annealing are higher than before annealing. The maximum increase of Lotgering Factors after annealing is ~0.20 with ~50% mechanical deformation.

4.2 GRAIN SIZE OF Bi2223 AND RECRYSTALLIZATION TREATMENTS

Fig. 21 shows the surface morphologies of Bi2223 tapes with various deformation ratios after annealing treatment. The effect of the deformation ratio on the grain size is shown in Fig. 22 in which the grain size is plotted versus the deformation ratio. It is clear that a higher deformation produces a smaller grain size. Fig. 22 also shows an approximately linear relation between the mechanical deformation and grain size after annealing. The average diameter of the Bi2223 grains D in μm can be expressed as:

$$D = -0.16R + 17.7 \quad (6)$$

where R is the deformation ratio which is defined in equation (1).

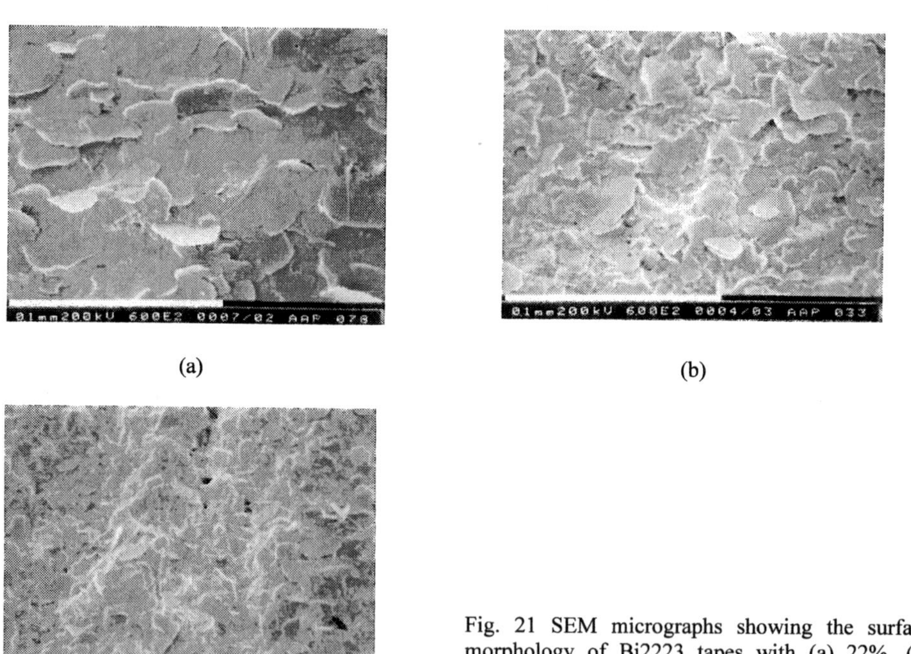

Fig. 21 SEM micrographs showing the surface morphology of Bi2223 tapes with (a) 22%, (b) 67%, and (c) 91% deformation after annealing.

This approximately linear relation between the grain size and deformation ratio does not follow the traditional theory for recrystallisation in metals, where the grain size decreases exponentially with the increasing deformation [38]. This observation implies different mechanisms of deformation and grain growth in metals and ceramic oxides. In metals, the final grain size is determined by the nucleation and grain growth processes. Higher plastic deformation introduces more energy in the deformed metal in the form of a high density of crystal defects, therefore increasing the nucleation rate during recrystallisation, and resulting in a smaller grain size. Mechanical deformation in ceramic oxides, on the other hand, mainly produces fracture of the plate-like grains, resulting in a very small grain size as shown in Fig. 18(b). The energy-storage mechanism in metals cannot be applied in ceramics. The nucleation-growth processes should also be different from that in metals. Therefore, the relations between the grain size and deformation are different for metals and ceramic superconductors.

In electronic materials including high T_c superconducting oxides, grain size and alignment are important because of their anisotropic properties. Therefore, understanding the relationships between grain size and processing parameters is an important topic. Coarse and textured grains are often desirable because of their favorable transport properties. From the present results, we suggest that multi-step small deformations be used before the final heat treatment in order to obtain a microstructure with relatively large and well-textured grains in preparation of Bi2223 superconductor tapes. However, the smaller grain size of Bi2223 oxides would create a number of flux pinning sites, thus increasing J_c in high field. It implies that the relation between J_c and grain size is not simple. Detailed research is being carried out to further understand this relationship.

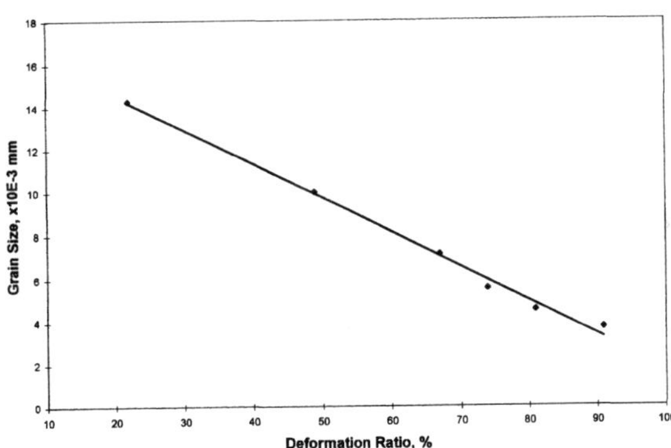

Fig. 22 Grain size of Bi2223 tapes after annealing versus mechanical deformation ratio.

5. FRACTURE BEHAVIORS AND GRAIN ALIGNMENT MECHANISMS OF POLYCRYSTALLINE Bi2223 OXIDES IN MECHANICAL DEFORMATION

Mechanical deformation with phase transformation treatment increases the grain alignment of Bi2223 superconductor tapes significantly while the intermediate deformation between the phase transformation and the recrystallization treatments may break the well-aligned Bi2223 grains formed in the phase transformation. However, the degree of texture on the surface of Bi2223 oxide cores is high, independent of the mechanical deformation ratio, [Fig. 20(a)]. This is because the silver sheath plays an important role in improving grain alignment of the core near the Ag-oxide interface, although the mechanisms of silver effects are different from the effects of different mechanical deformations [39]. The experimental results in Fig. 20 (a) demonstrate that a heavier deformation produces a deeper texture distribution and a higher degree of texture in the centre of the tapes. It implies that the mechanical deformation improves the grain alignment across the thickness of the tapes significantly. In order to understand the mechanisms of how the mechanical deformation induces the textured microstructure, the fracture behavior of polycrystalline Bi2223 oxide clad with silver sheets need to be understood.

As shown in Figs. 18(a) and 18(b), a small deformation broke the grains on the surface of the oxide core into relatively large pieces with rectangular-shape of ~20 μm × 10 μm, while a high deformation smashed the grains into much smaller pieces (2-3 μm or smaller). Fig. 23 shows an AFM micrograph that a piece of Bi2223 crystal (marked with C) has slid away from the original crystal. The sliding appears to take place along the a-b plane. It is believed that the fracture of the Bi2223 crystals often takes place along the basal plane due to the weak bonds in the Bi-O layers [40, 41]. With a low mechanical deformation, the rolling process mainly deformed the silver sheath as the plastic deformation ability of silver is high. Thus, the friction force caused by the silver flowing was transferred to the oxide surface. Since the friction force is likely to be higher than the compression force applied by the rollers, the shear force resulted from the friction plays a more important role than the compression force in the deformation processes. The crystals close to the interface with random orientation were not broken by the shear force, because the component on the a-b plane was not large enough in this case. Therefore, sliding along the a-b plane becomes the main feature of Bi2223 deformation under a low mechanical load, as shown in Fig. 23.

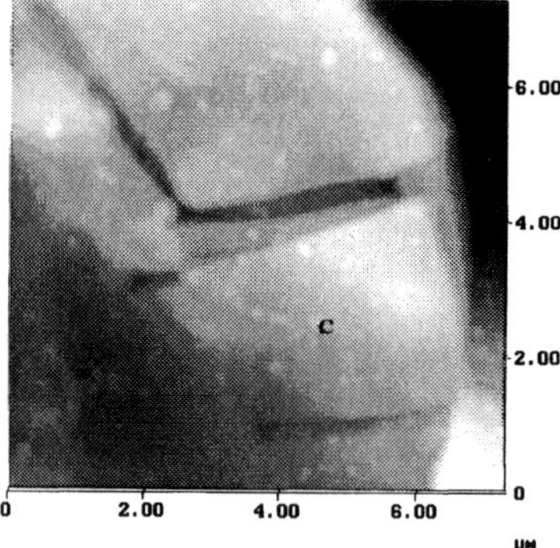

Fig. 23 AFM 2D surface morphology shows that a number of Bi2223 crystals fractured along a-b planes by frictional forces. The mechanical deformation was ~30%.

Figs. 24 and 25 show AFM 3D images of Bi2223 crystals with a higher deformation ratio (~50% thickness reduction). A bulk Bi2223 crystal (marked C in Fig. 24) shows multi-sliding steps along the a-b planes by the shear force. During this deformation process, the compression force provided by the rollers was relatively large. Therefore, the randomly oriented crystals close to the interface were also deformed. Fig. 25 shows a Bi2223 grain with its a-b plane almost perpendicular to the rolling plane. It can be seen that the multi-sliding action also took place along the a-b planes, evidence of the weak connections in the Bi2223 crystal. Fig. 26 shows SEM morphology of an oxide core with 50% mechanical deformation after a light polishing. Some crystals with a-b plane perpendicular to the rolling plane were fractured into smaller pieces, showing cracks developed along the a-b planes. Fig. 27 shows a schematic drawing illustrating the fracture mechanism of the randomly oriented Bi2223 crystals under a compression force. The sliding action took place along the a-b planes, resulting in multi-step feature which were developed on the edges of the Bi2223 grains, as shown in Fig. 24. With increasing mechanical deformation, the multi-sliding crystals slide further, and rotate their a-b plane toward the rolling plane, as shown in Fig. 27(c). The multi-step crystal structures were then further developed, resulting in a high degree of a-b plane texture.

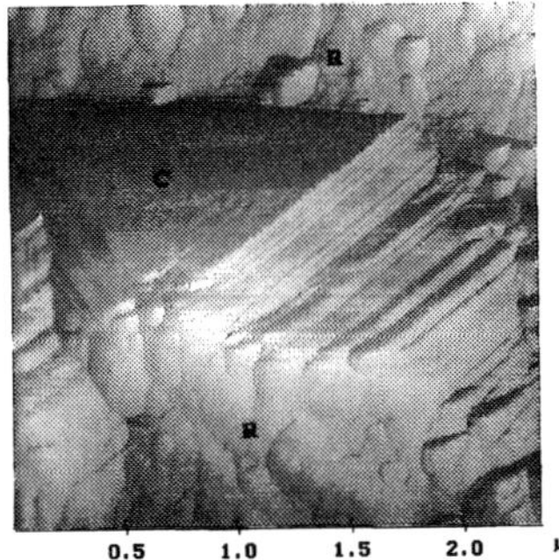

Fig. 24 AFM 3D top-view image showing the multi-sliding fracture feature of Bi2223 crystals produced by a compressive force in a tape with 50% mechanical deformation. The rolling plane and a-b plane are marked with "R" and "C", respectively.

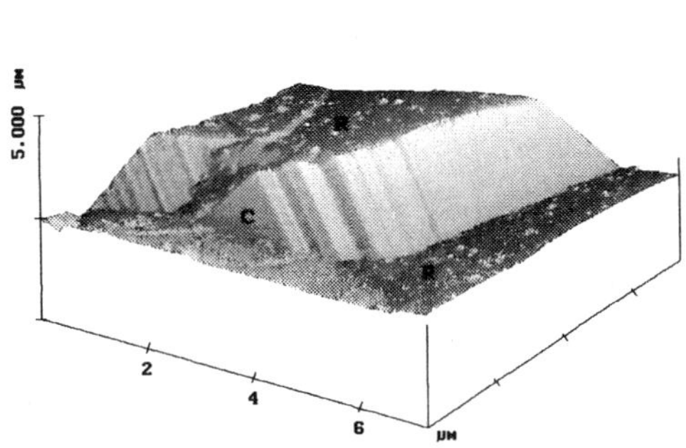

Fig. 25 AFM 3D image of crystals with the a-b plane perpendicular to the rolling plane. The multi-sliding was also produced by compressing force. The rolling plane and a-b plane were marked with "R" and "C", respectively.

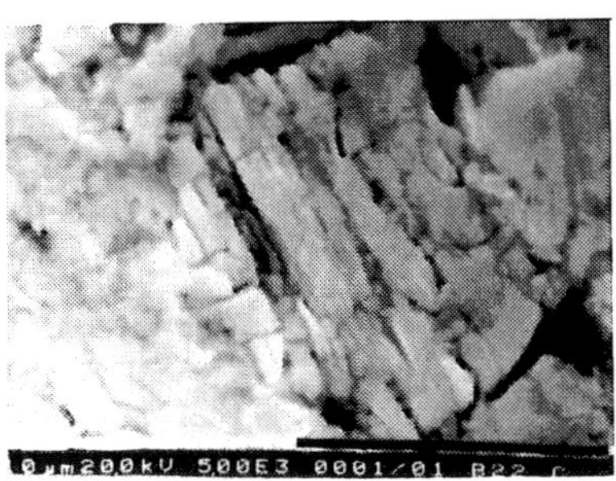

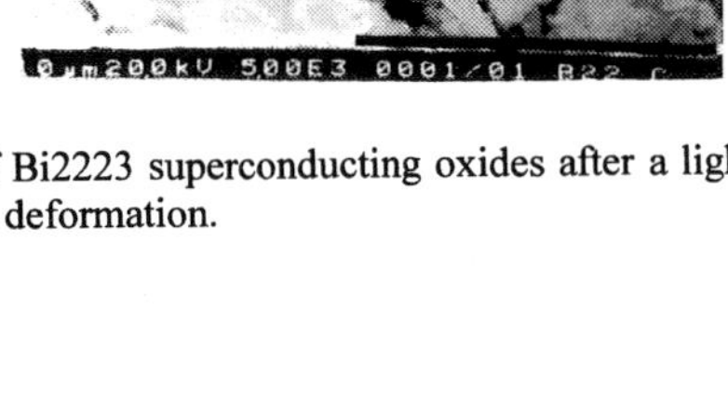

Fig. 26 SEM morphology of Bi2223 superconducting oxides after a light polishing. The specimen had 50% mechanical deformation.

Fig. 28 is 3D AFM image of Bi2223 oxide after a heavy mechanical deformation (80% thickness reduction). The contribution of the compression force became more important. The Bi2223 crystals in the interface areas were broken into small pieces, and brittle fracture feature can be seen on most areas. Multi-sliding also took place in a number of crystals after heavy deformation as shown in Fig. 28. In these crystals, the a-b planes perhaps had a large angle relating to the rolling plane. Cracks of ~10 nm wide can be seen on the edges of the deformed crystals (marked with A in Fig. 28). Fig. 29 shows plate-like Bi2223 crystals with large angles to each other. Multi-sliding features can also be seen after deformation.

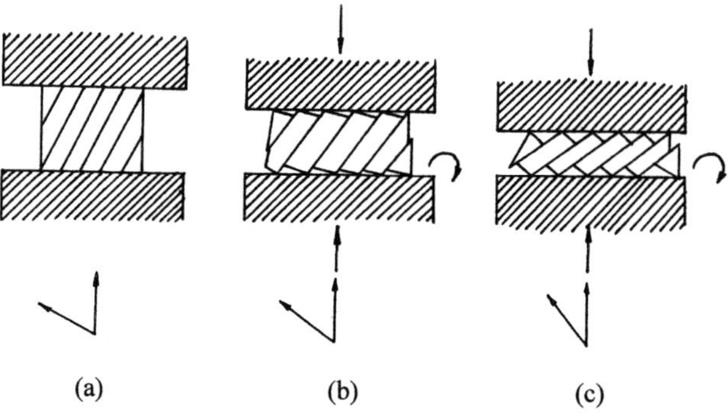

Fig. 27 Schematic drawings showing the multi-sliding fracture mechanisms for a randomly oriented Bi2223 crystal during the deformation processes.

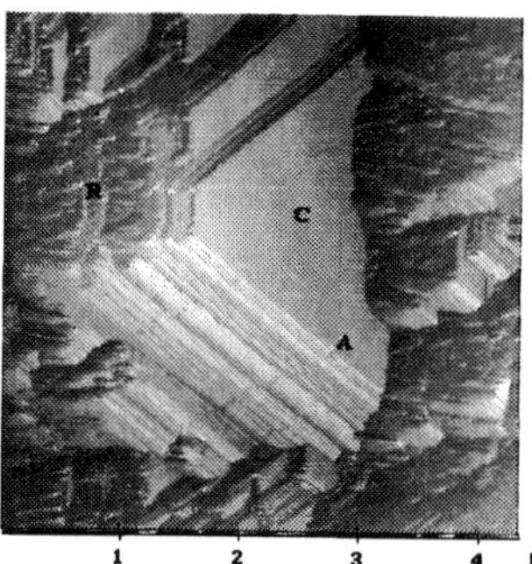

Fig. 28 AFM 3D top-view image showing the multi-sliding feature in Bi2223 crystals in a tape with 80% mechanical deformation. The a-b plane is marked with "C", the cracks are marked with "A", and the rolling plane is marked with "R".

Fig 29 AFM 3D top-view image showing the multi-sliding feature in the plate-like Bi2223 crystals with large angles relative to each other. The a-b plane is marked with "C", and the cracks are marked with "A".

After annealing at 833°C for only one hour, some heavily deformed Bi2223 crystals rapidly connected themselves and grew from ~2 μm to ~40 μm. Well-developed plate-like crystals formed after longer time annealing, as shown in Fig. 30. These results indicate that the recrystallisation and growth of the fractured crystals during annealing greatly improve the a-b plane alignment of the Bi2223 oxide.

The multi-step deformation morphology is a distinguished feature in the Bi2223 superconductors. This morphology also depends on the deformation ratio. With ~30% deformation, we see very little of this type of feature because the shear force on a-b planes was relative low and not enough to start the multi-sliding action. The multi-sliding of Bi2223 along the a-b plane can be observed with a higher deformation ratio (~50%), and increased with the increasing rolling force. Meanwhile, the high compression force also broke the Bi2223 grains into small pieces, as shown in Fig. 28.

Large amounts of microscopic observations indicated that this multi-step sliding feature resulted from mechanical deformation is quite different from the crystal growth morphology which is shown typically in Fig. 30. Firstly, the stacking plates after annealing are much thicker (in the order of 1 μm) than the sliding steps after deformation (~20 nm). Secondly, the plate-like crystals grew to different directions with some plates advancing much more than the others (Fig. 30), while the multi-steps produced by mechanical deformation always slid toward the same direction with approximately same step widths (Fig. 28). The fracture behavior of the polycrystalline Bi2223 clad with silver sheet can be summarized in Table 1.

It is clear that the fracture of Bi2223 crystals mainly takes place along their a-b planes and the fracture behaviors of Bi2223 oxide were different for low and high mechanical deformations. With a high mechanical deformation, the randomly oriented crystals in the tape cleave, slide along their a-b planes, rotate to the rolling plane direction, and form texture microstructure. Recrystallisation and grain growth during the subsequent thermal treatment produce microstructures with well-aligned plate-like crystals. With a low mechanical deformation, the dominant feature is the crystals with a-b planes parallel to the rolling plane sliding along their a-b plane. If the compressive force applied to the crystals has a component large enough along the a-b plane, multi-sliding deformation takes place.

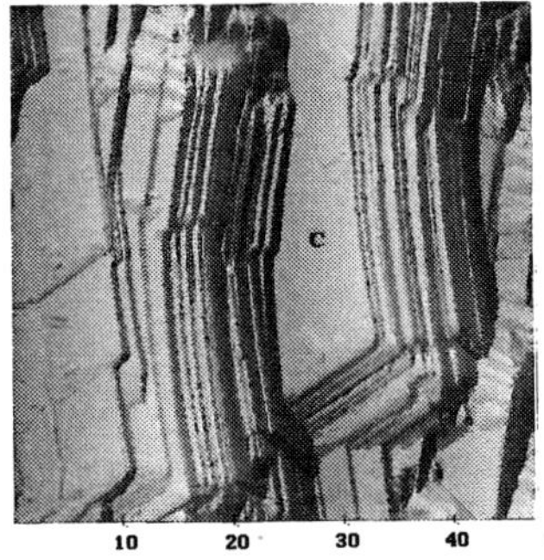

Fig. 30 AFM 3D top-view image showing the morphology of the Bi2223 crystals with 80% mechanical deformation after recrystallisation annealing at 833°C for 60 hours. The a-b plane is marked with "C".

Table 1. Summary of fracture behaviors in polycrystalline Bi2223 oxide.

Grain Orientation	a-b plane parallel to the rolling plane	a-b plane has a large angle to the rolling plane	a-b plane is approx. perpendicular to the rolling plane
Low Deformation (~30%)	The friction force from silver deformation drags the top crystals to move along the a-b plane. 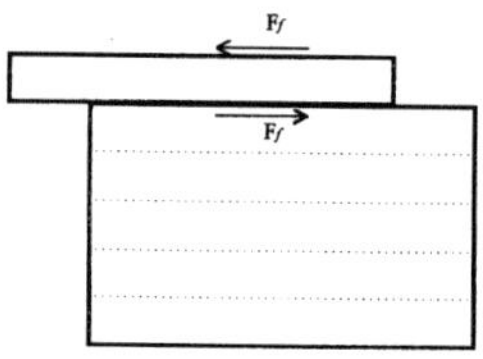	The component of friction force on the a-b plane is too small to fracture the crystals. Multi-sliding does not take place. 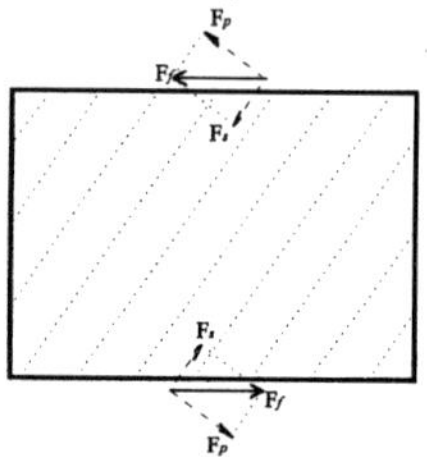	The component of friction force on the a-b planes equals zero. Fracture does not take place. 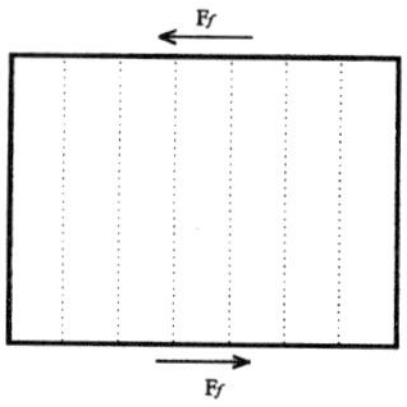
Medium Deformation (~50%)	Friction force moves the crystals along a-b plane and compression force breaks the crystals to small pieces. 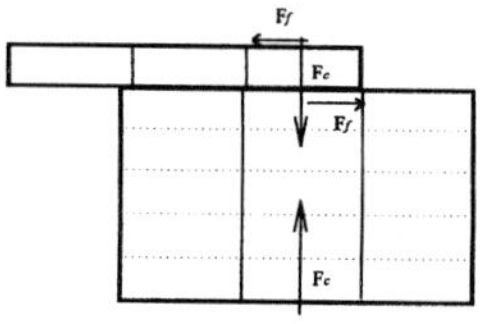	Compression force and friction force produce multi-sliding along a-b planes. 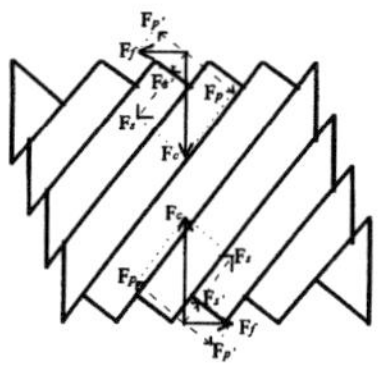	Multi-sliding along a-b planes takes place in some crystals due to compression force. 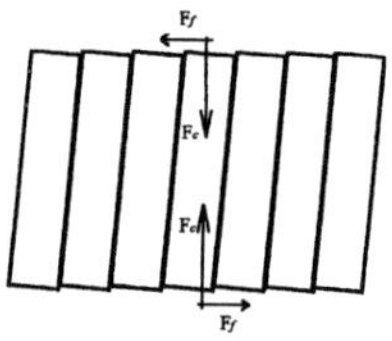
High Deformation (~80%)	Compression force breaks the crystals into small pieces. Friction force is small compared to compression force. 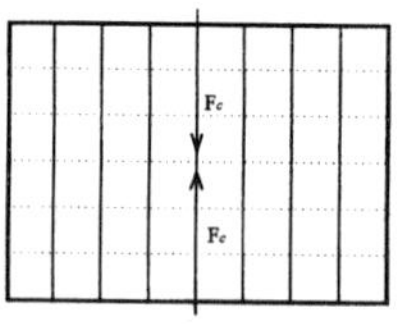	Compression force breaks the crystals, produces multi-sliding along a-b planes, and rotates a-b planes toward the rolling plane direction. 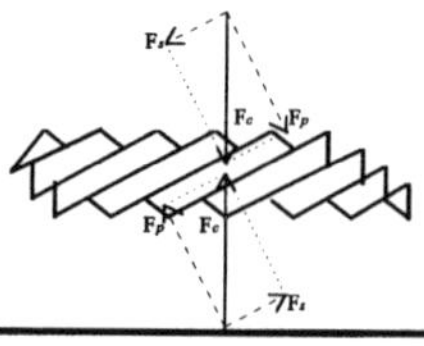	Crystals are fractured with multi-sliding action, and rotate toward the rolling plane direction. 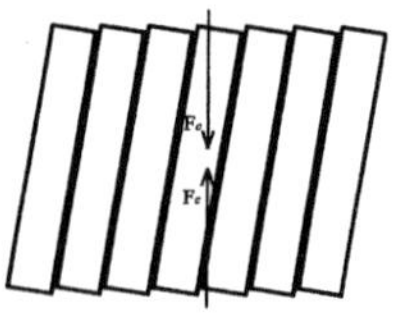

F_f — Friction force produce by silver flow
F_c — Compressive force
F_s — Shear force which is a component of F_c on a-b planes
$F_{s'}$ — Shear force which is a component of F_f on a-b planes
F_p — A component of F_c which is perpendicular to a-b planes
$F_{p'}$ — A component of F_f which is perpendicular to a-b planes

6. EFFECTS OF MECHANICAL DEFORMATION ON THE LATTICE DISTORTION AND GRAIN THICKNESS OF Bi2223 OXIDES

It has been evidenced that the textured microstructures of Bi2223 oxide in BSCCO superconductor tapes can also be greatly improved by recrystallisation treatments. The recrystallisation process is usually affected by the storage of the deformation energy in the material [42]. This energy influences the nucleation and grain growth processes during recrystallisation. Therefore, the mechanical deformation not only induces a textured microstructure by causing of the Bi2223 crystals sliding and rotation but it also plays a role in controlling the recrystallised textured microstructure.

Bi2223, as an oxide, is a brittle ceramic compound. It has been reported that the fracture toughness of bulk polycrystalline Bi2223 oxide is rather small, 0.52 ± 0.10 MPa $m^{1/2}$ [43]. Therefore, the deformation/fracture behavior of Bi2223 ceramics is certainly different from that of ductile metals. In this case, the special microstructural features of polycrystalline Bi2223 oxide under mechanical deformation need to be studied, thus establishing the relations between mechanical deformation, lattice distortion and the thickness of the plate-like crystals after deformation. The results are useful in designing and improving the mechanical deformation and heat treatment procedures in order to produce highly textured Bi2223 materials that have high critical current densities.

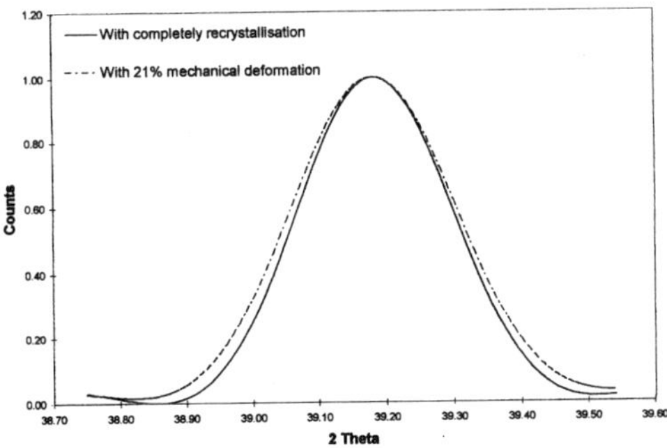

Fig. 31 The normalized $K\alpha_1$ X-ray diffraction profiles of the (*00<u>14</u>*) planes in the Bi2223 tape with a 21% thickness reduction.

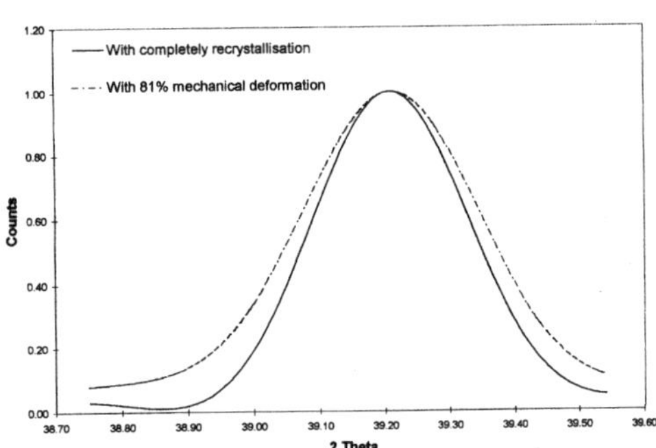

Fig. 32 The normalized $K\alpha_1$ X-ray diffraction profiles of the (*00<u>14</u>*) planes in the Bi2223 tape with an 81% thickness reduction.

Fig. 31 and 32 show the normalized $K\alpha_1$ X-ray diffraction profiles of the (*00<u>14</u>*) peak for the Bi2223 superconducting tapes with 21% and 81% thickness reductions, respectively. The two profiles were measured from the deformed (broken lines) and the annealed (solid lines) Bi2223 specimens, respectively. As the peak breadth of the deformed Bi2223 oxide is greater than that of the fully recrystallised Bi2223 oxide, the effect of mechanical deformation on the X-ray diffraction profile is obvious. The experimentally measured XRD profiles are affected by lattice distortion, crystal size, and instrumental broadening. The instrumental broadening effect can be removed by Fourier transformation. Fig. 33 shows the X-ray diffraction of the (*00<u>14</u>*) plane from the deformed tapes with 21% and 81% thickness reductions without the instrumental broadening effect. The breadth of the (*00<u>14</u>*) X-ray diffraction for the Bi2223 oxide core with 81% deformation is wider than that of the 21% deformation, indicating that the breadth of X-ray diffraction profile increases with increasing mechanical deformation.

In BSCCO thermo-mechanical treatments, the crystal grain size always decreases with increasing mechanical deformation. Figs. 18(a) and 18(b) are the surface morphology of Bi2223 samples after 21% and 81% deformation, showing that the crystals were fractured into smaller pieces when a higher mechanical deformation was applied, and that the mechanical deformation also affects the thickness of the plate-like Bi2223 crystals. Based on Fig. 31 and Fig. 18(a), we see that the broadening effect of the X-ray diffraction profile of a sample with 21% mechanical deformation is mainly caused by lattice distortion as the crystallite size varies very little after this light deformation. This gives a quantitative measurement of the broadening effect produced by lattice distortion resulting from mechanical deformation.

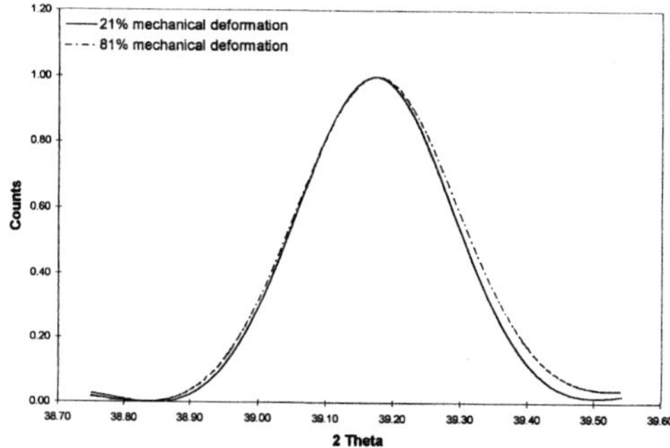

Fig. 33 The "pure" X-ray profiles of the (*00<u>14</u>*) peaks. Measured from the deformed tapes with 21% and 81% thickness reductions, without the instrumental broadening.

The lattice distortion and crystal grain size of cold rolled samples can be determined with precise X-ray diffraction and Fourier transformation [44-46]. For any diffraction of (*00l*) planes in Bi2223 oxide with distortion broadening and/or crystallite size broadening, the shape of the diffraction peak can be represented by a Fourier series:

$$P_{2\theta} = K\sum_{-\infty}^{+\infty} A_n \cos 2\pi n h_3 \tag{7}$$

where $P_{2\theta}$ is the distribution of power in unit length diffraction, A_n is the Fourier coefficient, K is a constant and $h_3 = 2a_3 \sin\theta/\lambda$. In equation (7), the Fourier coefficient A_n is a function of the (*00l*) plane index and it can be represented as $A_n(l)$. $A_n(l)$ is also the product of two independent Fourier coefficients that are related to the lattice distortion and crystalline size. Therefore, $A_n(l)$ can be expressed as:

$$\ln A_n(l) = \ln A_n^s - 2\pi^2 N_3^2 \langle \varepsilon^2 \rangle l^2 \tag{8}$$

where A_n^s is the Fourier coefficient for the crystallite size, N_3 is the average number of the cells per column, and ε is the strain in *c*-direction.

The strain ε can be obtained from the initial slope of the curve $\ln A_n(l)$ versus l^2. In order to calculate the slope, $\ln A_n(l)$ versus l^2 was plotted for $n = 1$, where n is the subscript number of the Fourier coefficient. The logarithmic Fourier coefficients $\ln A_n(l)$ versus l^2 for the Bi2223 superconducting tapes with 21% and 81% mechanical deformation are shown in Fig. 34. It shows an approximately linear relation with slopes of -0.0075 and -0.0059 and correlation coefficients, R, of 0.997 and 0.994 respectively. Extrapolating the lines of $n = 1$ to $n = 5$ for the Bi2223 superconducting tapes with 21% and 81% mechanical deformation to the $\ln A_n(l)$ axis, we have a series of intercepts $\ln A_n^S$. A plot of A_n^S versus n, allows the average number of cells per column, N_3, to be derived from the abscissa intercepts of the initial slopes of the two curves, as shown in Fig. 35. The results show that N_3 is ~67 for the tape with 21% deformation and ~7 for the tapes with 81% deformation. This means that the thickness of the plate-like crystals in Bi2223 tapes with 21% and 81% thickness reductions are assembled by approximately 67 and 7 Bi2223 unit cells along the c-axis direction, corresponding to a thickness of 248 nm and 26 nm, respectively. These results are in good agreement with those observed by AFM.

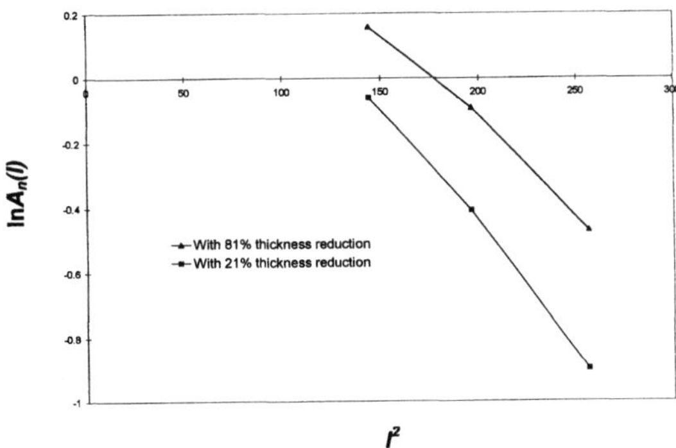

Fig. 34 The logarithmic Fourier coefficients [$\ln A_n(l)$] versus l^2 for the Bi2223 tapes with 21% and 81% thickness reductions.

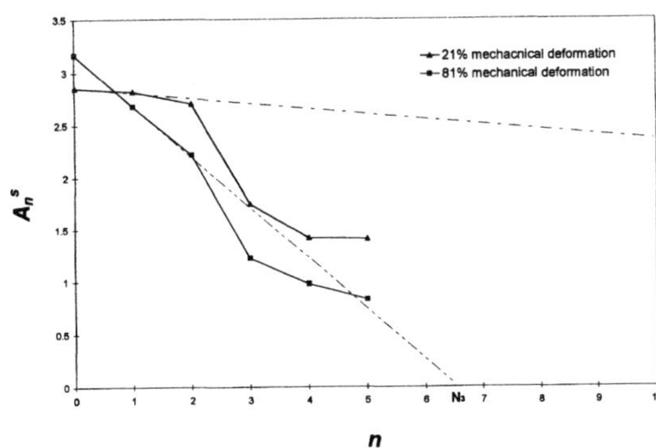

Fig. 35 Plot of the Fourier size coefficient A_n^s versus n for the Bi2223 tapes with 21% and 81% thickness reductions. The abscissa intercept of the initial slope is the mean column length number N_3.

Substituting -0.0075, 67, and -0.0059, 7 into following equation, which is deduced from the second item of equation (8):

$$\varepsilon = \left[-(slope)/\left(2\pi^2 N_3^2\right)\right]^{\frac{1}{2}} \tag{9}$$

where the slope is the gradient of the curve $\ln A_n(l)$ versus l^2 in Fig. 34, the lattice distortion can be calculated as 2.91×10^{-4} for the tape with 21% mechanical deformation and 2.47×10^{-3} for the tape with 81% deformation. This result indicates that the lattice distortion of Bi2223 crystals with high deformation is greater than those with low deformation. At the same time, with higher mechanical deformation, the thickness of the plate-like Bi2223 crystals is smaller than that with low mechanical deformation, indicating that different mechanical deformations produce not only different lattice distortions, but also different thicknesses of the plate-like Bi2223 crystals.

The lattice distortion introduced by mechanical deformation is stored in the Bi2223 crystals as deformation energy. However, the mechanical deformation not only produces lattice distortion in Bi2223 oxide, but also results in fracture of Bi2223 crystals. Fig. 36 shows a 3D AFM image, exhibiting the sliding zones produced by dislocation movement. These sliding planes (marked A) are perpendicular to the Bi2223 oxide columns (marked with B) fractured by the compressive force that is parallel to the *c*-axis of Bi2223 oxides. Fig. 37 shows several undulant Bi2223 oxide columns (marked with B), which resulted from cold deformation. The deformation mechanism can be explained by the analysis of the stress status as follows.

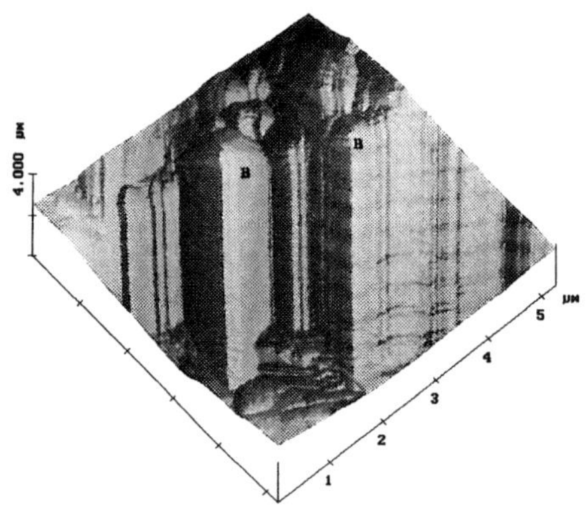

Fig. 36 AFM 3D image showing that the sliding zones were formed in the Bi2223 oxide columns after mechanical deformation.

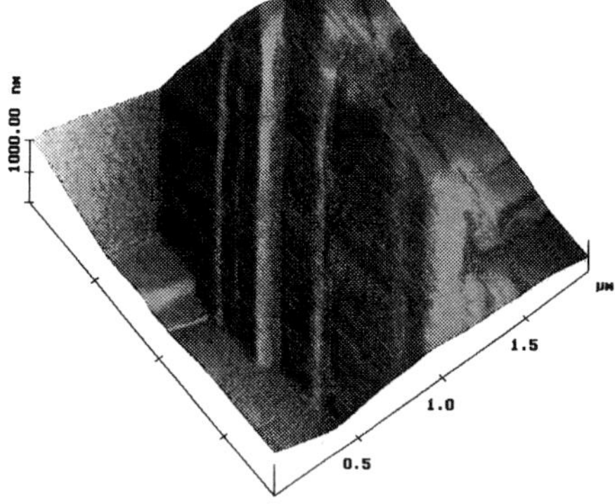

Fig. 37 AFM 3D image showing the undulant deformation zones formed in the Bi2223 crystals.

Fig. 38 is a schematic drawing of the analysis of the stress status in a crystal, showing that any compressive force loaded by the rollers can be divided into two components. The first component, F_s, is parallel to the *(00l)* planes of the Bi2223 oxide. The other component, F_p, is perpendicular to the *(00l)* planes. F_s is the shear force that makes the crystal cleave along *(00l)* plane, and F_p is the compressive force that breaks the plate-like Bi2223 crystals into columns. As the (*00l*) planes rotate towards the rolling plane direction, F_p increases to a maximum magnitude that equals the force applied by the rollers, while F_s decreases to zero.

For the crystals with (*00l*) orientation close to the rolling plane direction, the shear force component, F_s, is too small to start sliding along the (*00l*) planes even under a high deformation, but the compressive force component, F_p, is large enough to fracture the crystals into small columns along the (*100*) and (*010*) planes, as shown in Fig. 36. At the same time, high densities of dislocation are generated by the mechanical deformation. Although F_s is not large enough to cleave the Bi2223 crystals, it may move the dislocations in the crystals. Thus the movement of the dislocations produced the curve-shaped sliding zones observed in the Bi2223 crystals. Fig. 36 shows that these sliding zones are perpendicular to the (*100*) and (*010*) planes, indicating that dislocations slide along the a-b planes.

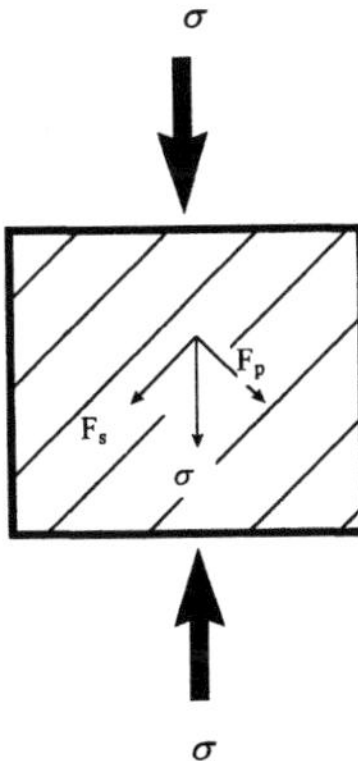

Fig. 38 Schematic drawing of the analysis of the stress status in Bi2223 crystals.

Although local micro-deformation was observed in the form of multi-step sliding, Bi2223 oxide is basically a brittle ceramic material. The energy obtained from the mechanical deformation process is mainly consumed in the fracturing of Bi2223 crystals. This means that only a small part of the energy can be stored in the Bi2223 crystals as the result of lattice distortion. Therefore, it is reasonable to believe that the energy stored in the oxide crystals is not a controlling factor to the recrystallisation kinetics, as is the case in metals. As shown in Fig. 22, the grain size of the recrystallised Bi2223 after annealing decreases linearly with increasing mechanical deformation. This relationship does not agree with the traditional theory of recrystallisation in metals, where the grain size decreases exponentially with increasing mechanical deformation. In metals, the recrystallisation grain size is determined by nucleation and grain growth processes that are in turn determined by the strain energy stored in the metal. The final gain size of Bi2223 is mainly determined by the grain size after deformation. Research was carried out to understand the role of lattice distortion in the recrystallisation processes of Bi2223 oxide.

7. TEXTURING KINETICS AND GRAIN GROWTH BEHAVIOR OF THE RECRYSTALLIZED Bi2223 OXIDE

The grain alignment of Bi2223 oxides in the BSCCO superconductor tapes is facilitated by the thermal treatment after deformation. As the description above, two types of processes occurred during annealing after mechanical deformation: (1) the phase transformation sintering in which some or all Bi2212 transformed into Bi2223, and (2) the recrystallisation annealing in which the deformed Bi2223 grains simply recrystallised. As the discussion in section 4.1, the phase transformation sintering is more effective in promoting grain alignment than the recrystallisation annealing.

For the recrystallisation texturing processes, there have been arguments concerning its physical origin. Opinions have been divided into two groups: one is the "oriented nucleation" hypothesis and the other is the "selective oriented growth" hypothesis [47]. There were some evidences of preferential grain growth along certain orientations on the surface of Bi2223 tapes clad with silver sheets as silver promotes the growth along the *(00l)* planes [39]. However, little work has been reported on the texturing and grain growth mechanisms of the deformed Bi2223 superconductor. Understanding these mechanisms is important in controlling the microstructure and the texture formation in the BSCCO high temperature superconductors.

In order to investigate the recrystallisation behaviour of Bi2223 superconducting oxide and to avoid other possible effects, pure Bi2223 phase was prepared. Mechanical deformation was performed by rolling these sintered wires into tapes with 80% thickness reduction by several rolling passes. The deformed specimens was annealed at 833°C in air for different duration, such as 1, 4, 12, 24, 60, and 100 hrs, and then quenched to the room temperature in air after annealing.

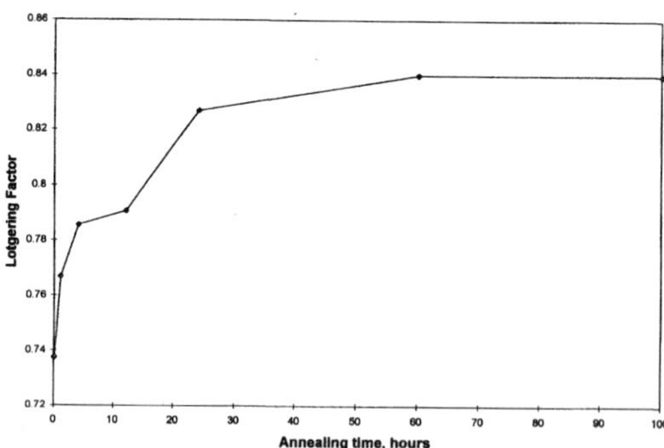

Fig. 39 The Lotgering Factor versus annealing time to show the grain alignment on the surface of the Bi2223 tapes with heavy mechanical deformation.

Fig. 39 plots the Lotgering factors of the grain alignment on the surface of a Bi2223 sample versus annealing time. The specimen had 80% thickness reduction before the annealing. It shows that the grain alignment increases with increasing annealing time. The kinetic curve also shows that the formation of the texture structure in the Bi2223 can be divided into two stages. In the first stage, the degree of grain alignment increased sharply with annealing time from 0 to 4 hrs, and then slowed down from 4 hrs to 12 hrs. In the second stage, the rate of grain alignment increased again between 12 to 24 hrs, and slowed down afterwards. These types of texturing kinetics implied that there were two different grain alignment mechanisms during the whole annealing process.

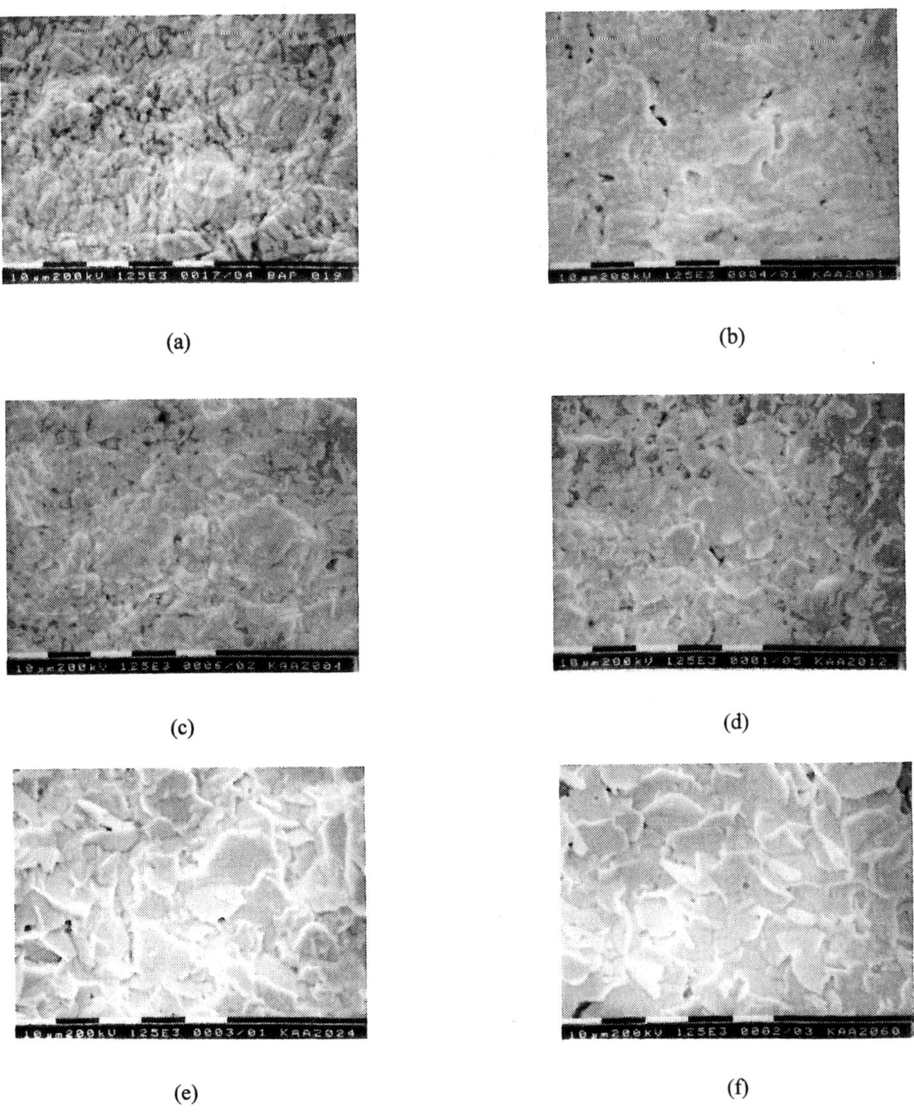

Fig. 40 SEM images showing the surface morphology of the deformed Bi2223 superconductor: (a) without annealing, (b) with 1 hour annealing, (c) with 4 hours annealing, (d) with 12 hours annealing, (e) with 24 hours annealing, and (f) with 60 hours annealing.

Fig. 40 shows the surface morphologies of the Bi2223 tapes after 80% deformation without annealing and with annealing for different periods of time. In Fig. 40(a), it can be seen that the Bi2223 crystals were heavily broken by the mechanical deformation. After annealing for only one hour, however, the sharp Bi2223 crystal edges became blunt and many fractured grains were reconnected, as shown in Fig. 40(b). This reflects the fast texturing process during the first stage of annealing. Figs. 40(c) and 40(d) shows the morphologies of the Bi2223 samples after 4 and 12 hrs annealing, respectively. It can be seen that the fractured grains were being connected and a few grains that are paralleled to the rolling plane grew much larger than before (~15 μm). It is believed that the texturing of Bi2223 is mainly contributed by reconnection of textured grains at this stage (0 – 12 hrs annealing). There was more grain growth during further annealing as shown in Figs. 40(e) and 40(f). The average grain size after 24 hrs annealing was much larger than after 12 hrs annealing, but some small grains could still be seen. After 60 hrs annealing, the growth of Bi2223 grains seemed to have reached saturation. The small grains formed by mechanical deformation disappeared. The microstructures of the Bi2223 were consistent with the texturing kinetics shown in Fig. 39, evidence of the different texturing mechanisms during the early and late stages of annealing

The two stages of texturing processes can be mathematically fitted as shown in Fig. 41. It can be expressed as:

$$F = 0.5a\left[1 + erf\left(\frac{t-b}{\sqrt{2}c}\right)\right] \quad (10)$$

where F is the Lotgering factor and t is annealing time in minutes. a, b and c are constants and equal to 0.789, -233.79 and 154.21, respectively, with correlation coefficient $r^2 = 0.994$. The error function in Equation (10) implies that the texturing process in this stage was controlled by diffusion processes. This is reasonable because the reconnection of the fractured crystals is likely to be controlled by mass transfer (diffusion). Therefore, it is proposed that the texturing processes in the first stage of annealing was dominated by reconnection of the fractured grains.

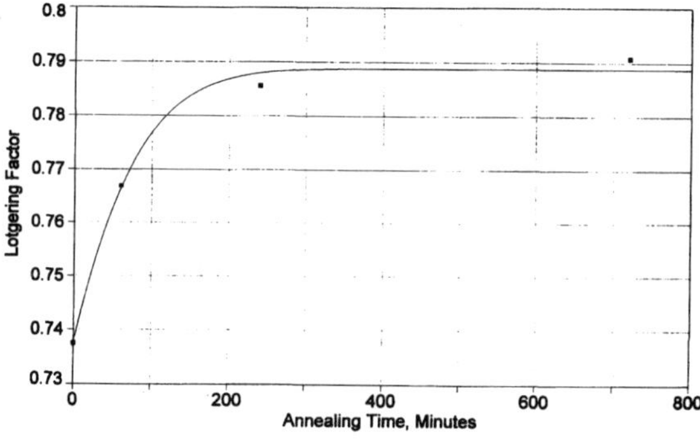

Fig. 41 Lotgering Factor versus annealing time in the first stage with curve fitting, showing the kinetics following an error function.

As the deception above, in ceramic deformation processes, almost all mechanical deformation energy was transferred to the fracture energy, which produced the small crystals in the Bi2223 tapes after deformation. The storage of deformation energy in the Bi2223 oxide lattice was very limited. Therefore, the free-surface energy in the Bi2223 oxides increased significantly after deformation. The recrystallisation processes of the deformed Bi2223 oxides were different from that of the deformed metals. The main driving force of recrystallisation in the deformed superconducting oxides was the reduction of the free-surface energy. In the first stage of annealing, the high surface energy held by the fractured crystals was quickly released, facilitating the reconnection of the textured grains. Due to the constraint of the silver sheets on the surface, only those grains with *(00l)* orientations can be well connected and grow [39]. Therefore, the texture degree increased rapidly in this stage, as shown in Fig. 39. Fig. 42 shows an AFM micrograph of a Bi2223 sample after mechanical deformation and 4 hrs annealing. It can be seen that some grains were reconnected (marked with A), consuming the fractured grains between them (marked with B). The reconnection processes were slowed down when most of the fractured grains were consumed, corresponding to the period of 4-12 hrs annealing as shown in Fig. 39.

Fig. 42 AFM image showing the growing crystals (marked with A) consuming the fractured grains (marked with B) nearby.

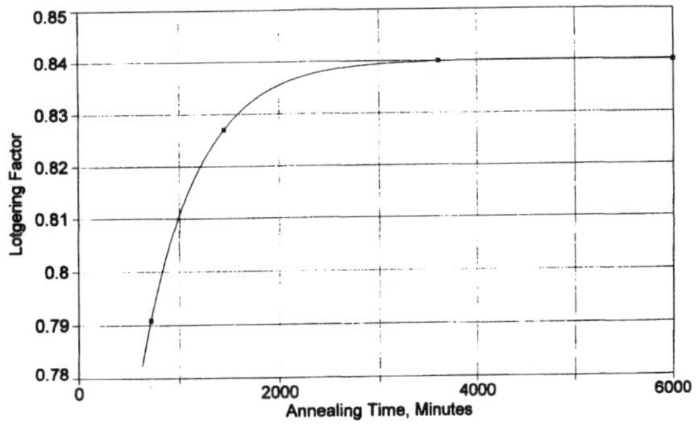

0.85
0.84
0.83
0.82
0.81
0.8
0.79
0.78
Lotgering Factor
0
2000
4000
6000
Annealing Time, Minutes

Fig. 43 Lotgering Factor versus annealing time during the second annealing stage, with curve fitting, showing the kinetics representing a recrystallisation process.

Fig. 43 plots the second part of the annealing processes (12 – 100 hrs). The texturing kinetics during this stage can be very well fitted with Equation (11), with a correlation coefficient $r^2 = 0.9999$.

$$F = 0.84\,[\,1 - 0.22\,exp\,(-1.84 \times 10^{-3}\,t)] \tag{11}$$

Equation (11) represents the late stage of typical recrystallisation kinetics for isothermal recrystallisation processes [47]. Typical recrystallisation processes include recovery, nucleation and growth. The late stage of recrystallisation is dominated by grain growth processes. The decrease in the recrystallisation kinetics is caused by the reduction of the reaction surface. The late stage of the texturing processes followed the same model as the recrystallisation processes. Therefore, it is considered that the texturing process of Bi2223 in this stage was controlled by the grain growth of the Bi2223 crystals. The grain size increased from ~4 μm after 12 hrs annealing to 15 – 20 μm after 60 hrs annealing as shown in Fig. 40, evidence of grain growth.

From microstructural observation, it was found that there were two growth mechanisms of Bi2223 crystals during the recrystallisation annealing. The first type can be called "ledge-growth", which corresponds to the grain growth along the *a-b* plane. The second type can be called "screw dislocation-growth", which corresponds to the growth along the *c* direction. Fig. 44 shows a flat Bi2223 crystal with ledges (marked with L) on the surface. Atoms are added to the ledge, thus the a-b planes propagate forward by lateral motion of the ledge. Fig. 45 shows a Bi2223 crystal growing with the "screw-dislocation-growth" mechanism. Fig. 46 shows the morphology of a well-grown crystal with the "ledge-growth" model. The ledge steps can still be seen on one side of the crystal. The mechanism of "ledge-growth" was schematically shown in Fig. 47. In fact, because the ledge and the underneath Bi2223 oxide have identical crystal structure and orientation, there was no difficulty in adding atoms onto the growing ledges. The growth rate depends on the crystal plane concerned, and the shape of the crystals is decided by the growth rates along different orientations (Fig. 48). A 3D AFM image shows the morphology of a few completely recrystallised crystals in Fig. 49.

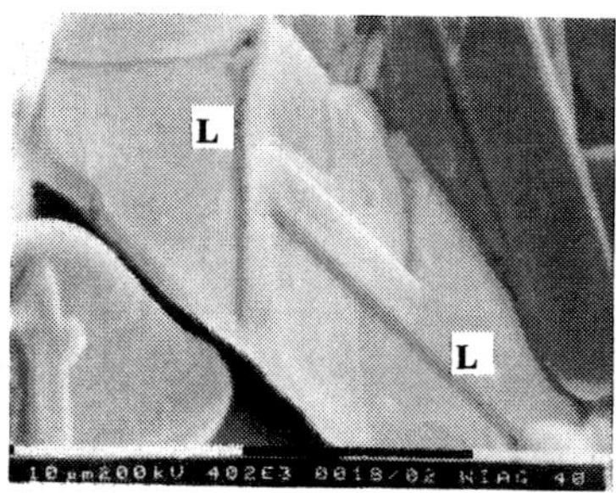

Fig. 44 A SEM image showing Bi2223 crystals growing with the "ledge-growth" mechanism.

Fig. 45 A SEM image showing a Bi2223 crystal growing with the "screw dislocation-growth" mechanism.

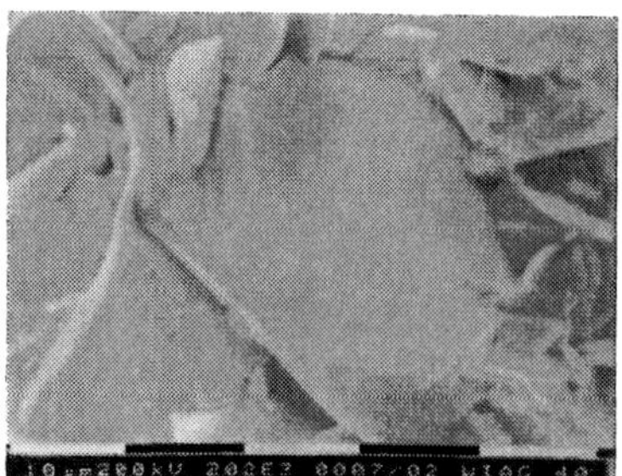

Fig. 46 A SEM image showing a completely grown Bi2223 crystal.

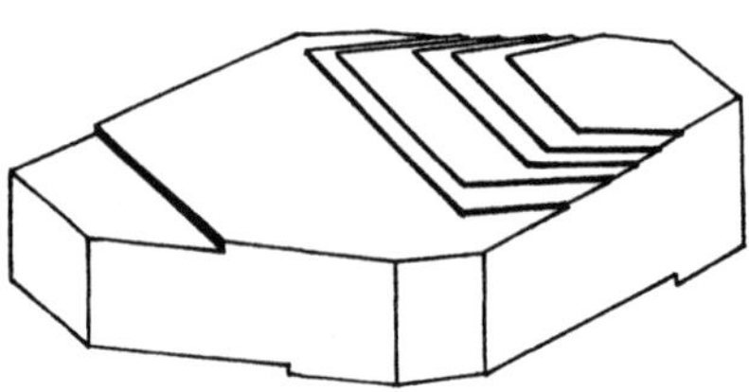

Fig 47 A schematic drawing showing the ledge-growth mechanism

The screw dislocation-growth of the deformed Bi2223 crystals was shown in Fig. 45 with a schematic drawing in Fig. 50. The step generated by each dislocation moves one plane along the *c* direction each time. The plate-like crystals grow thicker in this way. These two different growth mechanisms can also explain why Bi2223 crystals have the preferential growth direction along the *a-b* plane, producing the distinguished plate-like crystals.

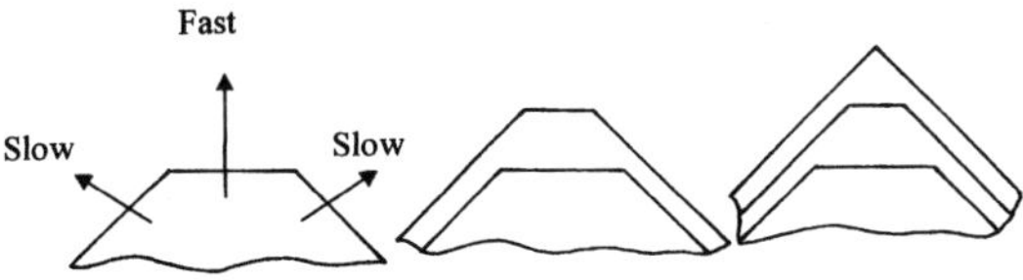

Fig. 48 A schematic drawing showing that the growth rates are different in different directions.

Fig. 49 AFM 3-D image showing the morphology of multiple layer plate-like crystals after 60 hours annealing.

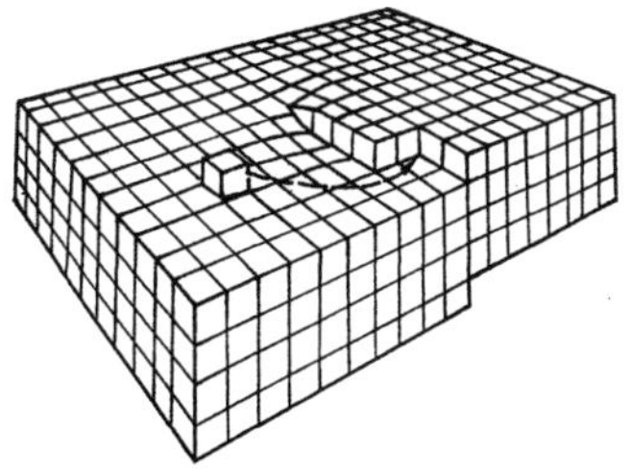

Fig. 50 A schematic drawing showing the screw dislocation-growth mechanism of Bi2223 oxide.

It can be summarized that the texturing processes in the recrystallization annealing consist of two stages - reconnection of the broken crystals and recrystallisation/growth of the crystals. In the first stage, the degree of grain alignment increased rapidly with increasing annealing time. This can be attributed to the rapid reconnection of the fractured Bi2223 crystals, which were laid roughly parallel to the rolling plane. In the second stage, grain growth dominated the processes, and the crystals in which the *a-b* planes were parallel to the rolling plane had the best opportunity to grow as the silver sheath restricted the growth of grains with other orientations. The recrystallisation of Bi2223 oxide was completed in ~60 hrs. There were two growth mechanisms of Bi2223 crystals during the recrystallisation annealing. One is the "ledge-growth" and the other is the "screw dislocation-growth". They represent the grain growth along the *a-b* plane direction and the *c* direction, respectively. These growth mechanisms resulted in the plate-like grains in the Bi2223 superconductor specimens.

8. CHARACTERISTICS OF MICRO-TEXTURE AND MESO-TEXTURE IN Bi2223 SUPERCONDUCTING TAPES SUBJECTED TO MECHANICAL DEFORMATION AND PHASE TRANSFORMATION.

It was found that J_c in the PIT processed tapes is highest at the edges of the tapes and lowest near at the center. This trend in J_c presents a parabolic relationship with respect to the transverse cross-section of the tape. It has been proposed that this relation strongly depends on the local crystallographic orientation distribution of the Bi2223 oxides [8, 11, 48]. However, the local three-dimensional crystallographic orientation distribution of Bi2223 crystals in the superconductor tapes has not been experimentally determined and theoretically analyzed. Understanding the relationship between the current paths and the crystallographic orientation resulted by mechanical deformation and phase transformation is one of the key issues in improving the J_c for power applications.

Several techniques, such as scanning electron microscopy (SEM), transmission electron microscopy (TEM), X-ray diffraction (XRD) etc. have been employed to evaluate the texture in Bi2223 superconductor tapes fabricated by PIT technique. However, SEM morphologies can only give the qualitative morphologies of the plate-like Bi2223 crystal distribution in the tape [49]. Although TEM can quantitatively describe the misorientation of the adjacent grains, the information is from a very limited area and cannot represent the texture behavior in a bulk specimen. XRD pole figure and Lotgering factor are generally used to determine the texturing behavior of bulk materials, but they only represent statistical information from a certain thickness underneath the material surface [34, 50]. It cannot express the local texture distributions in Bi2223 superconductor tapes. Therefore, the principle disadvantage of the traditional approach mentioned above is the lack of a direct link between the microstructure and crystallography in Bi2223 superconductor tapes.

Electron backscattered diffraction (EBSD) technique is a revolutionizing texture analysis method and has generated a whole new area of science named "microtexture" – the conjoining of microstructure with crystallographic, or micro-texture analysis [51, 52]. Its application in the texture distribution measurement of Bi2223 oxides in the tapes may reveal the relationships of the processing parameters, the microstructure (crystallographic orientation) and the current paths.

Fig. 51 shows maps of the crystallographic orientation distributions on the surface of a Bi2223 superconductor tape obtained by EBSD. It is represented by the combination of semi-transparent color orientation pixels and the polished surface morphology. In this case, the maps cover a region of 7038 μm^2 involving approximately 70 grains. Therefore, it can be considered that the results represent the local surface texturing behavior of the bulk Bi2223 superconductor.

Fig. 51(a) presents a map of Bi2223 grain orientation relative to the sample normal of the Bi2223 superconductor tape. The preponderance of red color pixels indicates that most of the Bi2223 grains in the measured area have their [001] axes nearly perpendicular to the rolling plane. Analysis of the orientation distribution reveals that 98% of the Bi2223 grains in the measured area have their [001] axes deviating from the sample normal in 5 degrees [Fig. 51(a)].

Fig. 51(b) shows the crystallographic orientation maps, with respect to the longitudinal direction, of the Bi2223 grains in the same measurement area as shown in Fig. 51(a). It shows that the [100] and [110] crystal orientations of Bi2223 dominate the surface with respect to the longitudinal direction. It indicates that 45% of the Bi2223 grains have their *a*-axes aligned closed to the rolling direction while an additional 45% of the Bi2223 grains are orientated closed to 45° with respect to the rolling direction. The latter grains therefore have an [110] orientation relative to the rolling direction after the mechanical deformation and thermal treatment. It is important to note however, that these grains have their *c*-axes, [001], aligned to within 5° of the normal of the rolling plane. The large amount of [110] alignment may be due to the intrinsic crystallographic relationship between Bi2223 and Bi2212. Since the properties of Bi2223 in the *a* and *b* axes are isotropic, the result in Fig. 51(c) showing the orientation distribution of the Bi2223 oxides along the transverse direction is the same as in Fig. 51(b). Therefore, it can be concluded that the surface of Bi2223 superconductor tapes mainly consists of $Bi2223_{a/b\text{-}axes}$ and $Bi2223_{[110]}$ orientations, in approximately equal proportions, with their *c*-axes ([001] orientation) perpendicular to the rolling plane.

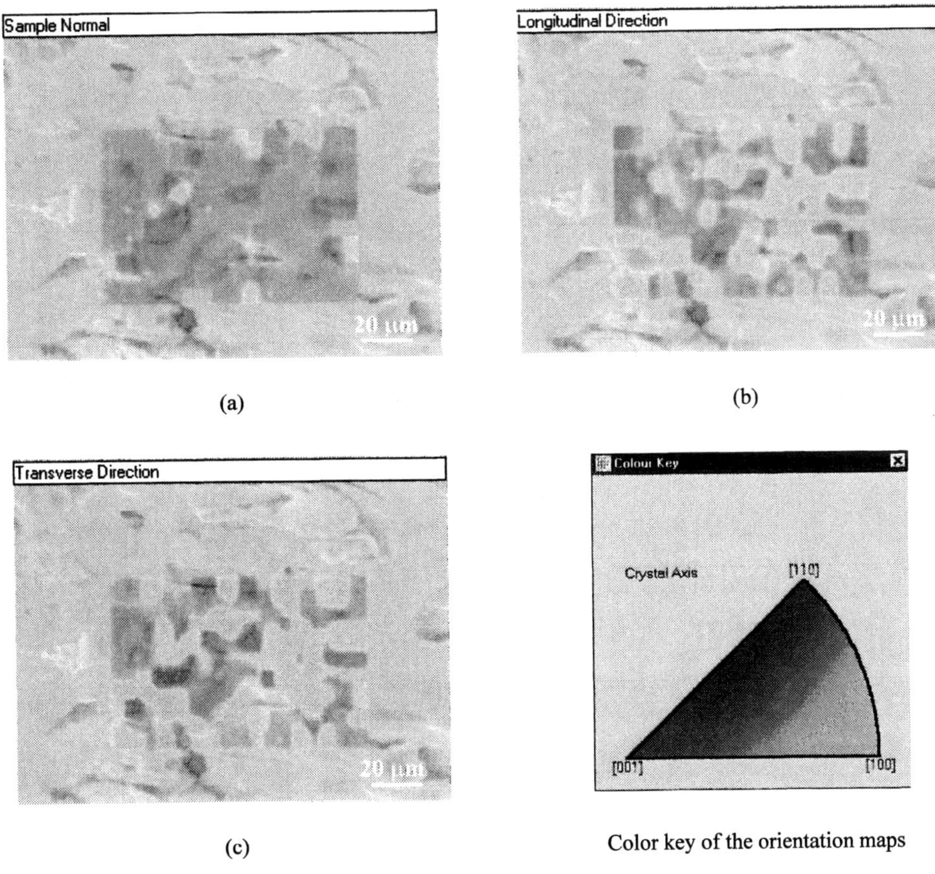

(a) (b)

(c) Color key of the orientation maps

Fig. 51 SEM micrograph and EBSD showing the orientation map in (a) the normal, (b) the longitudinal direction, and (c) the traverse direction of Bi2223 superconductor tapes.

Fig. 52 shows that the number of the grain boundaries versus the degrees of the grain boundary misorientations. It indicates that the low angle (0°<θ<7°) grain boundaries and 45° boundaries are frequently observed at this measured area of the sample. As the structure of Bi2223 crystals is tetragonal (a = b), the misorientation of a/b axes at 90° can be considered as grain boundaries without misorientation (θ = 0°). From Fig. 52, it can be observed that the grain boundaries with low-angle and 45° misorientation dominates at this measured area of the tapes. Fig. 53 shows the distribution map of the grain boundary misorientation at the surface of Bi2223 superconductor tapes and the color key is indicated in Fig. 52. It indicates that the grain boundaries with 45° misorientation are distributed in the middle zone of the measured area, along the rolling direction, while the low angle grain boundaries are gathered at the side zones of the measured area along the rolling direction.

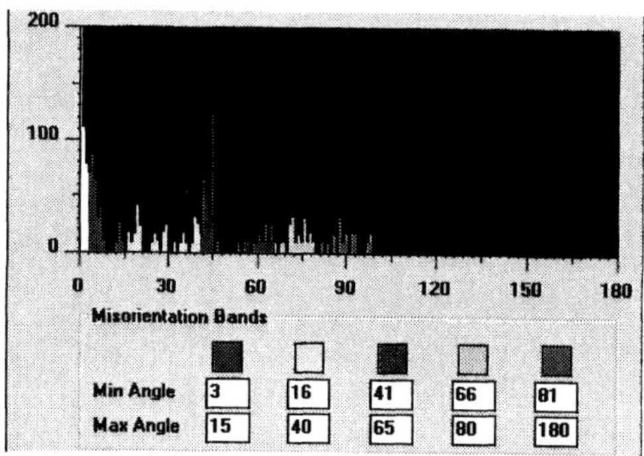

Fig. 52 A chart showing that the number of the grain boundaries versus the degree of the grain boundary misorientation.

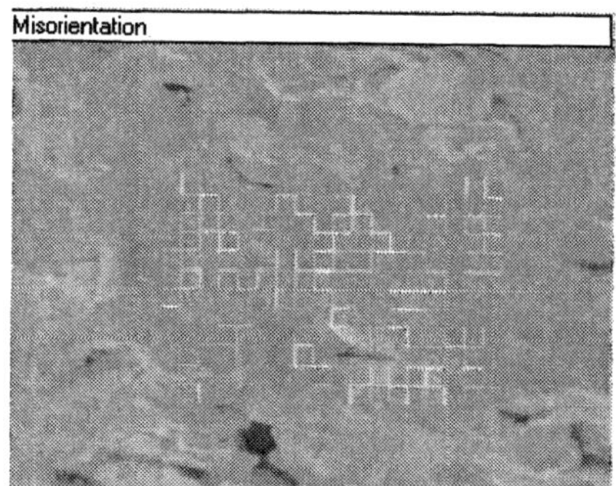

Fig. 53 SEM micrograph and EBSD showing the misorientation map of the Bi2223 grain boundaries.

Based on the analysis and the characteristics of the crystallographic orientation distribution, it was proposed that the a/b axes misorientation of the Bi2223 crystals on the surface might be the barrier of the current path in the Bi2223 superconductor tapes. However, the transverse microtxture and mesotexture distribution characteristics on the core surface of the Bi2223 tapes need to be clarified qualitatively.

The Rodrigues-Frank (RF) vectors, which reduce the four-parameter angle-axis pair expression for misorientation to three parameters suitable for plotting on a graph, is used for the quantitative representation of the texture in the Bi2223 tapes [53]. The vectors R1, R2 and R3 represents the product of the tangent of half rotation angle with ***u***, ***v*** and ***w*** coordinates of rotation axis, respectively [54]. The plot of RF vectors for the microtexture, which is a measure of individual grain orientation with respect to a reference frame, for the sample center is shown in Fig. 54. It shows that majority of the data points are distributed within ~±0.4 along the R3 axis while that along R1 and R2 are within ~±0.1. The distribution of RF vectors along R1 and R2 directions indicated that the basal planes of the majority grains at the center of tape are inclined with the angles less than 10° from the rolling plane. On the other hand, the distribution of RF vectors along R3 indicates that the grains have a/b axes oriented at up to 45° with respect to the rolling direction. The microtexture RF plot for the sample sides is shown in Fig. 55. From the plot, it can be observed that the data points with larger R3 values also have larger R1 and/or R2 values. This indicates that the platelet grains along the sample sides with the largest inclination angle to the rolling plane had the largest a/b axes misorientations.

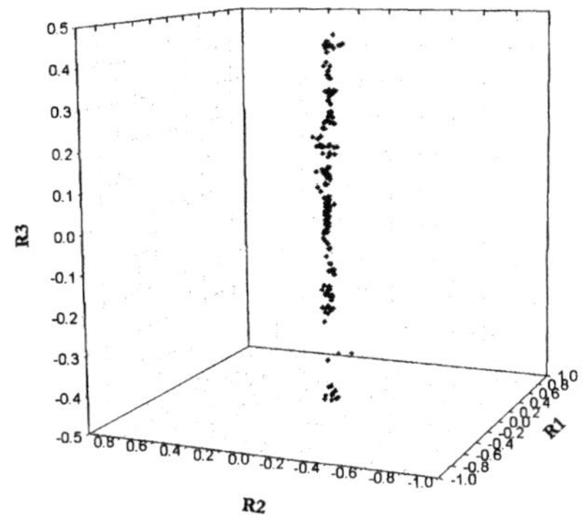

Fig. 54 Graph of microtexture RF vectors for sample center.

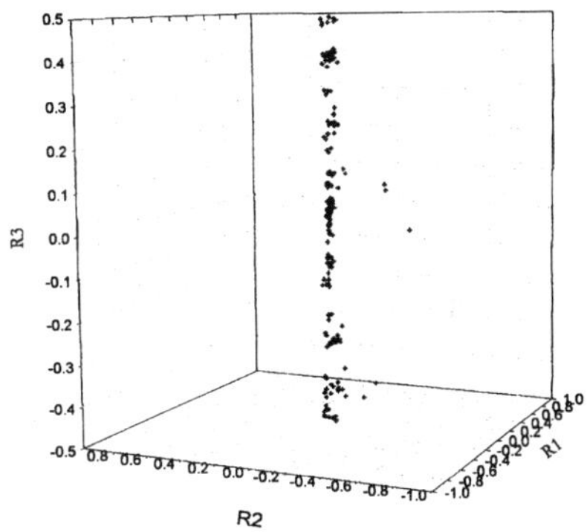

Fig. 55 Graph of microtexture RF vectors for sample sides.

The statistical results for the distribution of the orientation angle in sample center and sides, obtained from a few hundred individual areas in the tapes are plotted in Fig. 56. It shows that the peak distribution in orientation angle in the sample center occurred at 10°, 30° and 45°. In the sample sides, the peak distributions only occurred at 10° and 45°. The narrow distribution along R1 and R2 in Figs. 54 and 55 justifies such distribution plot. Through the computer simulation, it was suggested that a sample with randomly oriented grains would have a mono-peak distribution pattern with higher values in the larger angle region [55]. Comparing with the simulations, the results in this experiment show that biaxial texture exist in both of the sample center and side.

The RF plot for mesotexture, which is a measure of misorientation between grains, for the sample center and sides are shown in Figs. 57 and 58. Both exhibited that the vectors along R1 and R2 axes lie approximately between ±0.2 while the range in R3 is slightly higher at ±0.4. It shows that the range of misorientation in *c*-axis was large (±45°) while that in a/b axis was much smaller (±12°). It can be observed in both center and sides of the sample that there is clustering within the narrow region between ±0.1 R1 and ±0.1 R2 on the R1/R2 plane. This suggests that large quantity of low-angle mixed-axis tilt and twist boundaries within ±12° existed.

The rotation axis, in angle-axis pair representation of misorientation, is usually expressed in a unit vector with unit length. Fig. 59 shows a plot of the fraction of grain boundaries against the projection length of the rotation axes on the $(\boldsymbol{u}, \boldsymbol{v})$ plane. Unit projection length of $(\boldsymbol{u}, \boldsymbol{v})$ vectors indicates that the rotation axes lay completely on the $\boldsymbol{u}, \boldsymbol{v}$ plane while the rotation axis is parallel to *c*-axis when the length of (***u***, ***v***) equaled to zero. The plot also shows that approximately 70% of the grains had their rotation axes lying close to (***u***, ***v***) plane. This indicates that a greater proportion of the grains had rotation axes parallel to the R1 and R2 planes, suggesting that majority of the contiguous grains did not have parallel *c*-axis. More than 50% of grain boundaries, for both center and sides of the sample, had rotation axes within 3° from the (***u***, ***v***) plane. Their distributions were obtained for a better understanding of the misorientation characteristics.

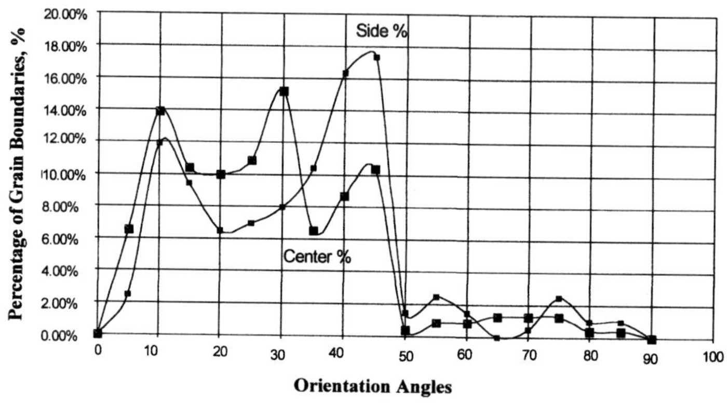

Fig. 56 The plots showing the orientation angle distribution in the Bi2223 superconductor tapes

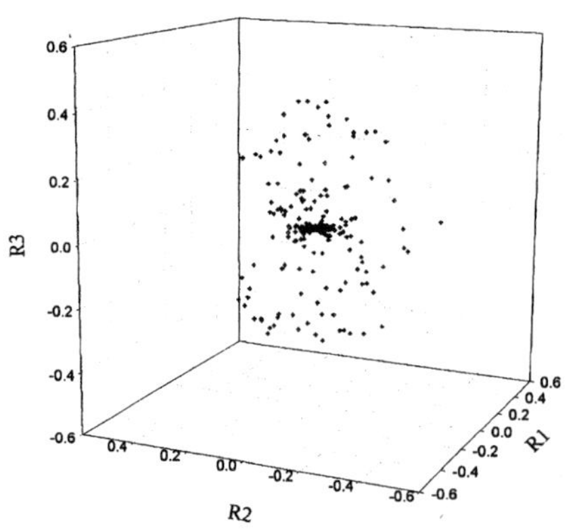

Fig. 57 Graph of Mesotexture RF Vectors for Sample Center

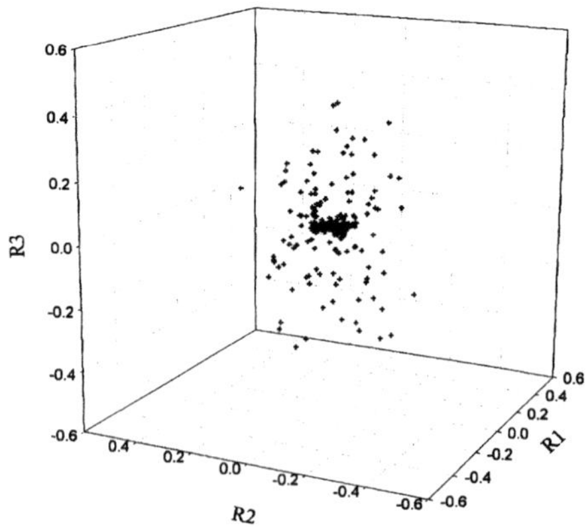

Fig. 58 Graph of Mesotexture RF Vectors for Sample Side

Fig. 60 shows the distribution of misorientation angles for the mixed-axis boundaries with rotation axis within 3° from the (***u***, ***v***) plane, both in the center and sides of the sample. It can be observed that the distribution of grain boundary misorientation peaked at 4°. This plot indicates that up to 80% of the mixed-axis grain boundaries had misorientation less than 10°. This result is in agreement with that obtained by the other research group [56], which indicated that about 40% of the boundaries were of CCSL Σ1 type. The grains analyzed may consist of a colony of sub-grains frequently observed in TEM [57]. These sub-grains are usually made up of Bi2223 platelet with sub-micron thickness connected by [001] twist boundaries. However, only the surface grains would contribute to the EBSD results since the pattern is formed from atoms not more than ~50 nm in depth. In addition, the Bi2223 crystals with *c*-axis oriented almost parallel to the surface of the tape would not contribute to the result, as the pixel size of 3 μm is larger than the average thickness of the platelet crystals.

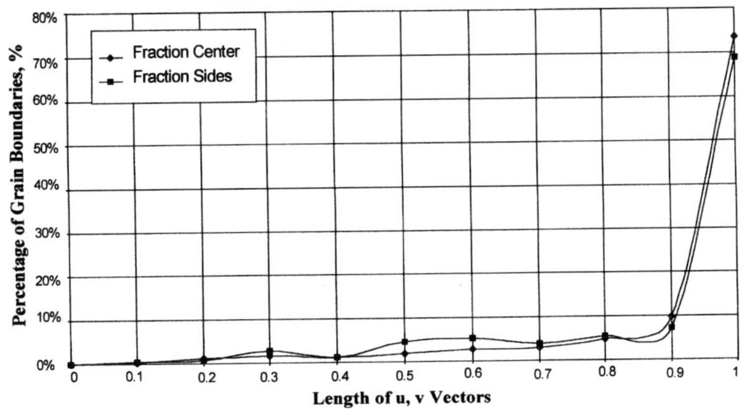

Fig. 59 The curves showing that fraction of grain boundaries against projected (u, v) length

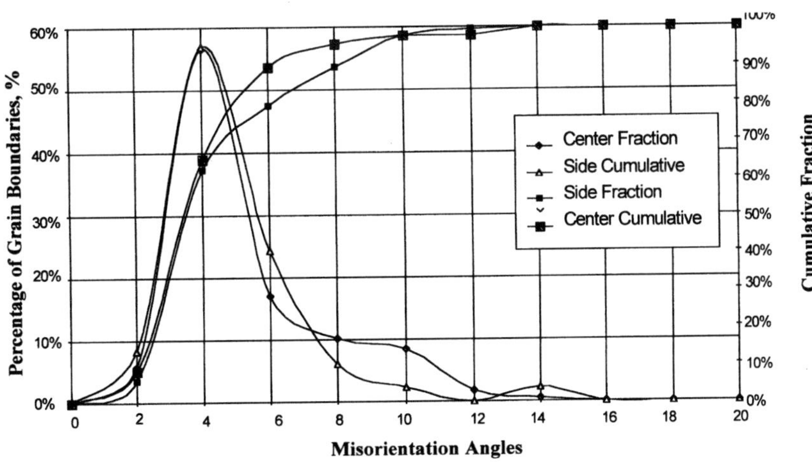

Fig. 60 The curves showing the misorientation angle distribution for both sample center and sides

It was reported that the J_c of PIT processed Bi2223 tapes is higher along the sides of the tape compared to the center where the J_c value correspond to the J_c of the bulk tapes [9,11]. There are a few mechanisms that can explain this. Experiments demonstrated that the presence of crack running perpendicular to the current path is one of mechanisms to lower the J_c [58-60] and the presence of intermittent high misorientation angle boundaries also acts as the bottlenecks to current flow [11, 59]. From Fig. 60, it can be observed that the fraction of higher angle boundaries (~10°) in tape center was slightly higher than that in the tape sides. This provides evidence that the J_c in the tape center is lower than that in the sides because of the inferior texture. However, the bulk J_c of the tape corresponds to the J_c of the tape center can be attributed to the presence of the intermittent cracks that disrupted current path in the sides.

From the texture measurement, the presence of a relatively large fraction of low-angle grain boundaries suggests that percolative current path exist in the sample. However, the low angle boundaries may not be the only current paths in the Bi2223 tapes as there were only 50% of low angle boundaries in the total boundaries analyzed. This would further reinforce that the intermittent occurrence of high angle boundaries, resulting in lower J_c attain. Higher J_c in the sides of the PIT tape could be due to the lower density of the higher angle boundaries.

Above results also indicate that the majority of the contiguous crystals did not have parallel basal plane as majority of the grain boundaries are of the mixed tilt and twist types of boundaries. From these, we believe that current flows along the basal planes of the crystals, in a manner similar to that described in "railway-switch" model [61]. It should be noted that the Bi2223 crystal platelets could either terminates edge-to-edge or edge-to-surface as illustrated in Fig. 61. Using EBSD technique, it is difficult to distinguish these two boundaries. This is because that both would give the same appearance after planarization of the sample. The evidence of current flowing along the *c*-axis would suggest that the majority of the boundaries analyzed were edge-to-surface type boundaries [9, 62]. Similar current transport in both parallel and perpendicular direction to the rolling plane was reported [7]. Such observation would suggest the dominance of edge-to-edge boundaries. We had found the characteristic misorientation angle of the grain boundaries to be 4°.

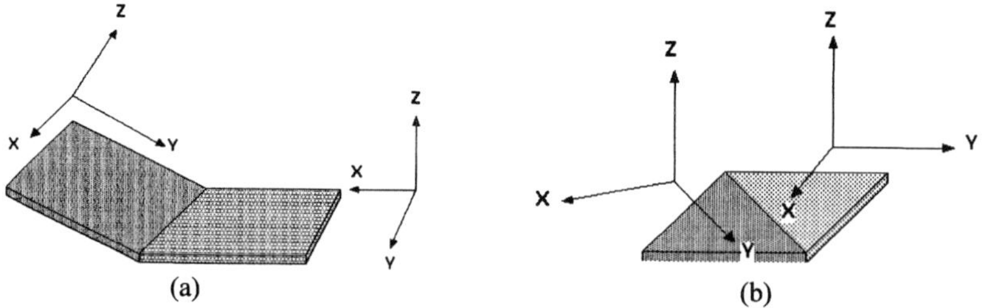

Fig. 61 Schematic view of (a) edge-to-edge and (b) edge-to-surface boundaries.

It is clear that the Bi2223 platelets are generally inclined at not more than 10° with respect to the rolling planes. These inclined grains generally have the largest a/b axes misorientation angle that generally increases with increasing inclination with respect to the rolling plane. In addition, it has been found that about 50% of the boundaries analyzed are mixed-axis boundaries of which approximately 80% have less than 10° misorientation. The dominating misorientation angle for both sample sides and center is the same in 4°. It evidences that the higher J_c at the sides of the PIT Bi2223 tapes is resulted from better texture, in terms of lower occurrence of the grain boundaries with intermittent high misorientation angle.

ACKNOWLEDGMENT

The authors would like to thank Professors S.X. Dou and H.K. Liu for their support and helpful discussion, and also appreciate the contributions made by T.T. Tan, H. Cooper and M. Bredehöft in this research.

REFERENCES

[1] D. Dimos, P. Chaudhari and J. Mannhart, and F.K. LeGoues, Phys. Rev. Lett. 61 (1988) 219.
[2] D. Dimos, P. Chaudhari and J. Mannhart, Phys. Rev. B 41 (1990) 4038.
[3] S.E. Babcock, MRS Bulletin August (1992) 20.
[4] S.E. Babcock and J.L. Vargas, Annu. Rev. Mater. Sci. 25 (1995) 193.
[5] S. Jin, J. E. Graebner, Mater. Sci. and Eng. B7 (1991) 243.
[6] D.C. Larbalestier, A. Gurevich, D.M. Feldmann and A. Polyanskii, Nature 414 (2001) 368.
[7] D.C. Larbalestier, Phsics Today June (1991) 74.
[8] D.C. Larbalestier, X.Y. Cai, Y. Feng, H. Edelman, A. Umezawa; Physica. C 221 (1994) 299.
[9] G. Grasso, B. Hensel, A. Jeremie and R. Flükiger, Physica C 241 (1995) 45.
[10] D.C. Larbalestier, X.Y. Cai, H. Edelman, M.B. Field, Y. Feng and J. Parrell, J. Met. 46 (12) (1994) 20.
[11] A.E. Pashitski, A. Gurevich, A.A. Polyanskii, D.C. Larbalestier, A. Goyal, E.D. Specht, D.M. Kroeger, J.A. DeLuca and J.E. Tkaczyk, Science 275 (1997) 367.
[12] A.E. Pashitski, A.A. Polyanskii, A. Gurevich, J.A. Parrell, D.C. Larbalestier, Physica C 246 (1996) 133.
[13] R. Hawsey and J. Daley, JOM August (1995) 56.
[14] M. Levovic, P. Krishnaraj, N. G. Eror and U. Balachandran, Physica C 242 (1995) 246.
[15] R. H. Arendt, M. F. Garbauskas, K. w. Lay, J. E. Tkaczyk, Physica C 194 (1992) 383.
[16] J.O. Willis, R.D. Ray II, D.S. Phillips, K.V. Salazar, J.F. Bingert, T.G. Holesinger, J.K. Bremser, and D.E. Peterson; Bi-2223/Ag-Sheathed Tape Processing Studies; 1995 TMS Annual Meeting, Las Vegas, NV., 13-16 February 1995.
[17] J.O. Willis, R.D. Ray II, J.F. Bingert, D.S. Phillips, And R.J. Beckman; Control of Processing Factors for High J_c Bi-2223/Ag Tapes; 1995 International Workshop on Superconductivity; Maui, Hawaii, 18-21 June 1995.
[18] Q.Y. Hu, R.M. Schalk, H.W. Weber, H.K. Liu, R.K. Wang, C. Czurda and S.X. Dou; J. Appl. Phys. 78 (1995) 1123.
[19] B. Hensel, G. Grasso and R. Flükiger; Physical Review B 51 (1995) 15456.

[20]L.N. Bulaevskii, L.L. Daemen, M.P. Maley and J.Y. Coulter; Physical Review B 48 (1993) 13798.
[21]W. Gao and J.B.Vander Sande; Physica C 181 (1991) 105.
[22]K. Heine, J. Tenbrink, M. Thoner, Appl. Phys. Lett. 55 (1989) 2441.
[23]Y. Yamada, B. Oberst, R. Flükiger, Supercond. Sci. Technol. 4 (1991) 165.
[24]Q. Li, K. Brodersen, H. A. Hjuler, T. Freltoft, Physica C 217 (1993) 360.
[25]K. Osamura, S.S. Oh and S. Ochiai, Supercond. Sci. Technol. 5 (1992) 1.
[26]G. Grasso, A. Perin and R. Flükiger, Physica C 250, (1995) 43
[27]T.H. Tiefel and S.Jin, J. Appl. Phys. 70 (10), 6510 (1991).
[28]J.A. Parrell, S.E. Dorris and D.C. Larbalestier, Physica C 231, 137 (1994).
[29]T. Uzumaki, K. Yamanaka, N. Kamehara and K. Niwa, Jnp. J. Appl. Phys. 28 (1989) L75.
[30]W. Wong-Ng, C.K. Chiang, S.W. Freiman, L.P. Cook and M.D. Hill, Am. Ceram. Soc. Bull. 71 (1992) 1261.
[31]N.Nijima and R. Gronsky, Appl. Phys. Lett. 58 (1991) 188.
[32]S.E. Dorris, B.C. Prorock, M.T. Lanagan, S. Sinha and R.B. Poeppel, Physica C 212 (1993) 66.
[33]S.X. Dou and H.K. Liu, Advances in High-T_c Superconductors, Materials Science Forum 137-139 (1993) 651.
[34]S. Li, M. Bredehöft, W. Gao, H.K. Liu, T. Chandra and S.X. Dou, Supercond. Sci. Technol. 11(1998) 1011.
[35]W. Gao and J.B. Vander Sande, Supercon. Sci. Technol. 5 (1992) 318.
[36]M. Merchant, J.S. Luo, V.A. Maroni, G.N. Riley and W.L. Carter, Appl. Phys. Lett. 65 (1994) 1039.
[37]S. Li, M. Bredehöft, Q.Y. Hu, H.K. Liu, S.X. Dou and W. Gao, Physica C 275 (1997) 259.
[38]R Reed-Hill, "Physical Metallurgy Principles", D Van Nostrand Company (Canada) Ltd. (1964) 195.
[39]S. Li, W. Gao, Q.Y. Hu, H.K. Liu and S.X. Dou, Physica C 276 (1997) 229.
[40]H.K. Liu, R.K. Wang and S.X. Dou, Physica C 229 (1994) 39.
[41]B. Raveau, C. Michel, M. Hervieu and D. Groult, Crystal Chemistry of High-T_c Superconductting Copper Oxides, Springer-Verlag Berlin Heidelberg (1991).
[42]R.W. Cahn and P. Haaden, Physical Metallurgy, Elsevier Science Publishers B.V. (1983) 1613.
[43]G. Lewis, J. Mat. Sci. Lett. 11, 321 (1992).
[45]B.W. Warren and B.L. Averbach, J. Appl. Phys. 21 (1950) 595.
[46]B.W. Warren and B.L. Averbach, J. Appl. Phys. 23 (1952) 497.
[47]R.W. Cahn and P. Haasen; Physical Metallurgy; North-Holland Physics Publishing, (1983) 1645.
[48]D.C. Larbalestier, X.Y. Cai, H. Edelman, M.B. Field, Y. Feng and J. Parrell, J. Met. 46 (1994) 20.
[49]G Grasso, A. Perin, B. Hensel and R Flükiger, Physica C 217 (1993) 335.
[50]J. M. Yoo and K. Mukherjee, Physica C 222 (1994) 241.
[51]D.J. Dingley, M. Longden, J. Weinber and J. Alderman, Scanning Electron Microscopy 1 (2), Scanning Electron Microscopy International Conference, Chicago (AMF O'Hare), IL 60666 USA, (1987) 451.

[52] D.J. Dingley, Scanning Elect. Micros. 11 (1984) 74.
[53] T.T. Tan, S. Li, H. Cooper, W. Gao, H.K. Liu and S.X. Dou, Supercond. Sci. Technol. 14 (2001) 471.
[54] V. Randle, Microtexture Determination and its applications, Institute of Materials (1992).
[55] A. Goyal, E.D. Specht, D.M. Kroeger and T.A. Mason, Appl. Phys. Lett. 68 (1996) 711.
[56] A. Goyal, E.D. Specht, D.M. Kroeger, T.A. Mason, D.J. Dingley, G.N. Riley Jr. and M.W. Rupich, Appl. Phys. Lett. 66 (1995) 2903.
[57] Y. Zhu, M. Suenaga and R.L. Sabatini, Appl. Phys. Lett. 65 (1994) 1832.
[58] J.A. Parrell, A.A. Polyanskii, A.E. Pashitski and D. C. Larbalestier, Supercond. Sci. Technol. 9 (1996) 393.
[59] X.Y. Cai, A. Polyanskii, Q. Li, G.N. Riley Jr. and D.C. Larbalestier, Nature 392 (1998) 906.
[60] D.C. Larbalestier, J.A. Anderson, S.E. Babcock, X.Y. Cai, S.E. Dorris, M. Feldmann, J. Jiang, Q. Li, J.A. Parrell, R. Parrella, M. Polak, A. Polyanskii, G.N. Riley Jr., M. Rupich and Y. Wu, Advances in Superconductivity XI, Springer Verlag, Tokyo (1999).
[61] B. Hensel, J.C. Grivel, A. Jeremie, A. Perin, A. Pollini and R. Flükiger Physica C 205 (1993) 329.
[62] B. Hensel, G. Grasso, D.P. Grindatto, H.U. Nissen and R. Flükiger Physica C 249 (1995) 247.

ATOMIC SCALE INVESTIGATION OF GRAIN BOUNDARIES AND DEFECTS IN Bi-2223/Ag TAPES BY SCANNING TRANSMISSION ELECTRON MICROSCOPY

K. Kishida [a] and N.D. Browning [b]

[a] Mechanical Properties Research Group, National Institute for Materials Science
1-2-1 Sengen, Tsukuba, Ibaraki 305-0047, JAPAN

[b] Department of Physics, University of Illinois at Chicago
845 West Taylor Street, Chicago, IL 60607-7059, USA

1. INTRODUCTION

It has been considered that both interfaces and defects in the Bi-2223/Ag tapes have a strong effect on the overall superconducting properties. In highly textured Bi-2223/Ag composite tapes, there are many different types of interfaces, including grain boundaries that occur between the superconductor phases and interfaces between the superconductor and Ag sheath. The role of the grain boundaries inside superconductor phases has been previously discussed according to the macroscopic configuration of the grains [1-4]. However, such models consider only the large-scale arrangement of the filaments, and the underlying mechanism controlling the properties has yet to be clarified.

To address the fundamental atomic scale effects at grain boundaries, transmission electron microscopy (TEM) is the most suitable method. Extensive studies using primarily high-resolution transmission electron microscopy (HRTEM) have revealed the atomic scale structure of the grain boundaries as well as the defects in the Bi-based superconductors [5-14]. However, one of the common disadvantages in the conventional HRTEM is that the interpretation of the HRTEM images is not straightforward but rather troublesome since iterative simulations of model structures are required until they match with the experimental images.

Recently it has been revealed that the Z-contrast imaging of the scanning transmission electron microscopy (STEM) provides an alternative means for the atomic scale characterization of defects such as dislocations, stacking faults, and interfaces in various materials such as semiconductors, intermetallics, ceramics and high-Tc superconductors [15-25]. The Z-contrast images exhibit an incoherent nature, i.e. no contrast reversal effect, and a compositional sensitivity that provides intuitively interpretable images of defects in materials [15-18]. These characteristics are advantageous for the study of defects and grain boundaries in high-Tc superconductors where the properties are determined by subtle changes in structure and composition.

In this article, we briefly mention the basics of the Z-contrast imaging technique in the STEM and describe the latest application of the technique to the Bi-based high-Tc superconducting tapes. This study is focused on the atomic scale characterization of the various grain boundaries observed to be numerous in the tapes as well as some defect structures found inside the superconductor phases.

2. SCANNING TRANSMISSION ELECTRON MICROSCOPY (STEM)

There are two different types of instruments capable for STEM experiments; one is a dedicated STEM, which is much closer to scanning electron microscopy (SEM), and the other a TEM with a scanning unit (TEM/STEM) [26, 27]. For both STEM instruments, the basic pre-specimen optics are designed to obtain a very fine probe on the surface of the specimen. One of the most important factors to obtain the atomic scale Z-contrast images is the size of the electron probe. Therefore, the use of a field emission gun (FEG) to generate a very fine probe with sufficiently high current, is essential for such high-resolution purposes. The microscope used in this study is a JEOL JEM-2010F TEM/STEM equipped with a Schottky field emission electron gun (FEG). The FEG together with the use of the ultra-high resolution pole-piece (Cs~0.5mm), the probe sizes under 0.14nm are obtainable in this microscope [20]. Figure 1 shows a schematic illustration of the STEM based on the TEM/STEM. Imaging basics are similar to those for the SEM. The probe is scanned over the specimen and signals collected by various detectors are displayed on synchronously scanning CRT monitors and can be stored in a computer as a function of the probe position. The difference for imaging in the STEM is that the electron passing through and interacting with a specimen are collected by various post-specimen detectors such as an annular dark-field (ADF) detector and an axial bright-field (BF) detector [26].

Z-contrast images (also known as high angle annular dark field (HAADF) images) are formed by collecting the high-angle scattering (40 - 100mrad at 200kV [19]) on an ADF detector [15-20]. Detecting the high-angle scattering intensities and integrating over a wide angular range effectively destroys coherent interference effects between adjacent atomic

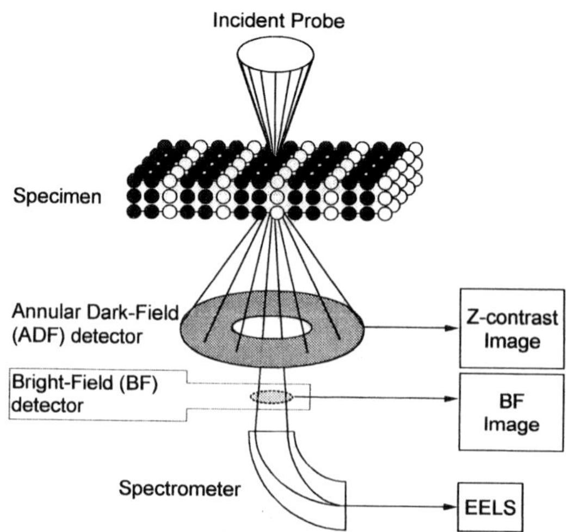

Figure 1: Schematic of the detector arrangement for STEM imaging and analysis.

columns in the specimen [15-18]. In addition, thermal vibrations reduce the coherence along each individual column to residual correlations between near neighbors [17]. Thus, each atom can be considered as an independent scatterer with a cross section that approaches a Z^2 dependence on atomic number with increase of the detector inner angle. This cross section forms an object function that is strongly peaked at the atom sites. In very thin crystal where there is no dynamical diffraction, the image can be simply interpreted as a convolution of the probe intensity with the object function [15]. Because of the small width of the object function (~0.02nm), the spatial resolution is limited only by the probe size of the microscope. Thus, for a crystalline material in a zone-axis orientation, an atomic resolution compositional map can be obtained when the spacing between adjacent atomic columns are greater than the probe size. The intensity depends on the average atomic number of the atoms in the columns in the Z-contrast images.

This result also holds true for thicker specimens. In this case, the experimental parameters cause dynamical diffraction effects to be manifested as a columnar channeling effect, thus maintaining the thin specimen description of the image as a simple convolution of the probe intensity and an object function, strongly peaked at the atom sites. The phase problem associated with the interpretation of conventional HRTEM images is therefore eliminated [15-18]. The effect of changing focus is also intuitively understandable as the focus control alters the probe intensity profile on the surface of the specimen. For defocus less than the optimum Scherzer condition, the probe broadens causing the individual columns not to be resolved. For higher defocus value, the probe narrows with the formation of more intense tail, causing sharper image feature but also with some artifacts [15]. The optimum focus condition therefore represents a compromise between high resolution (narrow probe profile) and the desire for a highly local image (no significant tails to the probe). This also corresponds to the optimum probe for microanalysis [28].

Samples used for the atomic resolution Z-contrast imaging are Bi-2223/Ag multifilament tapes with critical current densities (Jc) as high as $60kA/cm^2$ at 77K and self-field, which were provided by American Superconductor. Short and transverse cross section samples were prepared by gluing two tapes and two Si supporting blocks together using epoxy and mechanically grinding down to about 15μm thick. Finally, these samples were Ar ion milled at 4kV on a liquid nitrogen cooled stage. Since the spacing between the adjacent columns in the [110] projection of the Bi-2223 phase is more than 0.25nm, the probe size of about 0.16nm was used to increase the signal/noise (S/N) ratio in the Z-contrast imaging experiments.

3. Bi-2223 PHASE

The Bi-based superconductor phases are generally expressed by the formula $Bi_2Sr_2Ca_{n-1}Cu_nO_{2n+4+\delta}$ with n = 1 ($Bi_2Sr_2CuO_{6+\delta}$ or short notation: Bi-2201), n = 2 ($Bi_2Sr_2CaCu_2O_{8+\delta}$ or Bi-2212) and n = 3 ($(Bi,Pb)_2Sr_2Ca_2Cu_3O_{10+\delta}$ or Bi-2223) [11]. These three phases are formed by a conventional sintering method, while large-scale superconductor phases with $n \geq 4$ are not. Phases with large n have been formed only as thin films by a vapor deposition method or as inter-grown lamellae, less than a unit cell thick in conventionally sintered materials [11]. The critical temperature (Tc) increases with increasing the number (n) of CuO_2 layers up to n = 3; 20K, 80-90K and 110K for n = 1, 2 and 3, respectively. However, Tc values of phases with $n \geq 4$ are reported to be lower than that of n = 3 phase [29, 30]. This highest-Tc value of the Bi-2223 phase in this series of Bi-based superconductor is one of the advantages of using this phase in the production of superconducting tapes.

Figure 2 shows the typical Z-contrast images from bulk region of the Bi-2223 phase in the tapes. Incident beam directions are parallel to [110] and [010] in figs. 2a and 2b, respectively. Bi, Sr, Ca and Cu-O columns are clearly resolved in the [110] projected Z-contrast image (fig. 2a) with the strong Z-dependence of the contrast, i.e. the brightest spots in the image

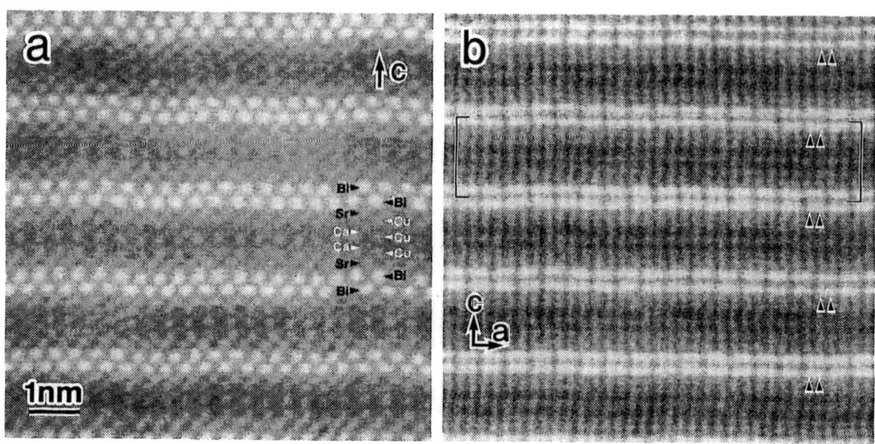

Figure 2: Z-contrast images of the bulk Bi-2223 phase in Bi-2223/Ag tapes in (a) the [110] and (b) the [010] projections. Arrowheads and brackets in (b) indicate the positions of Bi pairs and faulted stacking region, respectively.

corresponding to the Bi columns. In the [010] projection (fig. 2b), the pairing of Bi columns in the BiO double layer is clearly observed. The positions of the Bi pairs a half [001] apart are intrinsically shifted by 1/2 a unit cell toward the [100] direction. However, a faulted structure without the shift along [100] is sometimes observed as seen in the bracketed part of fig. 2b. The faulted structure was similarly observed in a conventional HRTEM study of the Bi-2212 phase [10, 13].

The structural modulations are well known to exist in the Bi-based superconductor phases [10, 11, 31-39]. Extensive studies have been carried out to analyze the modulated structure in both the Bi-2212 and the Bi-2223 by HRTEM [10, 11, 31-36]. These studies have revealed that all of the cations are displaced from the average atomic positions in both [010] and [001] direction. In the Bi-2223 phase with Pb addition, mainly two types of the modulated structure have been reported; Type-I modulation that is similar to those observed in the Bi-2212 phase, and Type II [10, 11, 33-36]. These two types of the structural modulation are also observed in atomic resolution Z-contrast images of the Bi-2223 phase projected along [100]. Figure 3 shows the Z-contrast images as well as the corresponding FFTs of the two typical modulated structures (type-I: fig. 3a, b and type-II: fig.3c, d). Positions of additional spots shown in the FFTs (marked by arrow heads) are consistent with the selected area diffraction patterns found in literatures [10, 11 33-36], further supporting that the observed modulated structures are the same as those observed by the conventional HRTEM.

It is important to note that not only the modulation of the atomic positions but also the contrast modulation along the BiO layers is clearly observed in the [100] projection of the Bi-2223 phase with the type-I modulation (fig. 3a). Since the intensity of each bright spot in the Z-contrast image reflects the average atomic number of the atoms in the columns, the observed contrast modulation may relate to the compositional variation in the BiO double layers. Previous powder neutron scattering experiments for the Bi-2212, which only possesses type-I modulation, have been carried out to determine the structural parameters of the modulated structure including the positions of oxygen atoms [37, 38]. According to the study by Yamamoto et al, the Bi and the O atoms are relatively well aligned along [100] in the Bi concentrate region, while they are not in the Bi-dilute region [37]. Also in the case of the Bi-2223, the similar variation of the atomic positions in the Bi-O layers would possibly occur. If this is the case, the Bi columns in the Bi-concentration region may show relatively higher scattering power and consequently the brighter contrast than those in the Bi-dilute region under the imaging condition of the Z-contrast technique. Thus, the contrast modulation would be introduced in the Bi-O double layers. In contrast, no apparent contrast modulation is seen in the type-II structure suggesting that the oxygen distribute rather uniformly in the double BiO layers (fig. 3b). Since the type-II modulation has been considered to relate closely to the Pb addition [10, 11, 33-36], the formation of these two modulated structure would be influenced by the compositional fluctuation of Pb in the filaments. However, the detailed mechanisms for the formation of these two modulated structure have not fully understood. Future STEM work in which we will combine the sub-nanometer scale Z-contrast imaging, the electronic structure investigation by Electron Energy Loss Spectroscopy (EELS) and the compositional analysis by both EELS and Energy Dispersive X-ray Spectroscopy (EDS), will be helpful in elucidating the mechanisms. The modulated structures shown in fig.3 correspond to two extreme cases. In many cases, mixed structures (or intermediate structures) of these two structures are observed (e.g. [100]-oriented grains in fig. 13).

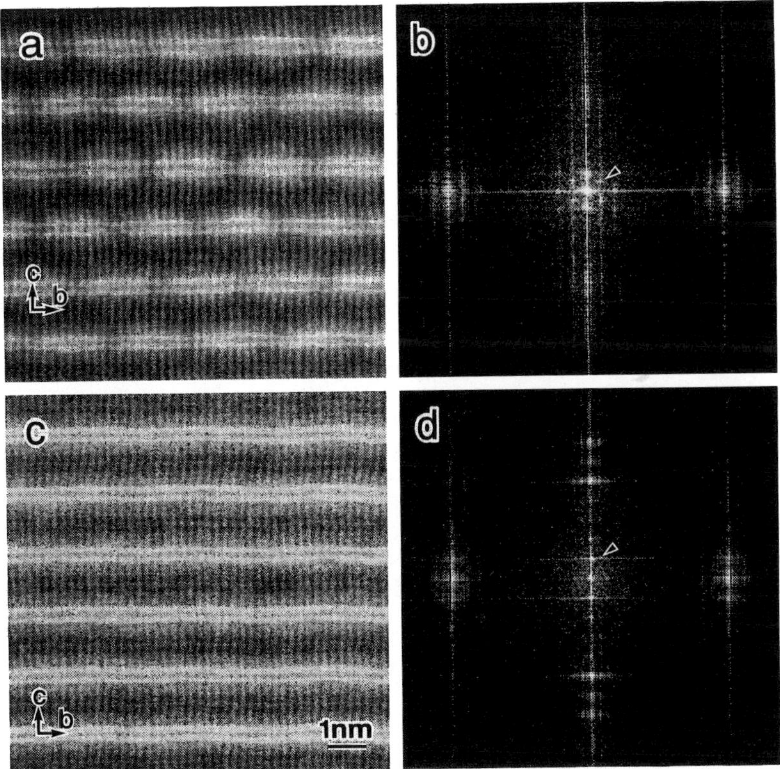

Figure 3: Z-contrast images and the corresponding FFTs of the two types of the modulated structures in the Bi-2223 phase: (a, b) type-I and (c, d) type-II modulations.

4. DISLOCATIONS, STACKING FAULTS and INTERGROWTH

Dislocations that separate the faulted from the not faulted regions are also occasionally observed. Figures 4a and 4b show Z-contrast images accompanied with the schematic illustrations of dislocations in the Bi-2223 phase. Double arrowheads in the figures show positions of Bi-column pairs. Burgers circuits are also indicated with black lines in figs. 4a and 4b. Taking into account the atomic arrangement of the BiO layers, both Burgers vectors are considered to be 1/2[110]. The intrinsic stacking sequence is observed on the right side of the dislocations shown in both figs. 4a and 4b. On the upper left side of the dislocation in fig. 4a, the faulted sequence, which is the same as the one shown in fig. 2b, is observed. The projection of the Burgers vector of the dislocation in fig. 4b is the same as that shown in fig. 4a. The sequence is preserved on both upper and lower side of the glide plane of the dislocation. This results in the formation of a stacking fault (SF) on the glide plane of left side of the dislocation. A strain effect on the image contrast is not observed in the Z-contrast image

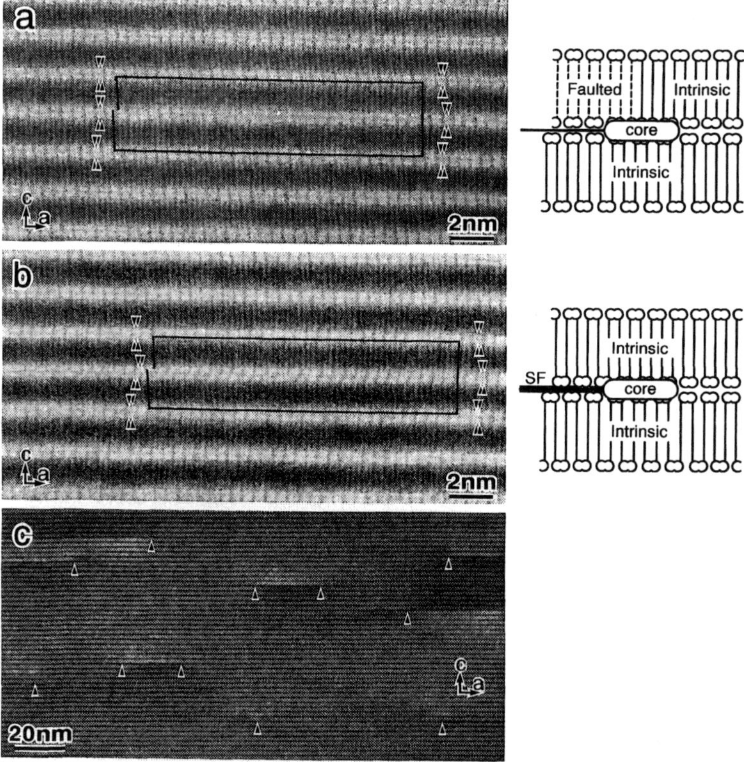

Figure 4: (a, b) Z-contrast images with schematic illustrations and (c) a bright field STEM image of dislocations and stacking faults in the Bi-2223 phase projected along the [010] orientation. Double arrowheads and black lines in (a) and (b) indicate the positions of Bi pairs and the Burgers circuits, respectively. Arrowheads in (c) indicate the positions of dislocations.

(figs. 4a and 4b) while strained regions associated with the dislocations and the SFs are clearly seen in the STEM-BF image, which contains the interference effect comparable to the conventional HRTEM image (fig. 4c). Arrowheads in fig. 4c indicate the positions of dislocations. The glide planes of these dislocations are always located in the middle of the double BiO layers.

These defects would be possible candidates for pinning sites of magnetic flux, especially when the magnetic field direction is parallel to the dislocation lines or stacking fault planes. Although the double BiO layers in the Bi base superconductor phases are considered to already play a role as intrinsic pinning sites in which the magnetic fluxes penetrate and stay preferentially [40], these defects would have strong additional effects on the flux pinning behavior when the density of the dislocations and stacking faults is high. In addition, the local changes of both the bond length and therefore the electronic structure are considered to occur under the strain field around the dislocations. This may be accompanied by either a rearrangement of the oxygen atoms or a change in the oxygen content in the BiO layers, which may be directly related to the density of holes in the CuO_2 layers. Thus, these dislocations and stacking faults may also change the superconducting properties such as the Tc value and the Jc value of the material.

Figure 5 shows an example of the intergrowth observed inside a grain. In the superconducting tapes studied here, the matrix phase is the Bi-2223 phase (numbered as “3” in the figure). As clearly seen in the fig. 5a, the Bi-2212 phase (marked as “2”) is observed in the matrix Bi-2223 phase. The amount of the inter-grown low-Tc phase is found to vary from place to place. In the lower middle of the fig. 5a, a transition region from the Bi-2223 to the Bi2212 is observed. Figure 5b is the image taken at higher magnification from the transition region (squared part in fig. 5a) showing the atomic scale configuration of the transition region. Three half unit cells of the Bi-2223 phase on the left side connects with a unit cell of the Bi-2212 phase and a half unit cell of the Bi-2245 (marked as “5”) phase producing steps at the transition region. These steps associated with the intergrowth may cause an additional effect on the vortex pinning behavior.

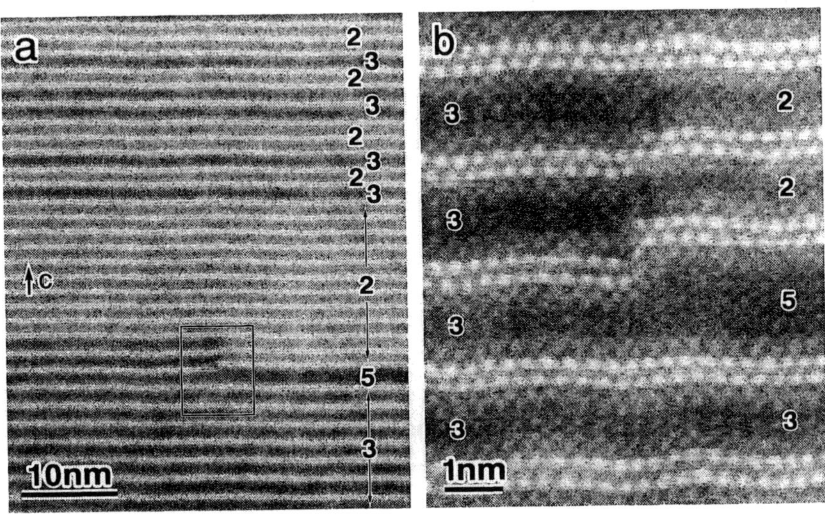

Figure 5: Intergrowth of the lower-Tc phases in the Bi-2223 phase.

5. GRAIN BOUNDARIES

In highly textured Bi-2223 composite tapes, grain boundaries are generally classified into three types: c-axis twist boundaries (TWBs), small-angle asymmetrical tilt boundaries (ATBs) (as termed as SCTILT boundaries by Hensel et al. [3]) and small-angle edge grain boundaries (EGBs) [4]. For simplicity, only the geometrical relationship between the basal planes of both grains is considered in classifying the grain boundaries in the present article. Schematic illustrations of these three boundaries are shown in fig. 6. The c-axis TWB connects two adjacent grains on their ab-basal planes having essentially perfect c-axis alignment. The other two kinds of boundaries can have a small tilt of the c-axis across the boundary as well as some rotation about the c-axis. One of the characteristics of the small-angle ATBs is that the basal planes of one grain are macroscopically parallel to the grain boundary plane. The small-angle EGBs are formed by joining thin edges of two grains.

From a viewpoint of morphology of the filaments, three models have been proposed to describe the current transport phenomena in the highly textured Bi-2223/Ag tapes [1-4]. In the brick-wall model, the grain structure is considered similar to a brick wall and the main paths for the current flow are assumed to be the c-axis TWBs [1, 2]. In fact, previous TEM studies of these materials have revealed that the c-axis TWB is one of the most dominant boundaries in the Bi-2223/Ag tapes [7-9, 14]. In addition, the critical current values across the boundaries may not exhibit an apparent angular dependence, if the properties of the boundary in the Bi-2223 tapes are similar to those in the Bi-2212 bicrystal system [13]. However, this model does not accurately describe the filament arrangement in the tapes. Taking into account a more realistic configuration of the filaments, Hensel et al. [3] have proposed the railway-switch model. In the framework of the railway-switch model, the small-angle ATBs are considered to be the fundamental elements for the current flow. However, data on the detailed microstructure of the small-angle asymmetrical tilt boundaries in the Bi-2223/Ag tapes as well as the relationship between tilt angle and the current value across the boundaries is not available. Recently, Riley et al. proposed the freeway model [4]. In the rotary junction mechanism of the freeway model, the small-angle EGBs are considered to have strongly linked regions where the supercurrent is assumed to flow selectively. In order to make this mechanism work, a redistribution of current across the c-axis TWBs within a colony is required. Thus, the c-axis TWBs are also considered to play an important role for the current transfer in the freeway model [4].

Figure 7 shows typical Z-contrast images of filaments close to the Ag interface of the tape. Since the ab-basal planes are parallel to the incident beam direction, each bright line in these figures corresponds to the double BiO layers at the magnification used. Neither large precipitates nor large amorphous phases at any type of boundaries are observed in the filaments located near the Ag interface of the tapes. As seen in fig. 7a, both the small-angle

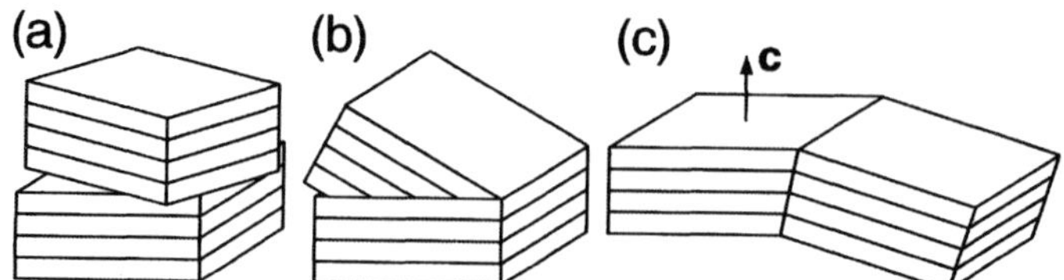

Figure 6: Schematics of three typical grain boundaries in Bi-2223/Ag tapes; (a) The c-axis twist boundary (TWB), (b) the small angle asymmetrical tilt boundary (ATB) and (c) the edge grain boundary (EGB).

ATB and the small-angle EGB regions can be formed continuously along a single boundary. Grains marked with asterisks in fig. 7b bend ~3.5° at the positions indicated by arrowheads, maintaining the layered structure of the superconductor phases. The inter-growth of the superconductor phase, which can be recognized by the different distance between the white lines, is observed to exist occasionally inside grains.

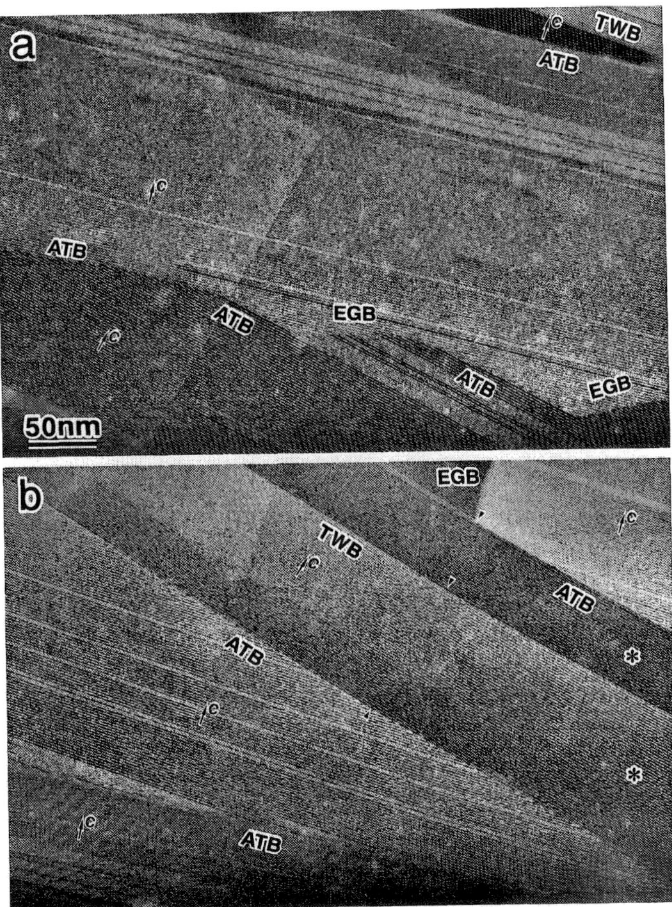

Figure 7: Z-contrast images of the filaments in Bi-2223/Ag tapes. TWB, ATB and EGB indicate the c-axis twist boundary, the small-angle asymmetrical tilt boundary and the edge grain boundary, respectively. Grains with asterisks in (b) bent about 3.5° at the positions indicated by arrowheads.

5.1. C-axis twist boundaries

The c-axis TWBs have been considered to play a dominant role in transporting the supercurrent within the framework of the brick-wall model [1, 2] and the freeway model [4]. Several c-axis TWBs imaged with the incident beam oriented parallel to the [110] direction on the upper side of the boundary are shown in fig. 8. In the lower half of figs. 8(a-f), several sets of two bright lines are clearly observed which correspond to the BiO double layers viewed end on. Arrows in fig. 8 indicate the positions of the boundaries. It is obvious from the figures that all of the boundaries are atomically flat and located in the middle of the double BiO layers. In addition, there are no amorphous layers observed on this type of the boundary. Such results are consistent with previous conventional HRTEM work on both the Bi-2212 [5, 12, 13] and the Bi-2223 systems [7, 8]. However, the great advantage of the Z-contrast technique is that the location and the nature of the boundary can be directly identified from the raw Z-contrast images without carrying out image simulation. In the majority of cases, local phase variation at the boundary is observed (figs. 8(b-f)) except for the boundary shown in fig. 8a at which the Bi-2223 phase is perfectly maintained. In figs. 8b and 8c, a half unit cell of the Bi-2234 phase and the Bi-2212 phase are observed only on one side of the boundary, respectively. A half unit cell of the Bi-2212 phase is formed on both sides of the boundary in fig. 8d. Thus, the TWBs are considered to be preferential sites for formations of the lower-Tc phases. Occasionally, the local formation of the lower-Tc phase such as the Bi-2201, the Bi-2212 and the Bi-2234 phases occur not only at the boundary but also near the boundary (figs. 8e and 8f) resulting in complicated phase variations near the boundary. Both the type of the phases and the number of the unit cells of the phases formed at/near the boundary is found to vary from boundary to boundary. In contrast to the case of the boundaries in the Bi-2212 materials, only lower-Tc phases are formed at the boundaries since the Bi-2223 phase has the highest Tc in the Bi-based superconductor system [29, 30]. Thus, the phase variation at/near the boundary could have a strong negative influence on the transport phenomena of the supercurrent across the boundary at liquid nitrogen temperature.

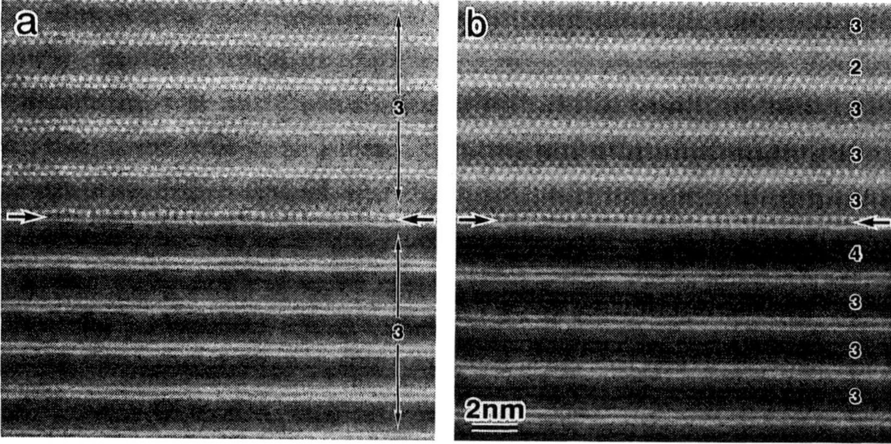

Figure 8: Z-contrast images of various c-axis twist boundaries in Bi-2223/Ag tapes. Positions of the grain boundaries are indicated by arrows. The numbers (n) in the figures represent the types of phases.

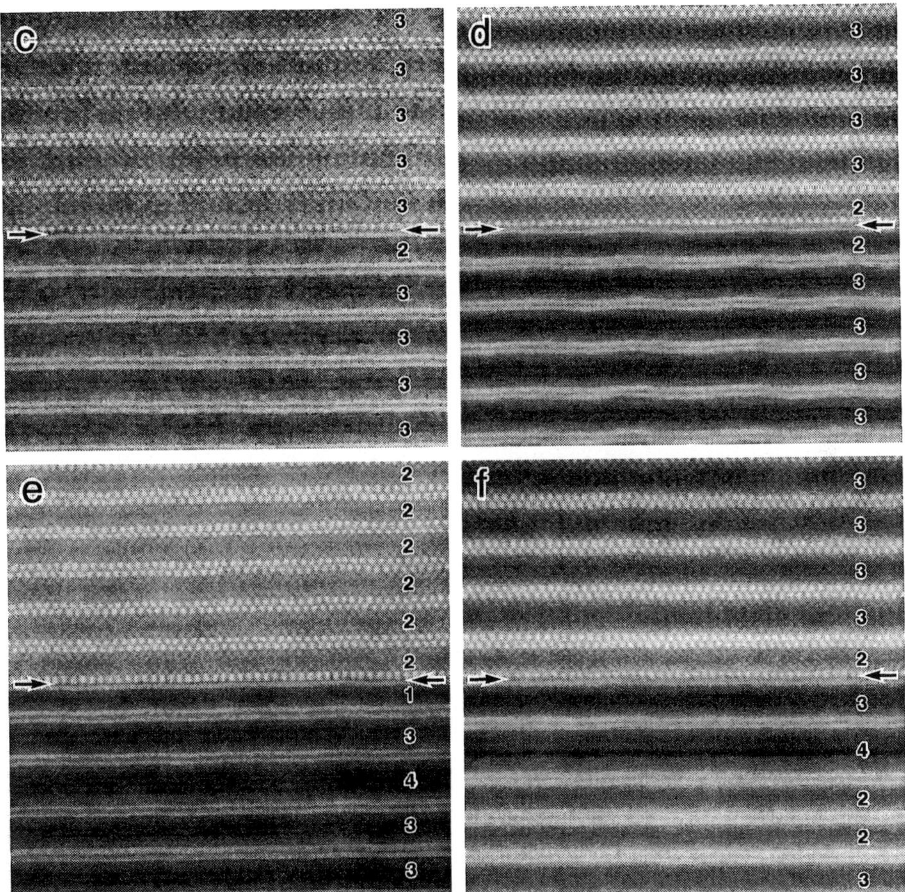

Figure 8: (continued).

Figure 9 shows an approximately 45° c-axis TWB in a composite tape projected along [110] of the upper half (fig. 9a) and [010] of the lower half (fig. 9b). Because of the slight orientation difference (~0.02°) between [110] of the upper grain and [010] of the lower grain, the lattice images for both grains were not obtained under the same imaging condition. This indicates that the high-resolution Z-contrast image is quite sensitive to the slight misalignment of the crystal orientation. In the upper grain, one unit cell of the Bi-2223 phase which is seen in left side of figs. 9a and 9b changes into three half unit cells of the Bi-2212 phase, eventually resulting in the formation of a step at the boundary as well as a half unit cell of the Bi-2234 phase in the lower grain. The formation of the lower-Tc phase resulting from the intergrowth may again have a strong negative influence on the current flow parallel to the basal plane of the superconductor phase.

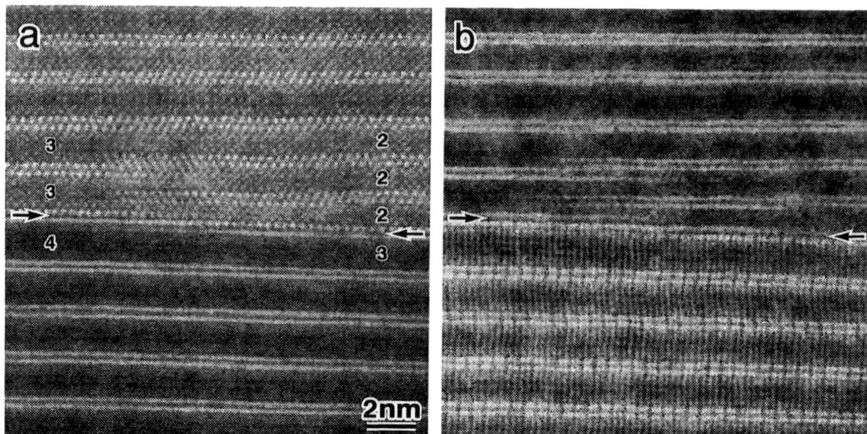

Figure 9: Local phase change along an approximately 45° c-axis twist boundary in a Bi-2223/Ag tape.

5.2. Small-angle asymmetrical tilt boundaries

Tilt angles between two adjacent grains are generally less than about 14° in filaments located near the surface of the highly textured tapes observed in this study. As seen in fig. 7a, small-angle ATB regions and EGB regions can exist along a single boundary. A characteristic of the small-angle ATB is that the basal planes of one grain are macroscopically parallel to the grain boundary plane [3]. Figures 10a and 10b show an 11° ATB taken from the same region of the boundary in the [110] projection for the lower grain (fig. 10a) and upper grain (fig. 10b). Slight orientations difference less than 0.1° prevent us obtaining a lattice image under the same conditions. In the grain where the basal planes are parallel to the boundary (upper side of the boundary in fig. 10b), periodic formation of the Bi-2212 and the Bi-2223 phase is clearly observed. On the other hand, four different phase; namely Bi-2201, Bi-2212, Bi-2223 and Bi-2234, are formed periodically along the lower half of the boundary in fig. 10a. Figure 10c shows a schematic diagram of the grain boundary constructed from figs. 10a and 10b. Solid lines and numbers in fig. 10c correspond to the BiO layers and the type of the phase, respectively. It is important to note that the periodic formation of the different phases along the boundary results in the boundary plane being in the middle of the double BiO layers, although there is slight distortion around the boundary. This observation of the periodic phase variation is in good agreement with the previous HRTEM works on both the Bi-2212 and the Bi-2223 textured bulk materials [5, 12].

The periodic phase variation on the inclined side of the boundary is a common characteristic for the small-angle asymmetrical tilt boundaries. Figure 11 shows the other examples of this type of boundary. In the case of a 4° ATB (fig. 11a), the same type of the phase variation is observed with longer periodicity on the inclined side while no phase variation is seen on the parallel side. In the center of the fig. 11a, an area with the darkest contrast is clearly seen (indicated by an arrowhead). Rearrangement of atoms occurs around this region, but the detailed structure is not clear from this image. The presence of such a large defect should change the local electronic structure and therefore superconducting properties. Future study of the electronic structure at the sub-nanometer scale by the EELS in the STEM will be helpful in understanding the effect of this defected region upon the current transport.

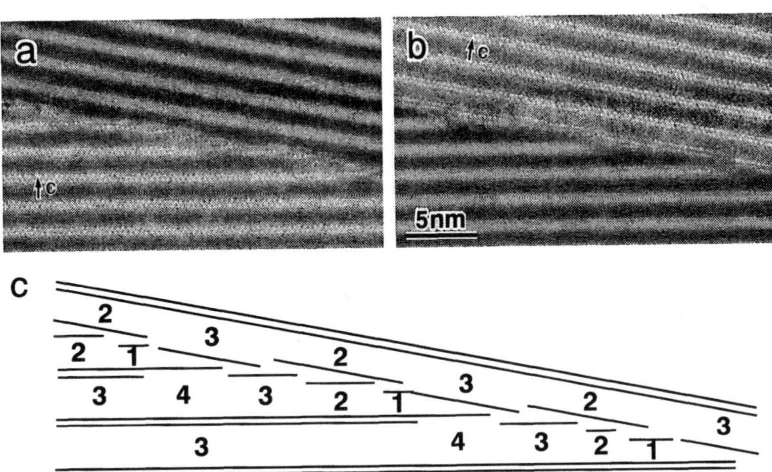

Figure 10: (a, b) Z-contrast images and (c) a schematic diagram of a small-angle (11°) asymmetrical tilt boundary in a Bi-2223/Ag tape. The black lines and the numbers (n) in the schematic diagram represent the BiO layers and the types of phases, respectively.

Defects of this kind are formed at the boundary where no phase variation is seen on the parallel side of the boundary. Along a 12° ATB (fig. 11b), the periodic formation of the Bi-2201, Bi-2212, Bi-2223 and Bi-2234 phases are also observed on the inclined side with shorter periodicity and only the Bi-2223 phase is seen on the parallel side. Dark and thick lines are clearly observed at the positions marked by arrowheads. In contrast, when the periodic phase change also exists on the parallel side of a small angle ATB, these defected regions are not formed even at a boundary with a relatively large (~ 14°) tilt angle. At the 14° ATB shown in fig. 11c, the Bi-2223 and the Bi-2234 phases are formed alternately on the parallel side of the boundary without forming large defects imaged as the dark contrast region in the inclined side of the boundary. Thus, all of the small angle (≤ 14°) ATBs in the highly textured tapes have the same basic structure with different periodicity according to the tilt angle and are always located in the middle of the double BiO layers.

The periodic formation of various phases along the boundary seems to effectively accommodate the mismatch between neighboring grains. Yan et al. [12] observed almost the same kind of phase modulation in textured bulk samples and considered that this type of the boundary may be favorable for transporting high Jc. They attributed the "pathway" for the current transport to the region that consists of the Bi-2212 and the Bi-2223 phases. In the present study, we also confirmed the presence of regions composed of two Bi-2223 phases along the small-angle ATBs. However, we have to consider the fact that the small-angle ATBs are always located in the middle of the double BiO layers. This means that each component with various combinations of the superconductor phases can be considered to be locally the same as the c-axis TWBs. Therefore, most of the boundary area may give rise to weak link behavior since the high-Tc Bi-2223 phase makes up only 1/4 of the inclined side of the all small-angle ATBs and about 1/2 of the parallel side of the most boundaries. Thus, we conclude that the current transportability of the small-angle ATBs may not be drastically high in comparison with that of the c-axis TWBs.

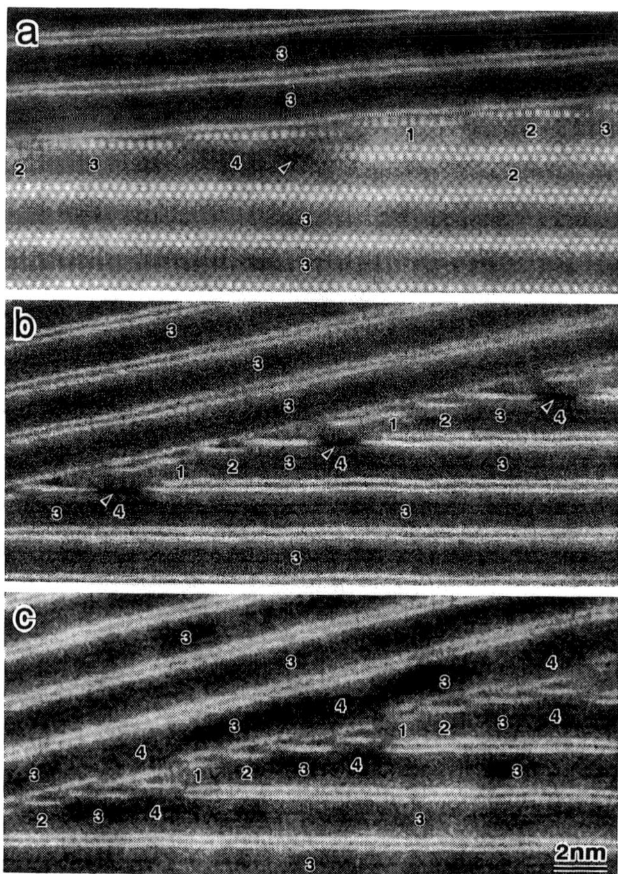

Figure 11: Z-contrast images of small-angle asymmetrical tilt boundaries with various tilt angles (a) 4°, (b) 12° and (c) 14°. Arrowheads in (a) and (b) indicate the nano-size defects in Bi-2234 (n = 4) phases.

5.3. Edge grain boundaries

As shown in fig. 7b, the superconductor filaments can tilt gradually at least up to ~3.5° maintaining the layered structure. However, in most cases, defected regions, where the layered structure is not well connected, are formed (indicated by small arrowheads in fig. 12a). Black lines drawn parallel to the ab planes in the middle of fig. 12a are guidelines to recognize the bending of the grain. This defected region can be considered as a sub-boundary having similar atomic scale structure to the EGB with low angle tilt. Except for the defected region, the layered structure of the superconductor phase is completely maintained. Dislocations either distributed randomly or ordered into a small-angle boundary may exist at such small-angle bended regions, any dislocations, however, are not imaged under the imaging condition used in this study. Figure 12b is an enlarged image of the defected region. An interlocking structure and slightly darker contrast of the BiO layers observed in the middle of the image, indicates that the grain boundary is inclined to the electron beam direction.

Figures 12c and 12d are examples of EGBs with a twist component. Tilt angles of the boundary seen in the middle of fig. 12c, the lower right of figs. 12c and 12d are 10°, 13.5° and 12°, respectively. When two grains meet at the boundary with almost the same tilt angle to the boundary, the layered structure is well preserved. On the other hand, when the two grains coincide asymmetrically at the boundary, each layer is not fully connected, resulting in the formation of the defected region (indicated by arrowheads). It is also important to note that no amorphous phases are observed at the EGB or even at triple points (fig. 12c).

The defected structure that is formed at asymmetrical regions can conceivably act as an obstacle in transporting current. However, since the layered structure of the superconductor phase is well connected at the symmetrical regions, there may be enough pathways for transport of large current, as assumed in the freeway model (at least when the tilt angle is relatively small) [4]. As seen in a boundary in fig. 7a, both the EGB regions and small-angle ATB regions can exist on a single boundary. Taking into consideration the fact that the small-angle ATBs have microscopically the same structure as the c-axis TWBs and most regions would exhibit weak link behavior, the large current would selectively pass through the EGB regions.

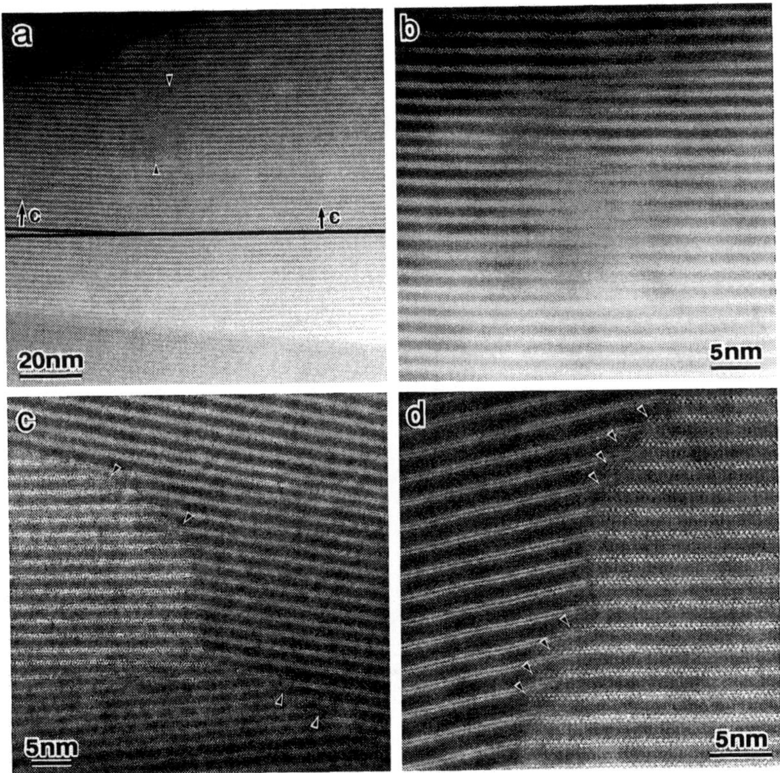

Figure 12: Z-contrast images of the edge grain boundaries in Bi-2223/Ag tapes. Tilt angles of each boundary are (a, b) 3°, (c) 10° (center) and 13.5 (lower) and (d) 11.5°. Arrowheads indicate the positions where the layered structure is disturbed.

5.4. Twin boundaries

The grain boundary formed between two grains with the orientation relationship of the 90°-rotation around the [001] axis often referred as the twin boundary [5, 11]. The twin boundary can be considered as special cases of the c-axis TWB or the EGB (more precisely the c-axis tilt boundary). Figure 13a shows an example of the TWB-type twin boundary in which the upper and the lower grains correspond to the [100] and the [010] projected images, respectively. Similarly to the case of the other c-axis TWBs shown in fig. 8, this type of the twin boundary is located in the middle of the double BiO layers. As seen in fig. 13a, the boundary is perfectly coherent and neither amorphous nor secondary phases are formed. Figure 13b shows an EGB-type twin boundary. Small arrowheads indicate the approximate location of the boundary region. Although the boundary location is shifted somehow, the layered structure is perfectly maintained across the boundary. Furthermore, the EGB-type

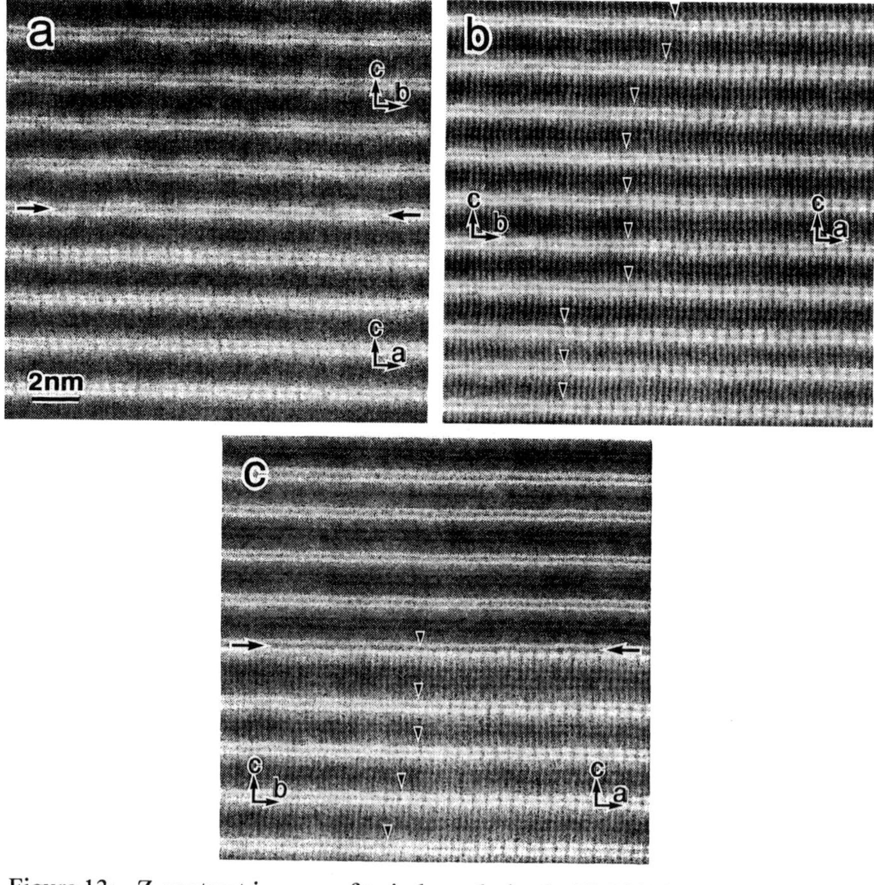

Figure 13: Z-contrast images of twin boundaries in Bi-2223/Ag tapes. Arrows and arrowheads indicate the positions of the c-axis TWBs and the EGB-type twin boundaries, respectively.

twin boundaries are clean and free from either amorphous or secondary phase, even at the triple point shown in the middle of fig. 13c. This suggests that the EGB-type twin boundary may just have a minor effect on the current transport properties parallel to the layered structure.

6. CONCLUDING REMARKS

The application of the STEM Z-contrast technique to the investigation of the atomic scale structural analysis of the Bi-2223/Ag tapes has been described in this article. The results shown here demonstrate the excellent capability of the Z-contrast imaging technique itself, which provides directly interpretable images suitable for atomic scale characterizations of crystal structure, grain boundaries and defects of the Bi-based superconductor. We have just focused on the structural features here, however, it is necessary to correlate these observations with the local variations of the electronic structure for further understandings of their effects to the superconducting properties. Such fine scale chemistry features can be analyzed by the EELS in the STEM. It is important to note that the Z-contrast technique allows us simultaneous operations of the EELS since the ADF detector does not interfere with the low angle scattering used for the EELS. This means that the Z-contrast image can be used to position the electron probe accurately over a particular feature for acquisition of a spectrum. Furthermore, the other analytical technique such as the EDS, CBED and the conventional HRTEM can be operative in the single microscope. Future works combining these techniques in the STEM will definitely be promising in elucidating the fundamental structure – property relationships.

ACKNOWLEDGEMENT

We would like to thank Mark Teplitsky of American Superconductor for providing the samples and for valuable discussions. This research is sponsored by NSF (DMR-9803021 and DMR 9601792). KK acknowledges the support from Research Fellowships of the Japan Society for the Promotion of Science for Young Scientists.

REFERENCES

[1] L.N. Bulaevskii, J.R. Clem, L.I. Glazman, and A.P. Malozemoff, Phys. Rev. B 45 (1992) 2545.
[2] L.N. Bulaevskii, L.L. Daemen, M.P. Maley, and J.Y. Coulter, Phys. Rev. B 48 (1992) 13798.
[3] B. Hensel, G. Grasso, and R. Flükiger, Phys. Rev. B 51 (1995) 15456.
[4] G.N. Riley Jr., A.P. Malozemoff, Q. Li, S. Fleshler, and T.G. Holesinger, JOM 49(10) (1997) 24.
[5] O. Eibl, Physica C 168 (1990) 215.
[6] O. Eibl, Physica C 168 (1990) 249.
[7] A. Umezawa, Y. Feng, H.S. Edelman, Y.E. High, D.C. Larbalestier, Y.S. Sung, and E.E. Hellstrom, Physica C 198 (1992) 261.
[8] A. Umezawa, Y. Feng, H.S. Edelman, T.C. Willis, J.A. Parrell, D.C. Larbalestier, G.N. Riley Jr., and W.L. Carter, Physica C 219 (1994) 378.
[9] H.K. Liu, R.K. Wang, and S.X. Dou, Physica C 229 (1994) 39.
[10] O. Eibl, Microsc. Res. Tech. 30 (1995) 218.
[11] S. Horiuchi and E. Takayama-Muromachi, in: H. Maeda and K. Togano (Ed.), Bismuth-based High-temperature Superconductors, Marcel Dekker, Inc., New York, 1996, p. 7.
[12] Y. Yan, M.A. Kirk, and J.E. Evetts, J. Mater. Res. 12 (1997) 3009.

[13] Y. Zhu, Q. Li, Y.N. Tsay, M. Suenaga, G.D. Gu and N. Koshizuka, Phys. Rev. B 57 (1998) 8601.
[14] T.G. Holesinger, J.F. Bingert, J.O. Willis, Q. Li, R.D. Parrella, M.D. Teplitsky, M.W. Rupich, and G.N. Riley Jr., IEEE Trans. Appl. Supercond. 9 (1999) 2440.
[15] S.J. Pennycook and D.E. Jesson, Ultramicroscopy 37 (1991) 14.
[16] D.E. Jesson and S.J. Pennycook, Proc. R. Soc. Lond. A 441 (1993) 261.
[17] D.E. Jesson and S.J. Pennycook, Proc. R. Soc. Lond. A 449 (1995) 273.
[18] P.D. Nellist and S.J. Pennycook, Ultramicroscopy 78 (1999) 111.
[19] E.M. James, N.D. Browning, A.W. Nicholls, M. Kawasaki, Y. Xin, and S. Stemmer, J. Electron Microsc. 47 (1998) 561.
[20] E.M. James and N.D. Browning, Ultramicroscopy 78 (1999) 125.
[21] N.D. Browning, J.P. Buban, P.D. Nellist, D.P. Norton, M.F. Chisholm, and S.J. Pennycook, Physica C 294 (1998) 183.
[22] Y. Xin, E.M. James, N.D. Browning, and S.J. Pennycook, J. Electron Microsc. 49 (2000) 231.
[23] N.D. Browning, I. Arslan, Y. Ito, E.M. James, R.F. Klie, P. Moeck, T. Topuria, and Y. Xin, J. Electron Microsc. 50 (2001) 205.
[24] K. Kishida and N.D. Browning, Physica C 351 (2001) 281.
[25] R.F. Klie, Y. Ito, S.Stemmer, and N.D. Browning, Ultramicroscopy 86 (2001) 289.
[26] N.D. Browning, I. Arslan, P. Moeck, and T. Topuria, phys. stat. sol. (b) 227 (2001) 229.
[27] R.J. Keyse, A.J. Garratt-Reed, P.J. Goodhew, and G.W. Lorimer, Introduction to Scanning Transmission Electron Microscopy, Springer-Verlag New York Inc., New York, 1998.
[28] G. Dusher, N.D. Browning, and S.J. Pennycook, phys. stat. sol. (a) 166 (1998) 327.
[29] H. Adachi, S. Kohiki, K. Setsune, T. Mitsuyu, and K. Wasa, Jpn. J. Appl. Phys. 27 (1988) L1883.
[30] T. Hatano and K. Nakamura, in: H. Maeda and K. Togano (Ed.), Bismuth-based High-temperature Superconductors, Marcel Dekker, Inc., New York, 1996, p. 545.
[31] S. Horiuchi, H. Maeda, Y. Tanaka, and Y Matsui, Jpn. J. Appl. Phys. 27 (1988) L1172.
[32] Y. Matsui, S. Takekawa, H. Nozaki, A. Umezono, E. Takayama-Muromachi, and S. Horiuchi, Jpn. J. Appl. Phys. 27 (1988) L1241.
[33] S. Ikeda, K. Aota, T. Hatano, and K. Ogawa, Jpn. J. Appl. Phys. 27 (1988) L2040.
[34] R. Ramesh, G. van Tendeloo, G. Thomas, S.M. Green, and H.L. Luo, Appl. Phys. Lett. 53 (1988) 2220.
[35] O. Eibl, Physica C 168 (1990) 215.
[36] R. Ramesh, K. Remschnig, J.M. Tarascon, and S.M. Green, J. Mater. Res. 6 (1991) 278.
[37] A. Yamamoto, M. Onoda, E. Takayama-Muromachi, and F. Izumi, Phys. Rev. B 42 (1990) 4228.
[38] Y. Gao, P. Coppens, D.E. Cox, and A.R. Moodenbaugh, Acta Cryst. A 49 (1993) 141.
[39] R.E. Gladyshevskii and R. Flükiger, Acta Cryst. B 52 (1996) 38.
[40] M. Tachiki and S. Takahashi, in: H. Maeda and K. Togano (Ed.), Bismuth-based High-temperature Superconductors, Marcel Dekker, Inc., New York, 1996, p. 153.

FABRICATION OF Bi-2223/Ag TAPES BY CONTROLLING PARTIAL MELTING TEMPEATURE AND COOLING RATE

S. K. Xia and E. T. Serra

CEPEL-Centro de Pesquisas de Energia Elétrica, P. O. Box: 68007,
CEP:21944-970, Rio de Janeiro, RJ, Brazil
E-mail: xia@cepel.br

I. INTRODUCTION

Owing to the large critical current density (J_c) at liquid nitrogen temperature and the feasibility of long length tape production, Ag sheathed $Bi_{2-x}Pb_xSr_2Ca_2Cu_3O_y$ (Bi-2223/Ag) tapes are still considered one of the most promising superconducting materials for large scale applications in the electric power industry.

The high cost of the Bi-2223/Ag tape fabrication is still the main impediment preventing Bi-2223/Ag from competing with conventional materials in the market [1]. The cost includes two parts: one is the cost related to the starting materials and tape processing and the other is associated with the tape performance, i.e. J_c. It has been addressed that the cost/performance must be reduced to as low as 10 \$/kA×m for a strong commercial interest [1,2]. The cost status for Bi-2223/Ag tapes, at present, is about 500 \$/kA×m [3]. Ag sheath is one of the parts that contribute to the high costs. Nevertheless, it is indispensable because it is permeable by oxygen that is necessary for 2223 transformation. Besides, Ag is not oxidized and almost has no reaction with the superconducting core at the tape processing temperature. In addition, Ag assists the conversion of 2212 to 2223 phases and promotes grain alignment [4,5]. Theoretically, one can increase the filling factor of superconducting core to reduce the consumption of Ag. However, There are technical difficulties in increasing the filling factor. First, the high core/Ag ratio increases the difficulty in mechanical deformation during the tape processing. Second, the final tape strength would be too low. Currently, the usual filling factor is around 25% to 35%. Thus, for reaching the competitive price, it is fundamental to improve J_c and reduce tape-processing costs.

At present the J_c of the Ag sheathed Bi-2223 long-length tapes, fabricated by several companies and many research groups, has reached as high as 20 to 40 kA/cm^2 [6-9]. For short tapes, J_c of 70.5 kA/cm^2 has been prepared [10]. However, these values are much more inferior to those of films with J_c of about 1 MA/cm^2 obtained from the same material [11] indicating that there is still large room for a significant improvement in J_c. The reason for the lower J_c in tapes is directly related to the conventional tape processing methods by which 2223 is formed by the conversion of 2212 incorporated with some secondary phases [12-15]. Although highly textured microstructure can be obtained, the ceramic core usually has lower density [16,17]. Void fractions > 15% are always observed in the Bi(Pb)-2223 filaments, even in the tapes with the higher J_c value. In addition, the tapes always contain more than 10% of secondary phases in the oxide core even for the better tapes. Another fact that effects J_c is the mechanical deformation induced cracks which are hardly been effectively healed by the conventional sintering process [18].

The melting process has been successfully used for processing Bi-2212 in order to achieve a higher superconducting core density, good grain connection and alignment [19]. Unfortunately, up to now, there is no any effective melting process method that can be applied for Bi-2223/Ag tapes [20].

Studies of the phase relations in the Bi-Pb-Sr-Ca-Cu-O system have been performed by several groups [20-30]. Phase equilibrium in large temperature range has been obtained. The published data agree well and are complementary. The result indicates that the 2223 phase decomposes to liquid and other cuprates phases at 880 °C in air depending on the Pb doping concentration. With doping of Pb, the decomposition temperature is reduced. According to Park et al. [31], This liquid phase has a composition closed to 2212. Thus 2223 phase is expected to be reversibly formed when annealed at lower temperature with incorporation of other cuprate phases [32]. Dou et al. have suggested a short period partial melting method [33-35]. This method was latterly modified by Hua et al [36,37]. They found that the partial melting method could improve the superconductivity properties in both J_c and J_c - H behavior. Nevertheless, differing from the 2212 tape processing, the excess quantity of the liquid phase is always detrimental for current transport [38,39].

Simplification of tape processing is another important aspect. It is not only the consideration of fabrication cost, but also the requirement of industry production. The simplifications include decreasing in the total processing time, reducing the sintering cycles and increasing in the sintering temperature range, i.e. the sintering processing window.

In the conventional Bi-2223/Ag tape preparation methods, tape processing usually takes hundreds of hours and several intermediate deformations [40,41]. Fast preparation of Bi-2223/Ag tapes has been studied by Dou et al. [42]. They developed a process that can reduce total sintered time to 30 h by using an intermediate quenching. The J_c and Bi-2223 volume fraction in the tapes are comparable to the tapes treated for 120 h. Regarding the sintering processing window, it is particularly important for production of long-length tapes since they occupy large space in furnaces and, therefore, inevitably are inhomogeneously heated during sintering process. The sintering processing window is usually narrower. When the tapes are sintered in air, the temperature window is just a few degrees [41]. Reducing oxygen partial pressure was found an effective method to increase the temperature window. In this case, however, the temperature window is still limited within 10 °C [43].

In our recent works, we proposed a new tape processing method denominated as continuos cooling sintering (CCS) [44-47]. In the CCS method, Bi-2223/Ag tapes are heated up to well above the 2223 decomposition temperatures then slowly cooled to lower temperatures. At the maximum temperatures, a large quantity of liquid phase can be formed and, upon the controlled cooling, the liquid phase re-transforms to 2223 with incorporation of other cuprate phases. The essential point of this process is to control the maximum temperature and the cooling rate. As the CCS process does not involve any lengthy annealing at any fixed temperature, the total sintering time reduces significantly. Another advantage of the CCS process is the large sintering processing window because the sintering temperatures are extended to the region above the 2223 single-phase region. In this chapter, an extensive discussion on the fabrication of Bi-2223/Ag tapes by partial melting, especially the CCS process will be given. The section II describes the CCS procedure. The optimization of the processing conditions and the experimental results of pressed (short) and rolled (1 meter long) tapes, prepared by the CCS process, will be presented in sections III to V. In section VI, the fundamental of partial melting process will be discussed and a summary will be given in section VII.

II. Bi-2223/Ag TAPE PROCESSING BY THE CCS METHOD

The powder-in-tube (PIT) technique has been used to process the Ag-sheathed Bi-2223 multi-filaments tapes in this work. Precursor powder from UNIQUEST LTD was used, which has a cation ratio of Bi:Pb:Sr:Ca:Cu = 1.84:0.35:1.91:2.05:3.06. Before filling into the silver tube, the precursor powder was pre-heated to 800 °C for 12 h to further reduce the residual carbon element in the precursor. The precursor treated in this way contains carbon less than 300 ppm. The powder was packed into silver tubes of 5.90 mm outer and 4.35 mm inner diameters in a glove box filled with N_2 gas. The protection from environment humidity is important to avoid the bubbles caused by the moisture. Tapes with 19 and 55 filaments have been produced in our group as shown in Fig. 1. For making the multi-filament tapes, tubes filled with

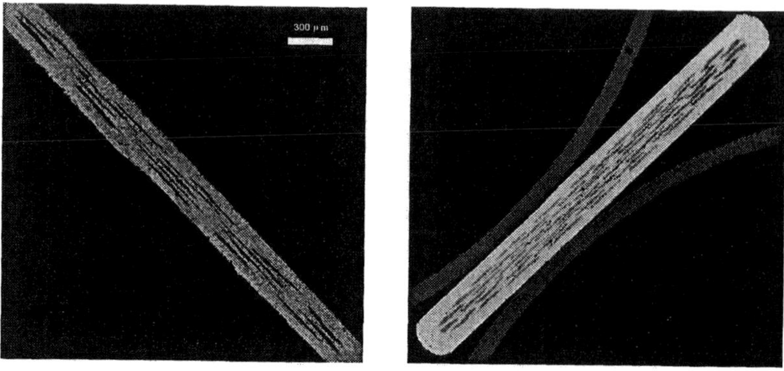

Fig.1. Transverse sections of the Bi-2223/Ag tapes with a)19-filaments and b) 55-filaments.

powder (composites) were drawn first into a wire with hexagonal cross section. Then, the desired number of wires were bundled and filled into a silver tubes of 5.90 mm outer and 4.35 mm inner diameters for 19-filament tapes and tubes of 10 mm outer and 8 mm inner diameters for 55 filament tapes. Then the multifilament wires were drawn into the diameter of 1.6 -1.2 mm and rolled into tapes of 0.23 to 0.35 mm thick.

Tapes with 4 cm and 1 m in length were prepared. For the short tapes (4 cm), the intermediate mechanical deformations between the sintering cycles were performed with uniaxial pressing using pressure of 15 tones. For the 1 meter long tapes, rolling was used for mechanical intermediate deformations with reduction of 20%.

The thermomechanical processing regimes used in our work are two sintering cycles (HT1 and HT2) and three sintering cycles (HT1, HT2 and HT3) with intermediate mechanical deformations, as shown in Fig. 2 and Fig. 3 respectively. As will be shown below, the two-cycle and three-cycle heat treatments do not have significantly difference regarding the final current transport properties for the short length tapes if pre-sintering is properly performed. However, three-cycle heat treatment would be useful for production of long-length tapes where bubbles appear frequently. Bubbles, which are harmful to the transport properties, are caused by relatively high carbon content or/and humidity in the precursor powders. Upon mechanical deformation, the bubbles can be removed and, in the next sintering cycles, the probability of bubbling is much reduced. Thus, one more intermediate mechanical deformation could diminish the effect of the bubble on the tapes. For the short-length tapes, small amount of gas can escape from the two ends of the tapes, therefore, the probability of bubbling in the short-length tape is much less than the long-length tapes.

The CCS method is applied in the last sintering cycle in both the two-cycle and three-cycle regimes. To simplifying the description, we call the sintering cycles before the last one *pre-sintering* and the last sintering cycle the *final sintering*. In the case of three-cycle sintering process, HT1 + HT2 are the pre-sintering and HT3 is the final

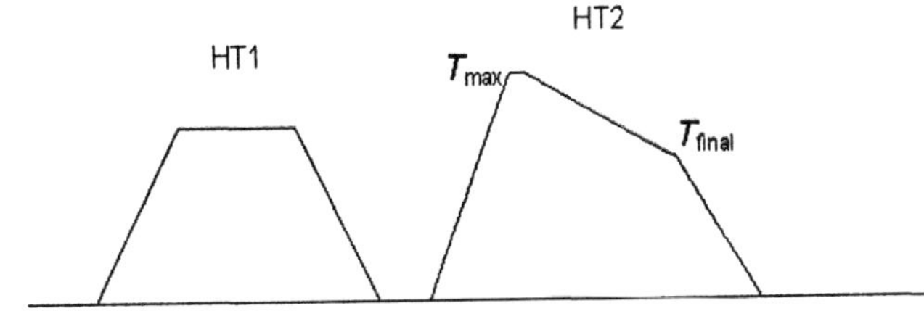

Fig. 2. Two-cycle sintering regime of the CCS process.

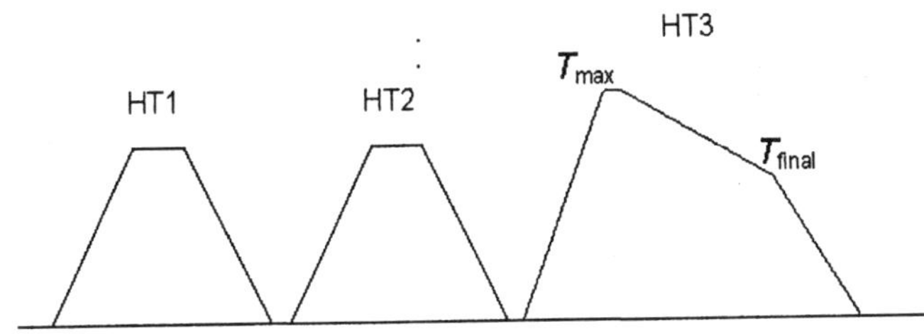

Fig. 3. Three-cycle sintering regime of the CCS process.

sintering cycles, while the HT1 is regarded as pre-sintering and HT2 is the final sintering in the case of two sintering cycles. In the final sintering cycle, where the CCS process is applied, tapes are heated to a maximum temperature T_{max} then, slowly cooled to temperature T_{final}, followed by a normal cooling to room temperature. All the heat treatments were performed under 1 atmosphere of a gas mixed with 7.5% oxygen balanced with nitrogen.

In the experiments, a tube furnace with three heating zones from LINDBERG was used for the tape sintering. The furnace was calibrated within ± 1 °C.

III. SHORT TAPES PREPARED BY THE CCS PROCESS

The structures and superconducting properties of the short-length tapes obtained by two-cycle sintering have been extensively studied and the results will be presented in sections III.1 to III.4. Since the heat treatment would be similar to the conventional 2223 tape sintering process if the maximum heating temperature goes down to the 2223 single-phase region, the comparison can be made between the partial melting process and conventional process by varying the T_{max}. In order to gain sufficient sintering time for the lower T_{max}, as usually required for conventional tape processing, the tapes heated to $T_{max} < 845$ °C are held at T_{max} for 10 h. The tapes heated to $T_{max} \geq 845$ °C, at which partial melting takes place, are held at T_{max} only for 10 min. In the two-cycle regime, all tapes are pre-sintered in HT1 at 830 °C for 30 h.

The comparative results of the tapes prepared by three-cycle sintering will be given in the section III.5.

III.1. CURRENT TRANSPORT PROPERTIES

The critical currents (I_c) of the tapes prepared by CCS method were measured using a four-probe method with a voltage criterion of 1μV/cm. J_c was obtained from the measured I_c and the overall transversal filament area. I_c versus applied magnetic field H was measured with the magnetic field parallel and perpendicular to the tape surface.

Fig. 4 shows the critical current density J_c, at 77 K and self-field, as a function of the maximum heating temperature T_{max} for tapes cooled to T_{final} of 790 °C and 770 °C. It reveals that the samples cooled to 770 °C exhibit better transport properties than those cooled to 790 °C. Although there is no significant change of J_c in the T_{max} range from 815 to 855 with T_{final} = 790, the J_c is indeed improved for the higher T_{max} heated samples with T_{final} = 770. The J_c reaches 30.5 kA/cm^2 for the sample heated to 855 °C. When T_{max} is higher than 855 °C or lower than 815 °C, J_c drops dramatically.

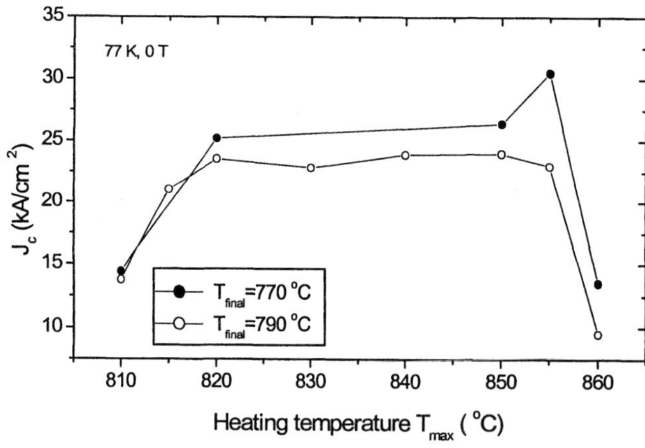

Fig. 4. J_c versus the maximum heating temperature T_{max} in the CCS process (T_{final} = 770 and 790 °C).

Fig. 5 shows the J_c, at 77 K and self-field, as a function of T_{final} for the tapes heated to T_{max} of 850 °C. It is evident that, at the cooling temperature T_{final} around 770 °C, the largest J_c can be obtained.

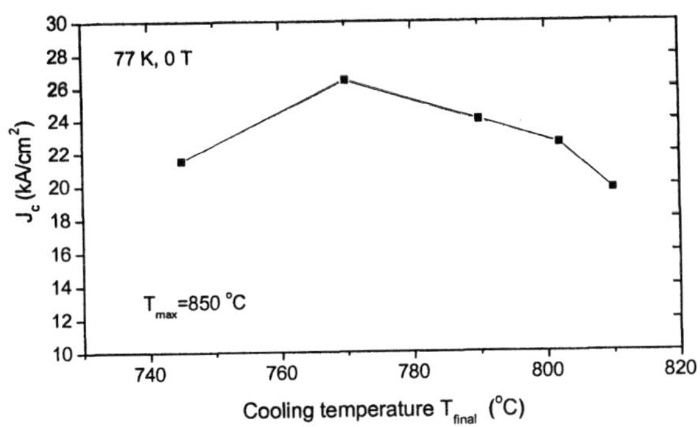

Fig. 5. J_c versus the cooling temperature T_{final} of the tapes heated to T_{max} = 850 °C in the CCS process.

Measurements of J_c under applied of magnetic fields can provide information about weak and strong links, which are related to grain connectivity, texturing and vortex pinning property [48]. Fig. 6 shows the normalized critical current density (J_c/J_{c0}) at 77 K as a function of applied magnetic field (H) for the samples heated to different T_{max} and cooled to T_{final} =770 °C. The fields are applied either perpendicular or parallel to the tape surface. It is clearly seen, from the curves of $H//ab$, that in the low field region the 850 °C heated sample drops more slowly than those heated to the lower temperatures and also than the 860 °C samples. Ekin et al. [49] suggested that at low magnetic field, a rapid drop in the J_c be attributed to Josephson weak links in the

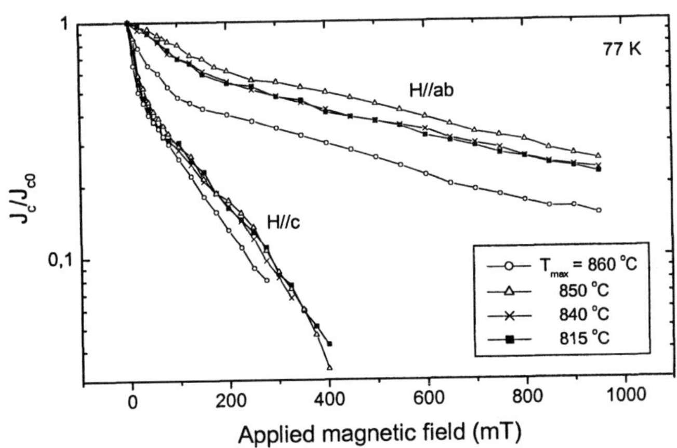

Fig. 6. The normalized J_c versus applied magnetic field of the Bi-2223/Ag tapes prepared by CCS method.

grain boundaries with poor grain connectivity and larger grain boundary angles. Thus, the better $J_c \sim H$ behaviour indicates an improved microstructure in the 850 °C heated sample. At the magnetic field of 950 mT, the 850 °C heated sample exhibit the (J_c/J_{c0}) of 26% against the values of around 21% for the 840 and 810 °C samples. In the case of *H//c*, there is no significant difference between samples heated to different T_{max}. This can be explained in the framework of the railway-switch model [50]. Based on this model, the field dependence with *H//c* is mainly the pinning-dependent, which is associated with intra-grain properties, such as structure, defects, composition, etc..

III.2. PHASE EVOLUTION

The phase content of the ceramic cores of the tapes was studied by X-ray diffraction (XRD) using Cu-Kα radiation. In order to study the melting behavior of the tapes, quenching has been performed at the different temperatures. All the samples for

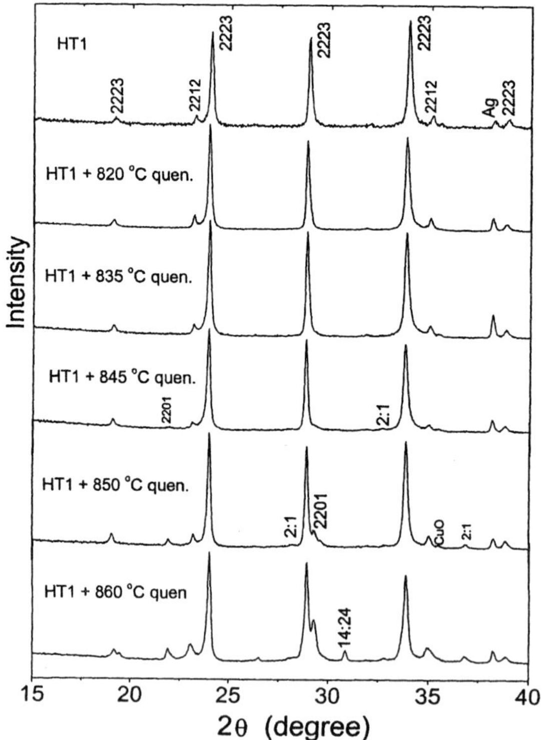

Fig. 7. XRD patterns of the Bi-2223/Ag tapes quenched at the indicated temperatures. The tapes heated to temperatures 845 °C were held at the quenching temperature for 10 min, while tapes heated to temperatures < 845 °C were held at the quenching temperatures for 10 h. Before quenching, the tapes had been sintered by HT1 (see Fig.2).

quenching experienced the pre-sintering heat treatments. The 2223 phase is dominant in the pre-sintered tapes, as shown in the Fig. 7. Two groups of samples are divided for quenching. In the group one, with quenching temperature < 845 °C, the tapes are heated to the quenching temperature and held at these temperatures for 10 h then quenched by quickly removing the samples from the furnace to air at room temperature. For the group two with quenching temperature ≥ 845 °C, the samples are heated to the quenching temperature and held at these temperatures for 10 min then quenched as those in the group one.

The XRD patterns of the quenched tapes reveal that, when tapes are quenched in the temperature range within 820 and 835 °C, the XRD patterns are similar to that of the pre-sintered tape. For the samples quenched at 845, 850 and 860 °C, several second phases appear beside 2212. The secondary phases are identified as $Bi_2Sr_2CuO_x$ (2201) and $(Sr,Ca)_2CuO_3$ (2:1 AEC). For the 860 °C quenched sample, beside these second phases, the $(Sr,Ca)_{14}Cu_{24}O_{41}$ (14:24 AEC) is also observed. The partial melting can be visually confirmed since some black liquid flows out from the two ends of the tapes. This was seen for the 845, 850 and 860 heated tapes. It is well known that the 2201 raises from the fast crystallization of the liquid phase even in the case of quenching [22]. Therefore, the amount of 2201 is proportional to the quantity of the liquid phase. It is evident that the contents of these secondary phases increase with increasing of heating temperature. From these results, one can concludes that temperatures ≥ 845 °C are above 2223 decomposition temperature.

Fig. 8 displays the XRD patterns of the tapes processed after complete sintering cycles. The tapes were heated to different T_{max} and cooled to T_{final} = 770 °C in the final sintering cycle. It can be seen that the secondary phases formed at the high temperatures, as shown by the quenched tapes in Fig. 7, are significantly transformed to 2223 phase. It is evident that control of the partial melting temperature is crucial for obtaining the high purity of ceramic cores. In the present experimental conditions, temperatures between 850 to 855 °C are optimized for obtaining high purity 2223 cores. When temperature is too high such as 860 °C, large quantity of secondary phases appears including 2201, cuprate secondary phases, and Pb-rich phases. Another interesting result is the considerable reduction of 2212 phase in the high temperature sintered tapes compared with those sintered at lower temperatures. It is known that the 2212 and 2223 have similar melting temperature [30]. At the 2223 decomposition temperature, 2212 also decomposes. The controlled cooling seems to only recover the 2223 phase but not 2212. This may be explained by the fact that the samples pass over the 2223 single-phase region during cooling. It is also noted from Fig. 8 that higher sintering temperature causes considerable residual Pb-rich phase in the tapes as compared with the lower T_{max} tapes. The Pb-rich phase shows a diffraction peak at 2θ ~ 30.5°, one of the principle peaks from Ca_2PbO_4. Such identification is consistent with the SEM observation, as shown below. For the very low T_{max}, for instance 810 °C, both Pb-rich and 2212 phases increase.

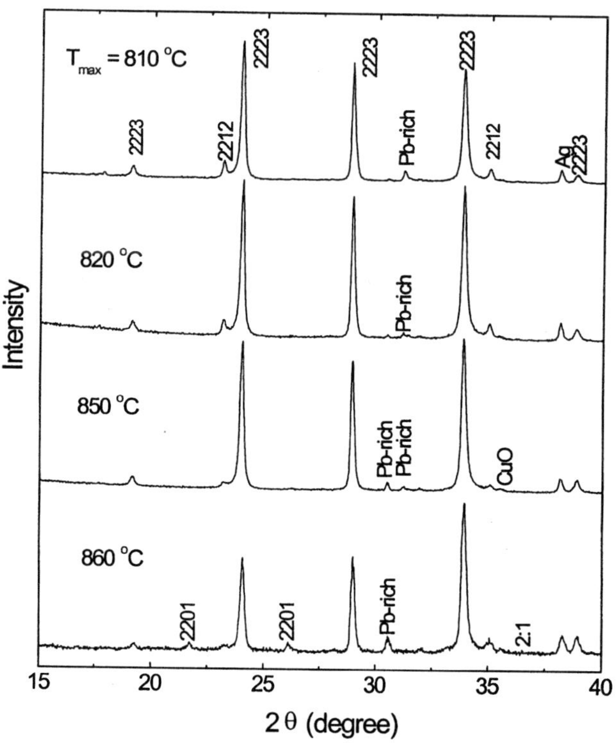

Fig. 8. XRD patterns of the Bi-2223/Ag tapes heated to different T_{max} and cooled to T_{final} of 770 °C.

The effect of T_{final} on the final phase content is shown in Fig. 9, where all the samples are heated to T_{max} of 850 °C. It indicates that the Ca_2PbO_4 (30.5°) and $(Pb,Bi)_3Sr_2Ca_2CuO_x$ (3221) (31.8°) increase with decreasing of T_{final}. This is the common feature of 2223 tape upon slow cooling [51]. On the other hand, annealing at lower temperatures (below 800 °C) can effectively depress the 2201 phase, which usually lies between the 2223 grains and, therefore, is extremely detrimental to the transport properties. The compromise of these two contributions leads to the maximum in the J_c curve in Fig. 5 at T_{final} of 770 °C.

From the above results, three temperature ranges can be divided. The middle temperature range (820 to 835 °C) is inside the 2223 single-phase region. Temperatures higher than 845 °C and lower than 810 °C are out of 2223 stability region. In the former case, 2212 and 2223 decompose forming a liquid phase and some other cuprates. In the latter case, the decomposition of 2223 into 2212 and Pb-rich phases takes place. It should be mentioned that, from our XRD results, it is difficult to determine the precise temperature boundaries that separate the regions.

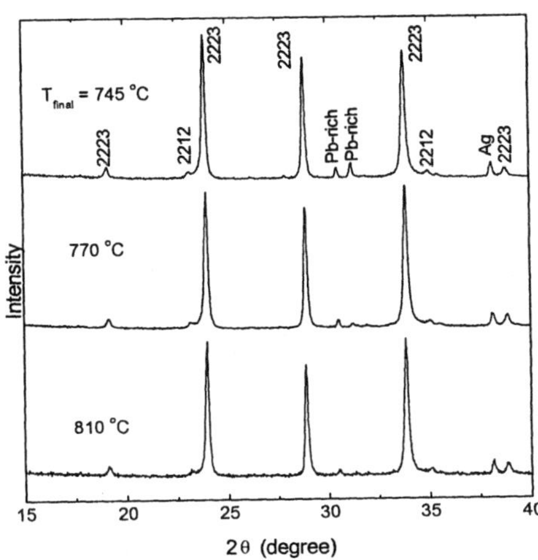

Fig. 9. XRD patterns of the Bi-2223/Ag tapes heated to T_{max} of 850 °C and cooled to the different T_{final}.

III.3. MICROSTRUCTURES

The microstructures of the tapes were studied by scanning electron microscope (SEM). The identification of different phases was done by energy dispersion X-ray spectroscopy (EDS).

Fig. 10 shows the microstructures of the Bi-2223/Ag tapes, where tapes quenched at 850 °C and 830 °C are also displayed for comparison. It is clearly seen that the 850 °C quenched sample contains large amount of secondary phases including 2201, 2:1 AEC and (Sr,Ca)CuO2 (1:1 AEC) while 830 °C quenched sample maintains higher purity of 2223. Namely, decomposition of 2223 takes place in the tape heating to 850 °C, which is consistent with the XRD observations in Fig. 6. Another strike feature is the better grain connection of the high T_{max} tapes, as can be noted by comparison of the morphology of the 850 °C quenched tape with the 830 °C quenched one. This is attributed to the large quantity of liquid phase as indicated by the 2201 phase in Fig. 10a).

The 2223 retransformation upon the slow cooling is evident in the photo Fig. 10d) for the 850 °C heated tape when compared with the same temperature quenched tape (in Fig. 10a). The secondary phases from the decomposition of 2223, such as 2201 and 2:1 AEC is hardly seen. Instead, some large size grains identified as Pb-rich phase appear. It is obvious that partially melted tapes maintain higher grain packing density and large grain size (Fig. 10(c) and (d)) in comparison with the low temperature heated samples (Fig. 10(e) and (f)). The porous morphology shown in Fig.

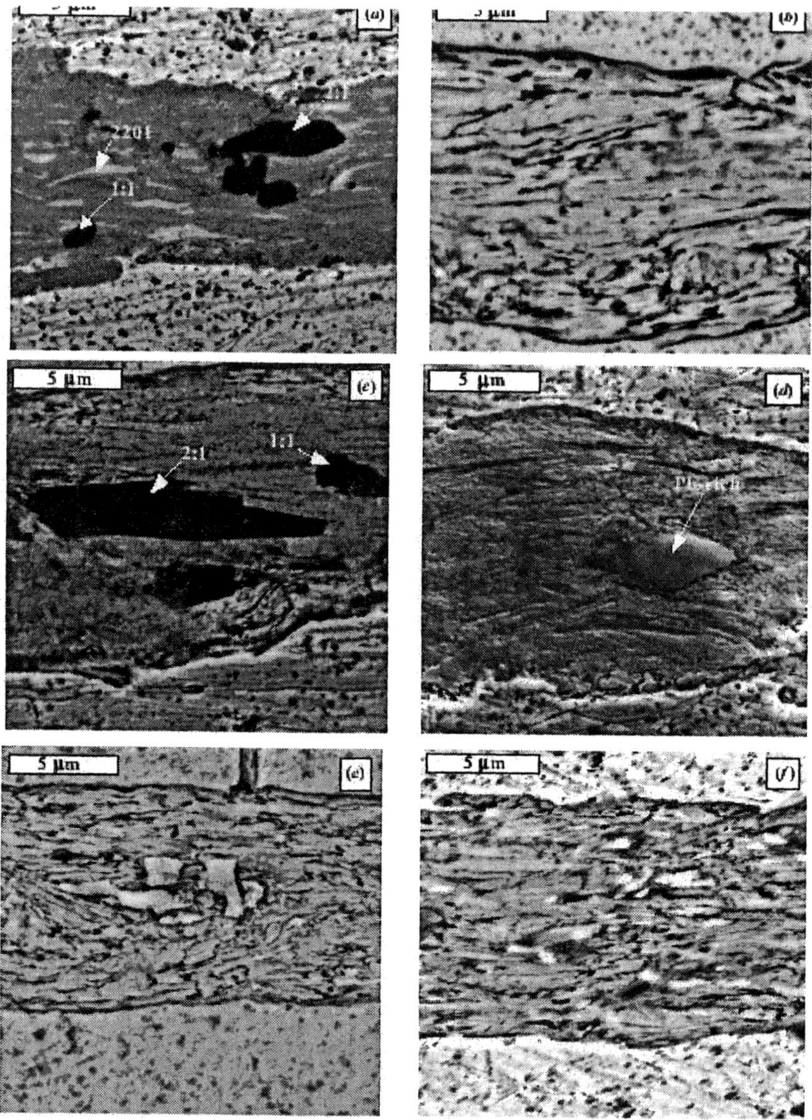

Fig. 10. SEM images showing the microstructure of the tapes a) quenched at 850 °C, b) quenched at 830 °C and the tapes prepared by the CCS process with T_{max} at c) 860 °C, d) 850 °C, e) 820 °C and f) 810 °C and with T_{final} of 770 °C.

10 (e) and (f) is the common feature for the 2223 tapes prepared by the conventional method. Beside the porosity, the low temperature heated tapes also show smaller grain size.

The SEM photos indicate once again that control of partial melting temperature is very important. For the samples heated to 850 °C, the main secondary phase is Ca_2PbO_4. However, when tapes are heated to 860 °C, in addition to the Pb-rich phase, there are also considerable other secondary phases such as 2201, 2:1 AEC, 1:1 AEC and 14:24 AEC etc..

III.4. EFFECT OF COOLING RATE

In the CCS process, the tapes are heated to T_{max} then slowly cooled down to T_{final} without any length annealing, therefore, one can expect a significant reduction of tape processing time. On the other hand, retransformation from liquid to 2223 requires a certain time, therefore, too fast cooling would lead to large quantity of residual secondary phases.

Fig. 11 shows the J_c (77 K and self-field) as function of the sintering time in HT2, which is the cooling time from T_{max} to T_{final} and is proportional to the cooling rate. The T_{max} and T_{final} were fixed to 850 °C and 770 °C, respectively. It shows that the tape sintered for 10 h has J_c comparable to those sintered for 67 h. When sintering time is shorter than 10 h, J_c reduces rapidly. Considering the pre-sintering time is 30 h, one can see that the total sintering time can be as short as about 40 h without significantly decrease in J_c.

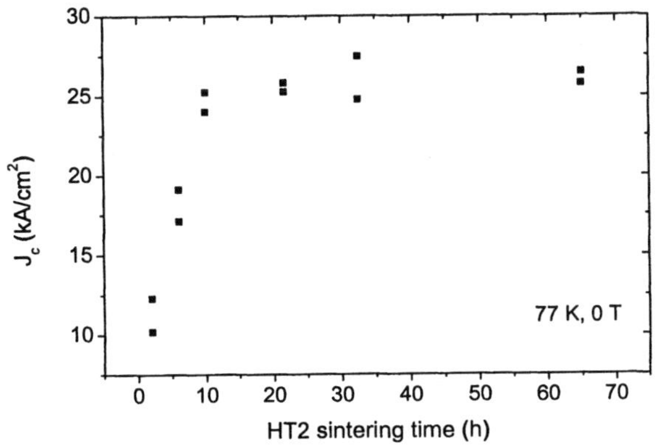

Fig. 11. J_c versus the HT2 sintering time, i.e. the cooling time from T_{max} =850 to T_{final} = 770 °C (see the two-cycle regime in Fig. 2).

The cooling rate dependence of $J_c/J_{c0} - H$ is shown in Fig. 12 with applied magnetic fields parallel and perpendicular to the tape surface. All these samples were heating to T_{max} = 850 °C and cooling to T_{final} = 770 °C. It reveals that the J_c decays more rapidly with increase in cooling rate in both H applied directions. Usually the Bi-

2223/Ag tapes shows J_c/J_{c0} of 20 to 30% under applied magnetic field of 1 T. This is shown for the tapes sintered for ≥10 h. However, when the sintering time is less than 10 h, the J_c/J_{c0} is much less than 20% at this field.

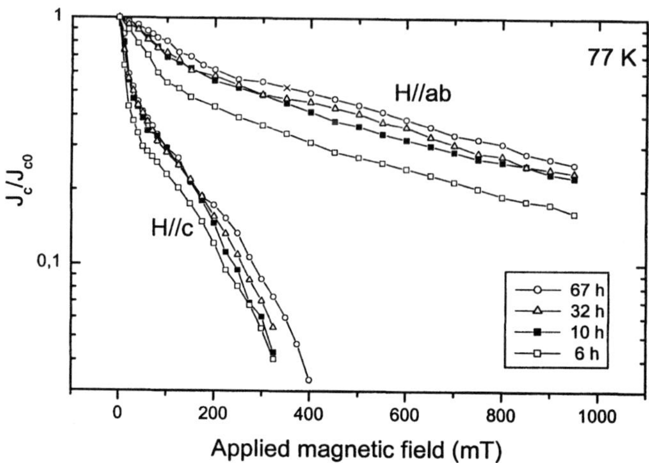

Fig. 12. J_c/J_{c0} of the tapes versus applied magnetic fields H parallel and perpendicular to tape surface for the tapes sintered with different HT2 time in the two-cycle process (see Fig.2), i.e. the cooling time from T_{max} = 850 to T_{final} = 770 °C.

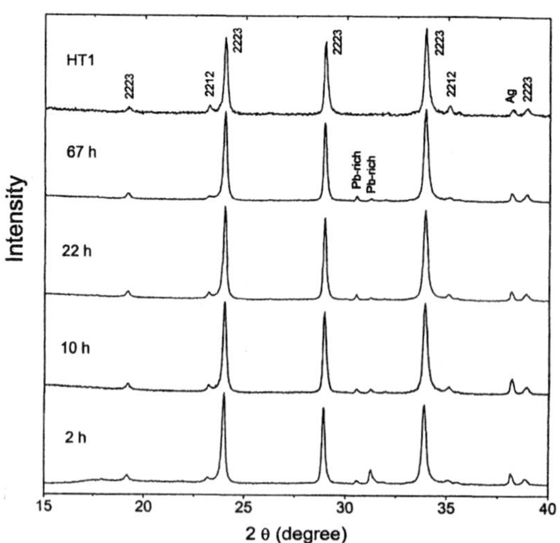

Fig. 13. XRD patterns of the tapes obtained with different HT2 time in the two cycle process (see Fig.2), i.e. the cooling time from T_{max} = 850 to T_{final} = 770 °C.

Fig. 13 shows the XRD patterns of the tapes with different HT2 cooling time. All the samples are heated to T_{max} = 850 °C and cooled to T_{final} = 770 °C. The XRD pattern of the pre-sintered tape is also displayed in the figure. Compared with the 850 °C quenching tape (see Fig. 7), significant 2223 retransformation occurred in the tape cooled even with rate 0.66 C/min, i.e. it was cooled from 850 to 770 °C with in only 2 h. However, compared with the slower cooled tapes with cooling rate less than 0.15 °C/min (cooling time longer than 10 h), the 2 h cooled tape contains larger quantity of Pb-riche phase, which is sufficient to depress the current transport properties. It is clearly seen that the Pb-rich phase decreases monotonously with decrease in cooling rate. Thus complete retransformation to 2223 indeed needs certain time to take place. As consequence of fast cooling, the secondary phases formed at higher temperature will transform to other secondary phases instead of 2223 phase, such as Pb-rich phases.

The effect of cooling rate on the microstructure of the tapes is shown by the SEM images in Fig. 14. Fig. 14 (a) shows the microstructure of the tape obtained at the end of pre-sintering. It can be seen that 2223 formation is nearly complete with a small

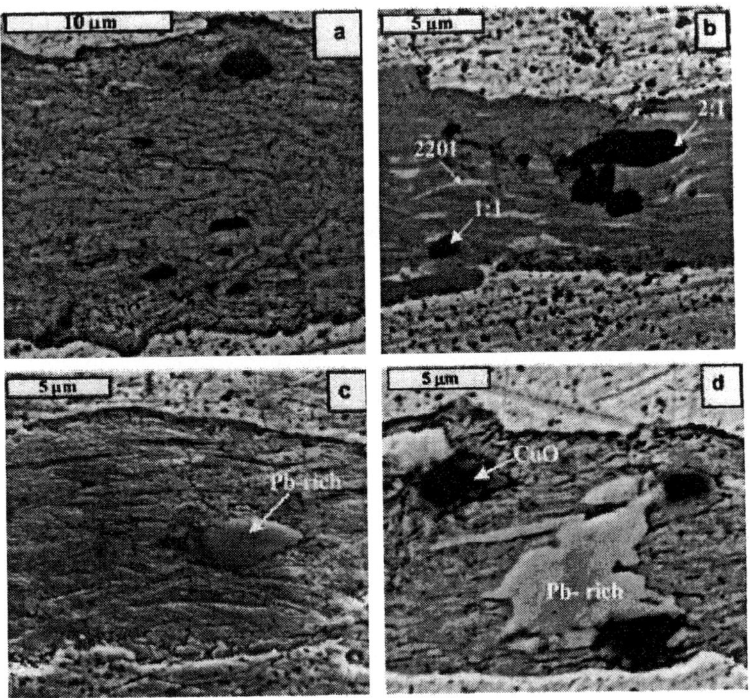

Fig. 14. SEM images of a) the tape sintered in HT1, b) the tape sintered in HT1 followed by heating to 850 °C, holding at 850 °C for 10 min, then quenchingin air, c) the tape sintered by HT1 + HT2 with cooling rate of 0.02 °C/min from 850 to 770 °C in HT2 and d) the tape sintered by HT1 + HT2 with cooling rate of 0.66 °C/min in HT2.

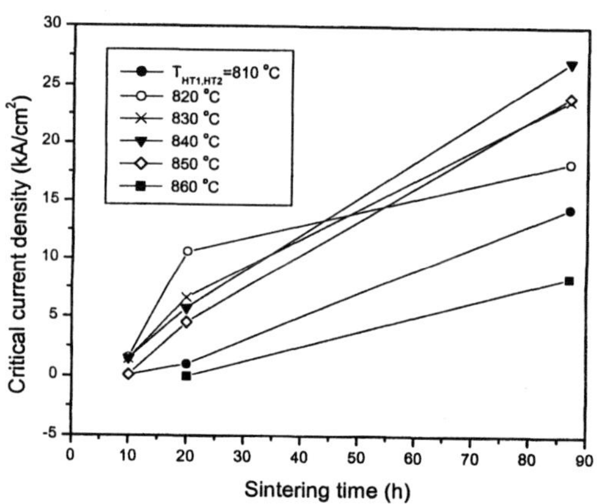

Fig. 15. J_c versus the sintering time in the three-cycle sintering process for different pre-sintering temperatures. The CCS process was applied in HT3 with T_{max} = 850 °C and T_{final} = 770 °C.

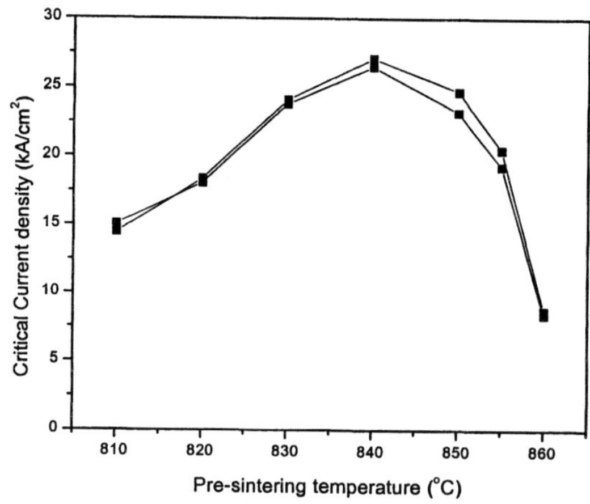

Fig. 16. J_c versus pre-sintering temperature in the three-cycle sintering process. The total pre-sintering time is 20 h. The CCS process was applied in HT3 with T_{max} = 850 °C and T_{final} = 770 °C.

Based on the above result, we know that 840 °C is near the upper limit of 2223 single-phase region and, therefore, higher reaction kinetics is expected. As result, the phase equilibrium can be reached within short time. When the pre-sintering temperature is lower, although the sintering temperature is inside the 2223 single-phase region, the lower reaction kinetics requires more time for accomplishing the 2223 phase transformation. When the pre-sintering temperature is above the 840 °C, tapes are sintered in a temperature region that is out of the 2223 single-phase region. Thus, beside the 2223 phase, there are other secondary phases in equilibrium with amount of secondary phases. When tapes are heated up to the T_{max} = 850 °C in HT2, large quantity of secondary phases appears due to the decomposition of 2223 displayed in Fig. 14 (b) that displays a tape held at 850 °C for 10 min then quenched in air. One sees that the 2201 (white color phase) grows along the 2223 *ab* plane and locates between the 2223 grains. Fig. 14(c) and (d) show the microstructures of the tapes cooled at the rates of 0.02 °C and 0.66 °C/min, respectively. These two tapes were heated to T_{max} = 850 °C and cooled to T_{final} = 770 °C. For the slowly cooled sample (Fig. 14 (c)), the high temperature formed secondary phases are significantly re-transformed to 2223. Beside small amount of residual Pb-rich phase, it is hardly to find other secondary phases by SEM. On the other hand, the fast cooled tape (Fig. 14 (d)) contains much more secondary phases including the larger size of Pb-rich phase and cuprates.

III.5. EFFECTS OF PRE-SINTERING TIME AND TEMPERATURE

In the two-cycle sintering process, pre-sintering at temperatures of 820, 830 and 840 °C has been tested and the results show no significant difference in the final J_c. We suggest that the pre-sintering serve for converting 2212 to 2223. No matter what sintering temperatures are used, the final tapes would show the similar properties if the pre-sintered tapes contain a dominant 2223. Thus in the two-cycle sintering process, 30 h sintering at 820 to 840 °C is long enough to convert majority 2212 to 2223. One expects that, at a lower sintering temperature, a longer sintering time is required for obtaining the same amount of 2223 than that of sintering at higher temperatures.

When sintering time is short, the phase equilibrium may not be reached and the pre-sintering temperature would have effect on the microstructure and properties even for the completely processed tapes. Such effect is studied for the tapes processed by three-cycle sintering with short and fixed sintering time. These experiments also aim to optimize the processing condition of the three cycle sintering, which may be useful for the long tape fabrication, as mentioned in section II. The tests were performed on the short-length tapes with 55-filaments.

In all experiments of the three-cycle process, the temperature of HT2 was maintained as equal as that of HT1. The HT1 and HT2 sintering times are fixed to 10 h for each. The HT3 regime was fixed as the follows: tapes are heated to 850 °C, held at this temperature for 10 min and then slowly cooled to 770 °C at the rate of 0.02 °C/min. The optimization was done by varying the pre-sintering temperatures.

Fig. 15 shows the J_c as a function of sintering cycle for different pre-sintering temperatures. The results indicate that there is a significant effect of the pre-sintering temperatures on the J_c of the tapes completely processed. The highest J_c was obtained for the tape with pre-sintered temperature of 840 °C. The comparison is clearer in Fig. 16. The J_c of the 840 °C pre-sintered tape is 27 kA/cm^2 that is similar to that of the tapes prepared by two-cycle sintering process.

2223. The results indicate that with an adequate pre-sintering temperature one can reduce significant processing temperature. The ideal pre-sintered temperature is that near the 2223 decomposition temperature.

The XRD patterns of the pre-sintered (HT1+HT2) tapes and the completely processed tapes are shown in the Fig. 17 and 18, respectively. It can be seen from Fig. 17 that, in the tape pre-sintered at 810 °C, only small amount of 2212 converts to 2223. With increasing in pre-sintering temperature, the 2223 fraction increases until 840 °C, at which almost all the 2212 converts to 2223, then, decreases when temperature is further increased. In the tape pre-sintered at 860 °C, there is almost no transformation from 2212 to 2223 during pre-sintering. It is also seen that, when the pre-sintering temperature is at and above 840 °C, the 2201 phase appears, which indicates existence of certain amount of liquid phase as mentioned above.

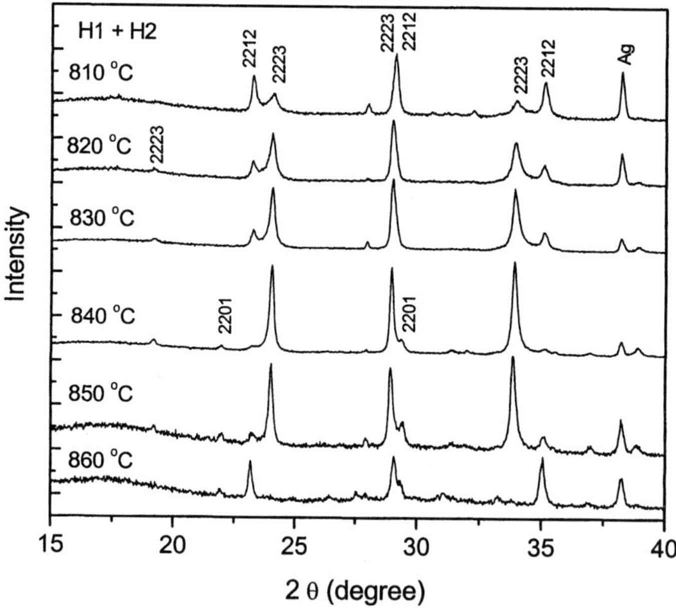

Fig. 17. X-ray diffraction patterns of the tapes pre-sintered through HT1+HT2 with different pre-sintering temperatures. The total pre-sintering time is 20 h.

For the completely processed tapes, (Fig. 18), the 2223 phase is dominant in all the tapes no matter what is the contents of 2223 phase in the pre-sintered tapes. Except the tapes pre-sintered at 810 and 860 °C, the 2223 phase contents in other tapes are similar. Combining with the results of J_c (Fig. 16), one sees that the 2223 phase content in a pre-sintered tape has a significant effect on the superconducting properties of the completely processed tapes. This may be explained as follows. If there is no significant 2223 conversion in the pre-sintering process, the conversion would takes place in the final sintering cycle. As the transformation involves variation of volume, the conversion will lead to cracking, poor grain alignment and connection. Thus, though the tapes contain high purity of 2223, these defects will affect the transport properties. On the other hand, if the 2223 phase is accomplished in the pre-sintered tape, these drawbacks can be significantly decreased.

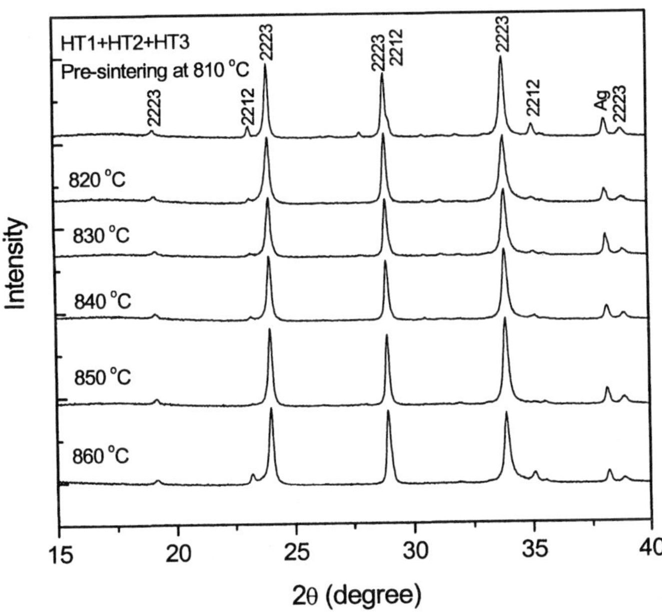

Fig. 18. X-ray diffraction patterns of the tapes after three-cycle sintering. The pre-sintering temperatures are indicated in the figure.

IV. ENHANCEMENT OF SINTERING PROCESSING WINDOW

While the precise temperature control can be usually obtained only in laboratories, a feasible industry production of Bi-2223/Ag tapes needs a tolerance in processing temperature. This is not only the requirement for decreasing the technical difficulties, but also for the reduction of the tape fabrication cost. Thus a broad sintering temperature window is an important parameter for large-scale tape production.

Using conventional tape preparation methods, i.e. tapes are sintered within the 2223 single-phase region, the sintering temperature window is narrow [41,43]. Although the 2223 has a quite large single-phase region, the effective sintering only concentrates the upper limit of this region because of the kinetic reason. It is only few degrees when sintering is performed in air and about 10 degree if sintering is done under an atmosphere of 7.5%O_2 balanced with an inert gas.

We have shown in Fig. 4 that the J_c is improved when a suitable partial melting temperature T_{max} is chosen. Nevertheless, tapes sintered within the 2223 region also maintain comparable high J_c values. Thus, within a tolerance range, one can consider the J_c obtained at some partial melting temperatures are comparable with those obtained at the temperatures within the 2223 single-phase region. Fig. 4 reveals that the J_c values maintain comparable from T_{max} = 815 to 855 °C with in J_c variation of 20%. In other words, the sintering processing window can be as large as 40 °C. Thus, the dramatically enhancement of the sintering processing window result from the fact that the higher J_c can be obtained not only in the 2223 single-phase region but also can be achieved at the temperatures above the 2223 decomposition temperatures.

V. ONE-METER LONG TAPES PREPARED BY THE CCS METHOD

The one-meter long tapes with 19 filaments are prepared by the CCS method. For heat treatment, tapes are wound on a cylinder made from alumina. The intermediate mechanical deformation, with a reduction per pass of about 20 % for all the tapes, is done by rolling from a roller with diameter of 100 mm.

Fig. 19 shows the critical current I_c as a function of the position along the tapes processed by two-cycle and three-cycle sintering. For the two-cycle process, the sintering temperature in HT1 is 830 °C and T_{max} and T_{final} in HT2 are 850 and 770 °C, respectively, with cooling rate of 0.02 °C/min from T_{max} to T_{final}. In the case of three-cycle process, the sintering temperature in HT1 and in HT2 is 840 °C, at which the sample is held for 10 h in each cycle. The regime of HT3 is the same as the final sintering cycle for the two-cycle process. For comparison, a tape prepared by two-step method is also displayed in the figure. The two-step method involves two-cycles of sintering and an intermediate mechanical deformation [52,53]. In the first sintering cycle, the tape is annealed at 830 °C for 30 h. For the second cycle, the tape is sintered at 835 °C for 20 h followed by a cooling at 1 °C/min until 800 °C at which the tape is annealed for 30 h then cooled to room temperature at rate of 0.5 °C/min. Thus, the two-step method represents a conventional tape processing method; i.e. the sintering does not involve decomposition of 2223 phase. The results indicate that the tape, processed by the two-cycle CCS, exhibits better transport properties than that prepared by three-cycle CCS process. The I_c is about 14 A along the tape for the former corresponding to a J_c of 13.5 kA/cm^2 against the J_c of 12 kA/cm^2 of the latter. The tape prepared by the two-step method shows the worst result with a J_c of 10.5 kA/cm^2.

These three tapes do not show bubbles after pre-sintering. Thus, it is not necessary to apply two times of intermediate rolling in order to reduce the bubbles as mentioned in the section II. More deformation would induce more cracks, especially the transversal cracks, a common drawback for the rolled tapes. The Fig. 19 clearly reveals that the partially melted tapes exhibit significantly higher J_c than the tape prepared by two-step process, a conventional method. Compared with the result of the short tapes (in Fig. 4), the CCS method has a more positive effect on the rolled tapes than on the pressed tapes. This may be explained by the fact that J_c is more sensitive to the transversal cracks, which are common in the rolled tapes, than the longitudinal cracks induced usually by pressing. Therefore, effective healing of the transversal cracks with the liquid phase, enhanced by the partial melting process, can improve significantly J_c.

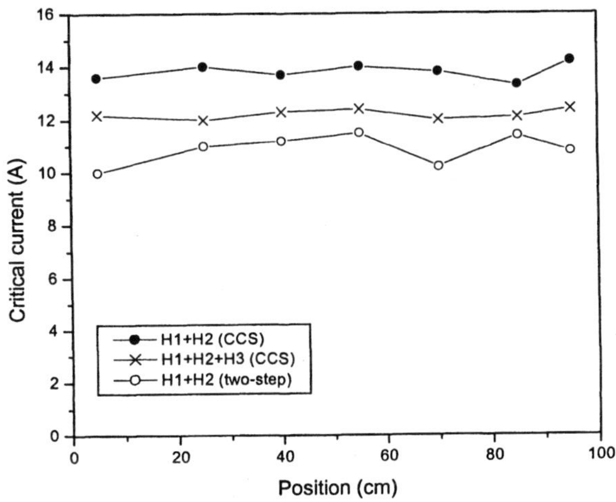

Fig. 19. Current densities of the one-meter long tapes prepared by the CCS two-cycle process, the CCS three-cycle process and the two-step sintering method.

VI. FUNDAMENTAL OF PARTIAL MELTING PROCESS

VI.1. MELTING EQUILIBRIUM OF Bi-2223 PHASE

The phase equilibrium of Bi-Sr-Ca-Cu-O system around 2223 single-phase region has been extensively studied. Fig. 20 displays the phase diagram obtained by Majewski [30] regarding the effect of Pb on the phase equilibria. It shows that the Bi(Pb)-2223 phase decomposition temperature depends sensitively on the Pb concentration. With absence of Ag, it can varies from 880 °C for the Pb-free 2223 phase [31] to 850 °C for the maximum Pb soluble concentration, which is about 15% of substitution for Bi. There is a temperature region immediately above 2223 decomposition temperature where 2223 phase is in equilibrium with several secondary phases, liquid, 2:1 AEC and 14:24 AEC. As shown in the Fig. 20, the 2223 single-phase region is also significantly Pb dependent. Immediately below the lower limit of

2223 single-phase region, 2223 is in equilibrium with 2212, Ca_2PbO_4, 3221, 2:1 AEC and 14:24 AEC.

The decomposition temperature of 2223 phase can be reduced to as much as 20 to 30 °C when Ag powder is added up to 50 mol% because of the eutectic reaction between Ag and the superconducting oxide [30]. In the Ag-clad Bi-2223 multifilament tapes, the decomposition temperature is 10 °C lower than that of the bulk material without presence of Ag [54]. Heat treatments under the reduced O_2 atmosphere leads to further reduction of the 2223 decomposition temperature. Although there is no 2223 diagram available concerned the effect of oxygen partial pressure, some experimental results [55, 56] indicate that the melting temperature can be lower to 30 °C with extremely diluted oxygen atmosphere. At the atmosphere of 7.5%O_2 balanced with N_2, the 2223 decomposition temperature can be reduced to 20 °C.

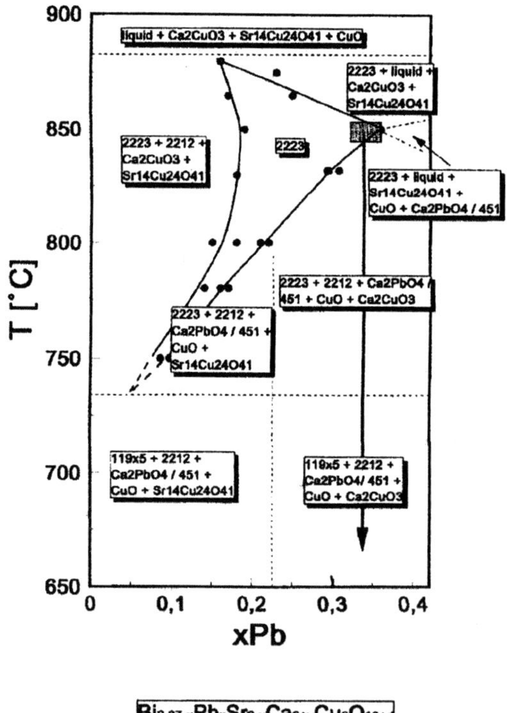

Fig. 20. The phase diagram of $Bi_{2.27-x}Pb_xSr_{2-y}Ca_{2+y}Cu_3O_{10+\delta}$ (from P. Majewski, Supercond. Sci. Technol., 10 (1997) 453)

It is known that, in the absence of Ag, there is almost no loss of Pb during sintering until temperature of 870 °C [57]. However, there is loss of lead when Ag is present even at lower temperatures. This is probably due to the formation of the low temperature melt with Ag. It has been shown that, the overall Pb concentration can reduce as much as 40% if a tape is sintered at 837 °C for 240 h [58]. Based on this result, under the tape processing condition of this work, the reduction of lead concentration of 20% is a reasonable estimation. Thus, according to the phase diagram of Fig. 20, the decomposition temperature of 2223 phase would be about 860 °C for a system with nominal composition $Bi_{1.84}Pb_{0.35}Sr_{1.91}Ca_{2.05}Cu_{3.06}O_{10+\delta}$. In the case of tape sheathed with Ag and under an atmosphere of 7.5%O_2 balanced with N_2 as in this work, the decomposition temperature can be estimated to about 830 °C.

VI.2. Bi-2223 PHASE FORMATION

In the conventional 2223 tape processing methods, tapes are usually processed in temperature range between about 835 to 845 °C in air [59,60] and 820 to 835 °C in the 7.5% oxygen partial pressure [43,53,61,62]. Based on the discussion in section 6.1, these are the temperatures closed to the upper limit of the 2223 single phase. According to the study of Strobel et al. [27], when the 2223 phase contains Ca little more than 2 and Cu little more than 3 in the formula $Bi_{1.6}Pb_{0.4}Sr_2Ca_{n-1}Cu_nO_x$, there is a temperature range, between 835 to 857 °C (in air and without addition of Ag), where 2223 is in equilibrium only with liquid phase. The upper limit of this temperature range is comparable to the 2223 decomposition temperature shown in the diagram in Fig. 20. Thus, for the conventional tape processing methods, sintering is actually performed in the 2223 single-phases region with small amount liquid phase.

Up to now the observed 2223 phase transformations are almost all through the conversion of 2212 incorporated with other secondary phases. Several 2223 formation routes have been proposed and the mostly accepted mechanisms are the intercalation process [63-67] and nucleation and growth process [68-73]. In the former case, 2223 is formed through the insertion of additional Ca and Cu-O into the 2212 lattice. The Ca and Cu-O are provided from pre-existing cuprates and 2201. In the latter case, the 2223 phase is formed by precipitation from the liquid phase. The result from the study of Park et al.[31] indicated that the liquid phase has a composition of Bi:Sr:Ca:Cu = 27.7:27.2:17.8:27.3, which is near 2212.

As the conventional tape processing is performed within the upper limit of 2223 single-phase region, the two 2223 formation mechanisms operate together [39]. The advantage of the conventional tape process is that it can lead to almost complete 2223 conversion as reported by many groups [41,52]. On the other hand, it shows serious drawbacks such as poor core density and unsatisfied texture causing by the following factors. Firstly, the ceramic core is formed from powder and it is difficult to obtain a core density near the full theoretic density without sufficient liquid. Secondly, the intercalation process is a solid state reaction and, thus, cracks and porosity would form during the 2223 formation, which involves change in volume and form of grains. This also affects the perfect grain alignment. Finally, mechanical deformation in tape processing induces great amount of cracks that not only reduce the core density but also affect the current transportation. In the case of rolling, the situation is even worse because rolling induce mainly the transverse cracks that may interrupt the flow current

[74]. With a small amount of liquid phase, these cracks cannot be effectively healed [18]. The results of density measurement indicate that even for the high J_c tapes, mass density of filaments is usually around 75% of its theoretical value [16]. Therefore, if tape processing is restricted within the 2223 single-phase region, it is impossible to significantly improve the J_c

Naturally, increasing the amount of liquid phase can be the way to solve the problems of the conventional tape processing. Unfortunately, direct formation 2223 phase from a melt seems to be impossible. From our knowledge, there is no any report in literature, which shows a minimum detective amount of 2223 phase formed from the liquid without containing 2212 or 2223 phases. The reason may be related to the strong elemental segregation in the liquid phase. However, our results in this work do indicate that, in a suitable partial melting temperature, 2223 can be formed from the liquid phase incorporated with other cuprates in a system which contains 2212 or/and 2223. Nevertheless, the excess liquid phase is always detrimental for the final properties. This may result from a significant phase segregation that takes place due to the higher atomic diffusion rate at high temperature. As consequence, there are both large quantity and large size of the cuprate secondary phases, which is hardly re-transformed totally to 2223 phase.

In order to obtain an improved microstructure, one needs to optimize the partial melting temperature. The objective is to increase the quantity of liquid phase meanwhile restricts the significant segregation of other secondary phases. With the control of the partial melting temperature, the retransformation of 2223 phase become possible when a subsequently sintering is performed in the 2223 single-phase region. In the present work, this is done through slow cooling. Optimization of cooling rate is also important since it can grantee enough time for 2223 recovery and, meanwhile, reduces the processing time.

VI.3. PARTIAL MELTING METHODS

Strictly speaking, the conventional tape processing also belongs to the partial melting method since a little amount of liquid phase exist during 2223 phase transformation as mentioned above. The partial melting indicated here refers to the phase mixture where 2223 is not only in equilibrium with liquid phase but also in equilibrium with other cuprates compounds, such as 2:1 AEC, 14:24 AEC etc. In the latter case, a significantly amount of liquid phase is produced.

There are few reports about the partial melting processing methods from the literature. Dou et al. first proposed the high T_c phase-formation-decomposition-recovery (PFDR) method and studied the effect of short partial melting on the 2223 superconducting properties [33-35]. This method requires three sintering cycles and two intermediate mechanical deformations. Tapes are sintered at 839 °C for 70 h in the first cycle, after which the 2223 phase becomes the main phase. In the second cycle, tapes are partially melted at 860 °C in air for 20 – 30 min., annealed at 839 °C for 60 h, and then cooled to room temperature rapidly. In the last sintering cycle, the tapes are annealed at 839 °C for 80 –100 h. Thus the total sintering time is about 210 to 230 h. Through the use of this method, they improved the flux pinning, i. e. the $J_c \sim H$ behavior, of the Bi-2223/Ag tapes. Following the same idea, Hua et al. [36,37] modified the PFDR method by using four sintering cycles instead of three and three intermediate mechanical deformations. The short partial melting is applied on the first sintering cycle after 70 h in air annealing and the total sintering time is about 280 h. They claimed that both J_c and $J_c \sim H$ behavior are improved. Dou et al. suggested that the improved the $J_c \sim H$ behavior has been attributed to the improvement of grain alignment and grain connectivity. In addition, the microstructure studies reveal that the PFDR processed samples contain highly dispersed, textured 2212 fine precipitates and a high density of find defects, which enhance flux pinning.

The CCS method discussed in this chapter is an alternative partial melting method in which tapes are sintered in a mixture of oxygen and nitrogen at PO_2 7.5%. Under such condition, the optimized partial melt temperature is reduced to 850 to 855 °C against 860 °C of the PFDR method. The great advantage of this method compared with the above mentioned methods is its simplicity in tape processing. The total sintering time can be reduced as short as 40 h with only two sintering cycles and one intermediate mechanical deformation.

VI.4. MECHANISM OF THE CCS PROCESS

For an ideal precursor, 2223 phase can be fast formed through the conversion of 2212 [59, 75] in a suitable sintering temperature range. Our results indicate that more than 90% of the 2223 phase can be formed within 30 h during the pre-sintering. Nevertheless, the J_c of the pre-sintered tapes is always poor as shown in the Fig. 15. This can be attributed to 1) the unhealed cracks induced by the tape deformation, volume change during phase transformation and the relative fast cooling [51]; 2) the residual secondary phase, especially the unconverted 2212, the main secondary phase after the pre-sintering; 3) the unsatisfied grain alignment and the poor density. As discussed above, the last one also arises from the phase transformation from 2212 to 2223.

Thus, the roles that the final sintering (the CCS process) comes to play are to eliminate or diminish these drawbacks. Firstly, the liquid phase formed during the melt process can effectively heal the cracks and porosity and therefore the core density can be improved. The liquid phase can also assist grain alignment, especially near the Ag sheath and promote 2223 transformation [4,5,13]. Secondly, 2212 phase content is reduced in the core. This may be due to that, at the partially melted temperature, 2212 also decomposes, and upon cooling these secondary phases re-transform to 2223 instead of 2212 because of the favorable overall 2223 composition.

The cooling rate between T_{max} and T_{final} is a critical parameter for obtaining high quality of Bi-2223/Ag tapes. The present results indicate that, although a reasonable high J_c can be obtained with 10 h sintering in the final sintering cycle, the slowly cooled tapes always exhibit better microstructure and superconducting properties. It is known that the formation of 2223 through precipitation from the liquid has a faster kinetics than that through the intercalation mechanism [39]. Therefore, the re-transformation of 2223 is expected to be a quick process, which has been seen from the present result. However, compared to 2201 phase precipitation, the 2223 formation rate is much lower. The 2201 precipitation becomes more significant when cooling rate is fast. In the case of quenching the 2201 phase will dominate the solidification product of the liquid phase. Fortunately, there is a temperature range where 2223 is in equilibrium with little amount of liquid phase and 2201 becomes unstable. Thus through a sufficient annealing at these temperature the 2201 phase can be diminished or eliminated. This temperatures as about 825 °C when tapes are sintered in air [52], which correspond to about 800 to 805 °C when tapes are sintered in a mixture of oxygen and nitrogen with PO_2 7.5%. Slow cooling guarantees an enough annealing time in this temperature range. The effective transformation of 2201 to 2223 is clearly seen by comparing Fig 8 with Fig. 7. Beside the 2201 phase, there are other secondary phases, which can be transformed to 2223 rapidly at higher temperatures inside 2223 single-phase region. Therefore, the continuous cooling has the advantage to depress variety of secondary phases when the tapes pass slowly over a wide temperature range.

The Pb contained phases, plumbates, are other principal residual phases of the CCS process. In the studied system, the plumbates exist in two forms: Ca_2PbO_4 and 3221. Which one becomes dominant in the ceramic cores depending on the O_2 partial pressure [76]. Generally, the amount of the residual plumbates depends on the quantity of the Pb being absorbed by the 2223 phase. According to the diagram in Fig.20, Pb has its maximum solubility in 2223 at about 850 °C in air without addition of Ag. Pb solubility reduces significantly when temperature is above or below 850 °C. With addition of Ag, this temperature reduces to about 835 °C [30]. In the CCS process, tapes are heated up to above 850 °C at which the Pb solubility in 2223 reduces significantly. As the phase diagram in Fig. 20 does not show any plumbate in the partial melt region, Pb dissolves into the liquid phase [30]. Although the tapes pass over the maximum Pb solubility temperature region (about 835 °C [30]) during cooling, the leads are not totally reversibly re-absorbed by the 2223 phase. This can be seen by comparing the tape heated to 850 °C to the tape heated to 820 °C as shown from XRD patterns in Fig. 8. For a suitable T_{final}, there seems to be no significant precipitation of the plumbate phases when tape are heated to temperatures below 835 °C. This can be explained by the diffusion kinetics. At the higher T_{max} (above the maximum solubility temperature), fast diffusion promotes the formation of large size plumbates. These large plumbates are hardly to be absorbed by the 2223 even in the 2223 single-phase region (Fig. 10). On the other hand, for the lower T_{max} below the maximum Pb solubility temperature, the slow diffusion makes the formation of plumbates difficult. However, if the tapes are sintered at the temperature range about 700 to 750 °C for a certain time, significant precipitation can take place as the result of large thermodynamic driving force. This is clearly seen from the XRD patterns for the tapes cooled to temperatures below to the 770 °C as shown in the Fig. 9.

The plumbates are detrimental to the transport current, especially the 3221 phase that lies along the 2223 grains increasing weak links. Large plumbate particles may even block up current path of superconducting filaments. Based on the above analysis, further modification of partial melting process may diminish the plumbates. Another way of effectively reduce the plumbates may be through a reduction of the overall Pb concentration in the precursor as the tapes prepared by partial melt process contain less Pb than those prepared by the convention method. We believe that J_c of Bi-2223/Ag tapes can be dramatically enhanced through the partial melt process if one can effectively depress the plumbates as well as the 2201 phase.

VII. SUMMARY

Although significantly high current densities have been obtained from the silver sheathed Bi-2223 tapes, the index of cost/performance is still much higher than the conventional materials. Therefore, the commercialization of the devises based on Bi-2223/Ag tapes is still in distance from the reality. The essential improvement depends on a significant enhancement of the critical current density. Besides, simplification of tape processing can also reduce the cost and facilitate large-scale fabrication. However, the drawbacks related to the conventional tape processing, such as the lower core density, poor grain connectivity and mechanical deformation induced cracks, make it difficult to further improve the critical current density without modifying the tape processing method.

In order to overcome these drawbacks, the partial melting process is proposed. In the partial melt state, large amount of liquid phase, which has a composition closed to 2212, is formed. Upon the subsequent annealing at the 2223 formation temperatures, the liquid phase can retransform to 2223 phase incorporating with other cuprate phases. It is essential to have the 2223 transformation that the partial melting system should contain substantial 2212 or 2223 phase. The liquid plays a role of healing the porosity and mechanical deformation induced cracks, improving the grain connectivity and promoting the grain alignment. Tapes prepared by the partial melting methods exhibit improved microstructure and superconductivity. Using the CCS method described in this chapter, one can also reduce significantly the sintering time and increase dramatically the sintering temperature window. We also show that partial melting has a more positive effect on the rolled tapes than on the pressed tapes since healing of the transversal cracks, produced mainly by rolling, is more essential for improving J_c.

Care must be taken in controlling the partial melting temperature and the subsequent low temperature annealing. Beside the 2223 phase, the main other secondary phases precipitated from the liquid phase are Bi-2201 and plumbates. Too high partial melt temperature leads to not only larger quantity, but also larger size of these secondary phases. The large size grains are hardly to react with the liquid and retransform completely to 2223. Too large grains may even interrupt transport current.

The cooling rate is also an important parameter. Generally, a slow cooling is always helpful for retransformation of 2223 phase. The 2201 phase has much faster precipitation kinetics than the 2223 phase from the liquid and can precipitate even during quenching. Thus, a slow cooling rate retards the 2201 formation. Beside, a slow cooling provides a sufficient time for a tape to pass the 2223 formation range and, therefore, to convert the liquid phase and other secondary phase to 2223. Slow cooling can also reduce the residual plumbate phases which has a different transformation temperatures from that of 2201. We believe that, if the residual secondary phases, especially the 2201 and plumbates are effectively depressed, significant enhancement of J_c will be obtained.

ACKNOWLEDGEMENTS

We thank F. Rizzo, M.B. Lisboa, A. Polasek and B. Malinkovic for stimulating discussions. The XRD and SEM measurements were done by L. A. Saléh, R. P. Silva and M. Simonson.

REFERENCES

[1] P. M. Grant, IEEE Trans. on Appl. Supercond., 7 (1997)112.
[2] D. C. Larbalestier, IEEE Trans. on Appl. Supercond., 7 (1997)90.
[3] R. D. Blaugher, The 1998 International Workshop on Superconductivity, (July 1999, Okinawa Japan) p13.
[4] T. Hikata, T. Nichikawa, H. Nukai, K. Sato and M. Hiotsuyanayi, Japan. J. Appl. Phys., 28 (1989) 288.
[5] S. X. Dou, H. K. Liu, J. Wang, M. H. Apperley, C. C. Sorrell, S. J. Gou, B. Loberg and K. E. Easterling, Phys. C, 172 (1990) 63.
[6] W. G. Wang, Z. Han, P. Skov-Hansen, J. Goul, M. D. Bentzon and P. Vase, IEEE Trans. on Appl. Supercond., 9 (1999)2613.
[7] T. Kaneko, T. Nikata, M. Ueyama, A.Mikumo, N. Ayai, S. Kobayashi, N. Saga, K. Hayashi, K. Ohmatsu and K. Sato, IEEE Trans. on Appl. Supercond., 9 (1997) 2465.
[8] A. P. Malozemoff, W. Carter, S. Fleshler, L. Fritzemeier, Q. Li, L. Masur, P. Miles, D. Parker, R. Parrela, E. Podtburg, G. N. Riley Jr, M. Rupich, J. Scudiere and W. Zhang, IEEE Trans. on Appl. Supercond., 9 (1997) 2469.
[9] B. Ficher, S. Kautz, M. Leghissa, H. W. Neumüller and T. Arndt, IEEE Trans. on Appl. Supercond., 9 (1997) 2480.
[10] M.W. Rupich, Q. Li, R.D. Parella, M. Teplitsky, E.R. Podtburg, W.L. Carter, S. Hancock, J. Marquardt, J.D. Schreiber, D.R. Parker, G.N. Riley Jr, L.J. Masur, P.K. Miles, Proc. Int. Workshop on Supercondutivity, Okinawa, (1998), p147.
[11] Y. Hakuraku and X. Mori, J. Appl. Phys., 73 (1992) 309.
[12] W. Zhang, E. E. Hellstrom and M. Polak, Supercond. Sci. Technol, 9 (1996) 971.

[13] Y. Yamada, B. Oberst and R. Flukiger, Supercond. Sci. Technol., 4 (1991) 165.
[14] S. X. Dou, H. K. Liu, M. Apperley, K. H. Song and C. C. Sorrell, Supercond. Sci. Technol., 3 (1990) 138.
[15] S. X. Dou, Y. C. Guo and H. K. Liu, Phys. C, 194 (1992) 343.
[16] J. Jiang, X. Y. Cai, J. G. Chandler, M. O. Rikel, E. E. Hellstrom, R. D. Parrella, Y. Dingan, Q. Li, M. W. Rupich, G. N. Riley Jr., and D. C. Larbalestier, IEEE Trans. Appl. Supercond., 11 (2001) 3561.
[17] J. W. Anderson, X. Y. Cai, M. Feldmann, A. Polyanskii, J. Jing, J. A. Parrell, K. R. Marken, S. Hong and D. C. Larbalestier, Supercond. Sci. Technol., 12 (1999) 617
[18] X. Y. Cai, A. Polyanskii, Q. Li, G. N. Riley jr and D. C. Larbalestie, Nature, 392 (1998) 906.
[19] H. Miao, H. Kitaguchi, H. Kumakura, K. Togano and T. Hasegawa, Physica C, 301 (1998) 116.
[20] B. Hong, J. Hahn, and T. O. Mason, J. Am. Ceram. Soc., 73 (1990)1965.
[21] R. Müller, T. Schweizer, P. Bohac, R. O. Suzuki and L. J. Gauckler, Phys. C, 203 (1996) 299.
[22] K. Schulze, P. Majewski, B. Hettich and G. Petzow, Z Metallkde., 81 (1990) 836
[23] B. Hong, and T. O. Mason, J. Am. Ceram. Soc., 74 (1991)1045.
[24] T. G. Holesinger, D. J. Miller, L. S. Chumbley, M. J. Kramer and K. W. Dennis, Phys. C, 201 (1992)109.
[25] M. R. de Guire, N. P. Bansal, D. E. Farrell, V. Finan, C. J. Kim, B. J. Hill and C. J. Allen, Phys. C, 179 (1992)333.
[26] C. L. Lee, J. J. Chen, W. J. Wen, T. P. Wen, T. P. Perng, J. M. Wu, T. B. Wu, T. S. Tin, R. S. Liu and P. T. Wu, J. Mater. Res., 5 (1990) 1403.
[27] P. Strobel, J. C. Tolédano, D. Morin, J. Schneck, G. Vacquier, Omonnereau, J. Primot and T. Fournier, Phys. C, 204 (1992) 27.
[28] P. Majewski, B. Hettich, H. Jaeger, K. Schulze, Adv. Mater., 3 (1991) 67.
[29] W. Wong-Ng, L. P. Cook and F. Jiang, J. Am. Ceram. Soc., 81 (1998)1829.
[30] P. Majewski, Supercond. Sci. Techonol., 10 (1997) 453.
[31] C. Park, W. Wong-Ng, L. P. Cook, R. L. Snyder, P. V. S. S. Sastry and A. R. West, Phys. C, 304 (1998) 265.
[32] T. Hatano, K. Aota, S. Ikeda, K. Nakamura and K. Ogawa, Japan. J. Appl. Phys., 27 (1988) L2055.
[33] Y. C. Guo, H. K. Liu and S. X. Dou, Phys. C, 200 (1992) 147.
[34] S. X. Dou, Y. C. Guo, H. K. Liu, Appl. Phys. Lett., 60 (1992) 2929.
[35] Y. C. Guo, H. K. Liu and S. X. Dou, Appl. Supercond., 1 (1993) 25.
[36] L. Hua, Y. Z. Wang, C. Zhang, Z. Q. Yang, G. W. Qiao, Y. C. Chuang, J. M. Yoo, H. S. Chung and H. D. Kim, Phys. C, 282-287 (1997) 2601.
[37] L. Hua, J. M. Yoo, J. Ko, H. D. Kim, H. S. Chung and G. W. Qiao, Supercond. Sci. Technol., 12 (1999) 153.
[38] S. X. Dou, H. K. Liu, Y. C. Guo, D. L. Shi, M. D. Sumption and E. D. Collings, Proc. Iss'92 (Kobe, Japan, Nov, 1992).
[39] J. Jiang and J. S. Abell, Supercond. Sci. Technol., 11 (1998) 705.

[40] S. X. Dou, H. K. Liu, Supercond. Sci. Technol., 6 (1993) 297.
[41] R. Flukiger, G. Grasso, J. C. Grivel, F. Marti, M. Dhallé and Y. Hang, Supercond. Sci. Technol., 10 (1997) A68.
[42] S. X. Dou, R. Zheng, B. Ye, Y. C. Guo, J. Horvat, H. K. Liu, T. Beales, X. F. Yang and M. Apperley, Supercond. Sci. Technol., 11 (1998) 915.
[42] S. X. Dou, R. Zheng, B. Ye, Y. C. Guo, J. Horvat, H. K. Liu, T. Beales, X. F. Yang and M. Apperley, Supercond. Sci. Technol., 11 (1998) 915.
[43] Y. L. Liu, W. G. Wang, H. F. Poulsen and P. Vase, Supercond. Sci. Technol., 12 (1999) 376.
[44] S. K. Xia, M. B. Lisboa, E. T. Serra and F. Rizzo, Phys. C, 354 (2001) 463.
[45] S. K. Xia, M. B. Lisboa, E. T. Serra and F. Rizzo, IEEE Trans. Appl. Supercond., 11 (2001) 2971.
[46] S. K. Xia, M. B. Lisboa, E. T. Serra and F. Rizzo, Supercond. Sci. Technol., 14 (2001) 103.
[47] S. K. Xia, E. T. Serra and F. Rizzo, Phys. C, 361 (2001) 175.
[48] J. Horvat, Y. C. Guo and S. X. Dou, Phys. C, 271 (1996) 59.
[49] J. W. Ekin, T. M. Larson, A. M. Hermann, Z. Z. Sheng, K. Togano and H. Kumakura, Phys. C, 160 (1989) 489.
[50] B. Hensel, G. Grasso and R. Flukiger, Phys. Rev., B15 (1995) 15456.
[51] J. A. Parrell, D. C. Larbalestier, G. N. Riley Jr, Q. Li, R. D. Parrella and M. Teplitsky, Appl. Phys. Latt., 69 (1996) 2519.
[52] S. X. Dou, R. Zheng, X. F. Fu, Y. C. Guo, J. Horvat, H. K. Liu, T. Beales and M. Apperley, IEEE Trans. on Appl. Super., 9 (1999)2436.
[53] S. K. Xia, A. Polasek, M. B. Lisboa, E. T. Serra and F. Rizzo, 1999 Applied Superconductivity (Inst. Phys. Conf. Ser. 167) ed. X Obradors, F. Sandiumenge and J Fontcuberta (Bristol: Institute of Physics Publishing) p479.
[54] S. X. Dou, H. K. Liu, J. Wang, M. H. Apperley, C. C Sorrell, S. J. Guo, B. Loberg, K. E. Esterling, Phys. C, 172 (1990) 63.
[55] W. Zhu, P. S. Nicholson, J. Appl. Phys., 73 (1993) 8423.
[56] T. G. Holesinger, J. F. Bingert, J. O. Willis, V. A. Maroni, A. K. Fisher and K. T. Wu, J. Mater. Res., 12 (1997) 3046.
[57] S. Kaesche, P. Majewski and F. Aldinger, Z. Metallkde., 87 (1996) 587.
[58] F. Marti, G. Grasso, J. C. Grivel and R. Flükiger, Supercond. Sci. Technol., 11 (1998) 485.
[59] H. K. Liu, R. Zeng, X. K. Fu and S. X. Dou, Phys. C, 325 (1999) 70.
[60] G. Grasso, A. Jeremie, R. Flükiger, 1995, Supercond. Sci. Technol., 8 (1995) 827.
[61] M. Lelovic, T. Deis, N. G. Eror, U. Balachandran and P. Haldar, Supercond. Sci. Technol., 9 (1996) 965.
[62] W. L. Carter, G. N. Riley, Jr., A. Otto, D. R. Parker, C. J. Christopherson, L. J. Masur and D. Buczek, IEEE. Trans. Appl. Supercond., 5 (1995) 1145.
[63] J. S. Luo, N. Merchant, V. A. Maroni, D. M. Gruen, B. S. Tani, W. L. Carter, G. N. Riley, IEEE. Trans. Appl. Supercond., 1 (1993) 101.

[64] Y. L. Wang, W. Bain, Y. Zhu, Z. X. Cai, D. O. Welch, R. L. Sabatini, L. Suenaga and T. R. Thurston, Appl. Phys. Lett., 69 (1996) 580.
[65] Q. Feng, H. Zhang, S. Feng, X. Zhu, Z. liu and L. Xue, Solid State Commun., 78 (1991) 609.
[66] D. Shi, M. Boley, J. G. Chen, M. Xu, K. Vandervoot, Y X. Liao and A. Zangvil, Appl. Phys. Lett., 55 (1998) 699.
[67] Z. X. Cai, Y. Zhu, and D. O. Welch, Phys. Rev., B 52 (1995) 13035.
[68] Y. L. Chen and R. Stevens, J. Am. Ceramic Soc., 17 (1992) 1150.
[69] A. Ono, Japan. J. Appl. Phys., 27 (1988) L2276.
[70] J. Tsuchiya, H. Endo, N. Kijima, A. Sumiyama, M. Mizuno and Y. Oguri, Japan. J. Appl. Phys., 28 (1989) L1918.
[71] Q. Y. Hu, H. K. Liu and S. X. Dou, Phys. C, 250 (1995) 7.
[72] J. C. Grivel and R. Flükiger, Supercond. Sci Technol., 9 (1996) 555.
[73] P. E. D. Morgan, R. M. Housley, J. R. Porter and J. J. Ratto, Phys. C, 176(1991) 279.
[74] Q. Li, K. Broderson, H. A. Hjuler and T. Freltoft, Phys. C, 217 (1993) 360.
[75] J. Müller, O. Eibl, B. Fischer and P. Herzog, Supercond. Sci. Technol., 11(1998) 238.
[76] J. Müller, J. H. Albering, B. Fischer, A. Kautz and P. Herzog, Supercond. Sci. Technol., 11(1998) 111.

STUDIES ON THE ANOMALOUS MAGNETIZATION PEAK OF $Bi_2Sr_2CaCu_2O_{8+\delta}$ AND $(Bi,Pb)_2Sr_2CaCu_2O_{8+\delta}$ SINGLE CRYSTALS

Y. P. Sun

Key Laboratory of Internal Friction and Defects in Solids, Institute of Solid State Physics, Chinese Academy of Sciences, Hefei 230031, P. R. China and National High Magnetic Field Laboratory, Hefei 230031, P. R. China

1. INTRODUCTION

The anomalous magnetization, i.e. peak effect of the magnetization, where the magnetization increases anomalously with increasing magnetic field in the mixed state when an applied magnetic field is parallel to the c-axis direction of crystals, was first reported in textured ceramic and single crystals of $YBa_2Cu_3O_{7-\delta}$ [1,2,3] and $Bi_2Sr_2CaCu_2O_{8+\delta}$ (Bi-2212) [4,5,6,7]. Similar peak effect has been observed in many other superconducting systems such as $La_{2-x}Sr_xCuO_{4-\delta}$ [8], $Nd_{1.85}Ce_{0.15}CuO_{4-\delta}$ [9,10,11], $YBa_2Cu_4O_{8-\delta}$ [12], $Tl_2Ba_2CuO_{6+\delta}$ [13,14,15] and $HgBaCa_{n-1}Cu_nO_x$ (Hg-12(n-1)n, n = 2, 3) [16]. Particularly the peak effect in Bi-2212 single crystals has the characteristic that the peak is very sharp and the peak field, H_{2p}, is low (from 100 G to 1kG) and almost temperature independent. On the contrary, H_{2p} in other superconducting systems mentioned above shows a strong temperature dependence and the peak shape is broader.

As to the origin of the peak effect in Bi-2212 crystals, several interpretations including surface barrier effect [13], matching effect [5], a crossover from the surface barrier to bulk pinning [4], dynamic effect [6], 2D-3D dimensional crossover in the vortex structure [17] and vortex softening [18] have been proposed. However, what is the exact origin of the anomalous magnetization peak of Bi-2212 superconductors is still an open question so far.

Following, a brief description of some aspects as to the crystal growth of Bi-2212 and Pb-doped $(Bi,Pb)_2Sr_2CaCu_2O_{8+\delta}$ single crystals [(Bi,Pb)-2212]] single crystals and experimental technique will be presented in Section 2. In Section 3, the effect of the surface state and shape of Bi-2212 single crystals on the anomalous magnetization peak will be discussed. In Section 4, a detailed analysis of anomalous magnetization peak of (Bi,Pb)-2212 single crystals with decreasing anisotropy will be described. Section 5 explores dynamic magnetic moment relaxation properties in the vicinity of the anomalous magnetization peak of (Bi,Pb)-2212 single crystals will be described. Finally, a summary is presented in Section 6.

2. SAMPLE PREPARATION AND EXPERIMENATL TECHNIQUE

Bi-2212 and (Bi,Pb)-2212 single crystals were grown by the self-flux method using Bi_2O_3 as flux. The nominal starting composition of two kinds of single crystals was Bi : Sr : Ca : Cu = 2.4 : 2 : 1 : 2 and Bi: Pb: Sr: Ca: Cu = 1.6 : 0.8 : 2 : 1 : 2 , respectively. The mixture of high-purity Bi_2O_3 , $SrCO_3$, $CaCO_3$, CuO, PbO powders was well ground and presintered for two times at 800 – 820°C for 48 h. After prereaction, the mixture was well reground and put into Al_2O_3 crucible. The powder was melted at 1020 °C for 6h and then cooled quickly to 900°C. After that, the temperature was decreased very slowly at a rate of 0.6°C/h to 830°C. It was found that the crystal growth was performed in this slow cooling process. Finally, temperature was cooled to room temperature by turning off the power. (Bi,Pb)-2212 crystal growth was carried out in a sealed Al_2O_3 crucible to avoid the volatilization of Pb during the growth process of single crystals. The as-grown crystals are black and shiny platelets with typical dimension of $6 \times 3 \times 0.03$ mm^3 and $1 \times 0.5 \times 0.03$ mm^3 for Bi-2212 and (Bi,Pb)-2212 single crystals, respectively. The single crystals were post-annealed at 500 ^{0}C in air for 5 days to ensure sample homogeneity. Single crystal x-ray diffraction shows only the (00*l*) peaks with c = 3.0799 nm and 3.0784 nm for Bi-2212 and (Bi,Pb)-2212 single crystals, respectively. No secondary phases were observed even when the diffraction intensity is plotted logarithmically. The actual composition of Bi-2212 and (Bi,Pb)-2212 single crystals was determined via a large area energy-dispersive x-ray (EDX) analysis with typical crystal composition of Bi : Sr : Ca : Cu = 2.1 : 1.7 : 1 : 2 and Bi: Pb: Sr: Ca: Cu = 1.8 : 0.8 : 2.1 : 1.1 : 2, respectively. Magnetic susceptibility M (T), magnetization M (H) measurements with an applied field H parallel to c-axis of single crystals were carried out using a Quantum Design MPMS or μ-metal shielded $MPMS_2$ SQUID magnetometers. The irreversibility temperature T_{irr} is determined according to the magnetization measurement M (T) under different applied fields with H // c-axis in so-called zero-field-cooling (ZFC) and field-cooling (FC) process using their superimposing point. The resistivity was measured by a dc four-probe technique. The electrical contacts were made by evaporating silver pads onto the surface of the crystals soon after the crystals were annealed. Leads were attached to silver pads using silver paste. The contact resistivity was less than 1 Ω.

3. ANOMALOUS MAGNETIZATION PEAK OF BI-2212 SINGLE CRYSTALS

3.1 The effect of the surface state of Bi-2212 single crystal on the anomalous magnetization peak H_{2p}

For Bi-2212 single crystals grown by flux method, some flux lying on the surface of the crystal can usually be observed by optical microscopy. In fact, even for Bi-2212 crystals grown by floating zone method, some impurities are also found on the surface of crystals. Fig.1 (a) shows that there are some column fluxes with diameter of 10 ~ 25 μm besides spot flux on the surface of Bi-2212 single crystal. These fluxes usually can't be detected by x-ray diffraction due to its limited resolution. To check the possible effect of these impurities on the anomalous magnetization peak H_{2p}, the visible fluxes were removed from the surface of the single crystal by cleaving using adhesive tape. Using this method, no flux was observed and no surface damage occurred on the new crystal surface as shown in Fig.1 (b).

Isothermal magnetization data were obtained with zero-field-cooled (ZFC) from above T_c to the set temperature. Half magnetic hysteresis M (H) loops for the Bi-2212 single crystal with T_c = 88 K before (marked as as-surface) and after (marked as surface-1) its surfaces are cleaved at 28 K is plotted in Fig.2. It shows that there is an obvious anomalous magnetization peak at H_{2p} = 750 G for as-surface single crystal. For surface-1 single crystal, though the width of M (H) loop is dramatically reduced related to as-surface single crystal, a magnetiza-

(A)

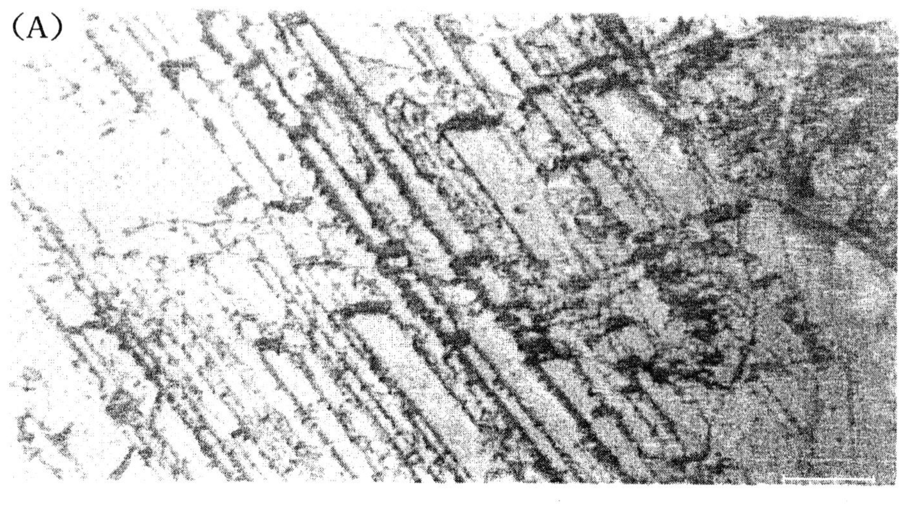

(B)

Fig.1. (a) Optical microscopy picture of the surface (as-surface) of Bi-2212 single crystal, Bar indicates 100 μm. (b) Optical microscopy picture of the new surface (surface –1) when the surface flux in (a) was removed

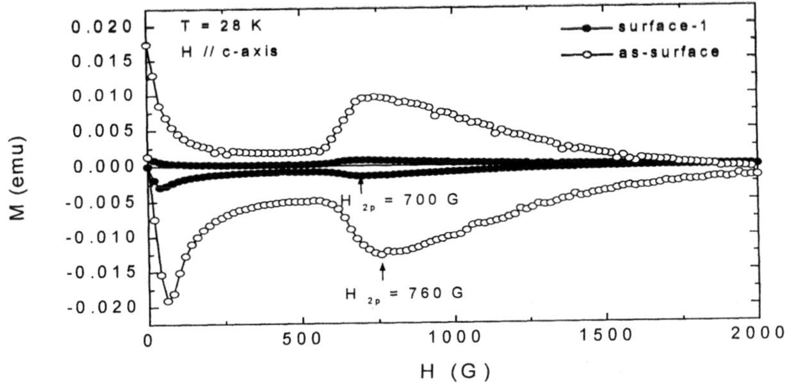

Fig.2. Magnetic hysteresis loops M (H) for $T_c = 88$ K Bi-2212 single crystal before (as-surface) and after (surface-1) the surface was removed at T = 28 K.

tion peak at H_{2p} = 700 G can be still observed. That is to say, the anomalous magnetization peak still exists when the surface layers are cleaved. This result indicates that the surface state of the Bi-2212 single crystal has only an influence on the width of M (H) loops because the surface state may contribute to the surface barrier and flux pinning of single crystals. However, the surface state do not results in the disappearance of H_{2p}, which is different from the result obtained in spiral-grown Bi-2212 crystals by Wang et al.[19], which the anomalous magnetization peak almost completely disappears after removing growth spiral patterns from the crystal surface. Therefore, the origin of the anomalous magnetization peak H_{2p} of Bi-2212 single crystal can't be attributed to its surface state.

3.2 The effect of the shape of Bi-2212 single crystal on the anomalous magnetizatn peak H_{2p}

As to the origin of H_{2p} of Bi-2212 single crystals, Kopylov et al. argue that it originates from the surface barrier effect [13], which is closely related to the shape of single crystals. As we know, for circle single crystal, the surface barrier, in particular, Bean-Livingston's surface

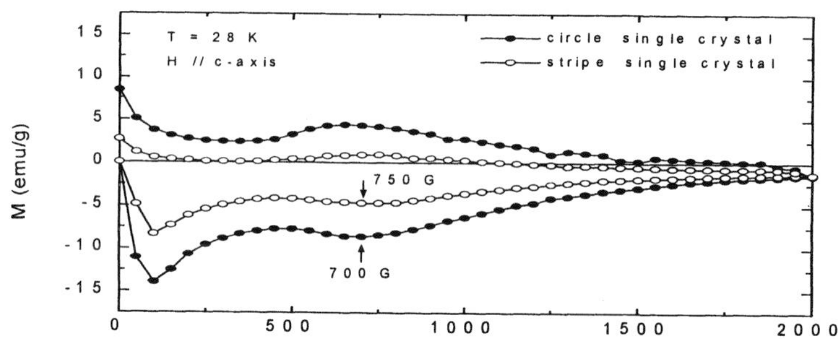

Fig.3. Half magnetic hysteresis loops M (H) for $T_c = 88$ K Bi-2212 single crystal with stripe and circle shape at T = 28 K.

barrier [20] stemming from the mirror image force of vortices due to a platelet single crystal when the applied magnetic field is aligned perpendicular to the broad surface of the platelet single crystal, can be dramatically suppressed. In order to investigate the surface barrier effect of Bi-2212 single crystals on the anomalous magnetization peak H_{2p}, the stripe single crystal was cut into circle single crystal with the diameter of 1.7 mm and M (H) curves of two types of single crystals were measured at T = 28 K as shown in Fig. 3. It indicates that the anomalous magnetization peak still exists at $H_{2p} = 750$ G for the circle single crystal whose M (H) loop becomes more symmetry relative to the H-axis compared with the M (H) loop of the stripe single crystal with a dimension of $2.25 \times 0.2 \times 0.03$. This result indicates that the origin of H_{2p} of Bi-2212 single crystal should not be attributed to the surface barrier effect.

4. ANOMALOUS MAGNETIZATION PEAK OF (Bi,Pb)-2212 SINGLE CRYSTALS

As to the origin of the peak effect in Bi-2212 single crystals, though all kinds of models have been presented, there has no consensus about its origin and it also remains unclear to what extent a single underlying mechanism might be responsible in all high temperature superconductors (HTS). For Bi-2212 single crystals, one of the difficulties in analyzing the nature of H_{2p} is its narrow temperature window with a weak temperature dependence, which usually appears in the temperature range from 0.2 to 0.4 T_c, compared with other superconducting systems [8, 9, 10, 11,12,13,14,15,16]. The narrow temperature window of H_{2p} seems to be related to the strong anisotropy of Bi-2212 superconductors with a typical anisotropies parameter $\gamma > 100$. [21, 22] Therefore, it is very necessary to reduce the anisotropy of the Bi-2212 single crystal and investigate its anomalous magnetization peak H_{2p} behavior, which will supply a more comprehensive understanding for the origin of anomalous magnetization peak of Bi-2212 superconductors. On the other hand, in all kinds existed explanations with respect to the origin of H_{2p} of Bi-2212 single crystals, the dimensional crossover model [17] seems to be most plausible because it is supported by the observation of the flux line lattice by neutron diffraction [23] and the field distribution as detected by μSR experiment [24]. The 2D-3D dimensional crossover in the vortex structure is expected to occur at $H_{cr} = \phi_0/(\gamma s)^2$ as Josephson coupling between CuO_2 layers is the dominant coupling mechanism (i.e. $\gamma s < \lambda_J$) , where s is the layer spacing, γ an anisotropic parameter, λ_J Josephson length [25]. Hence, in order to test whether the dimensional crossover model can be as the origin of the peak effect, it is also very meaningful to study how the anisotropy of the samples affects the peak effect.

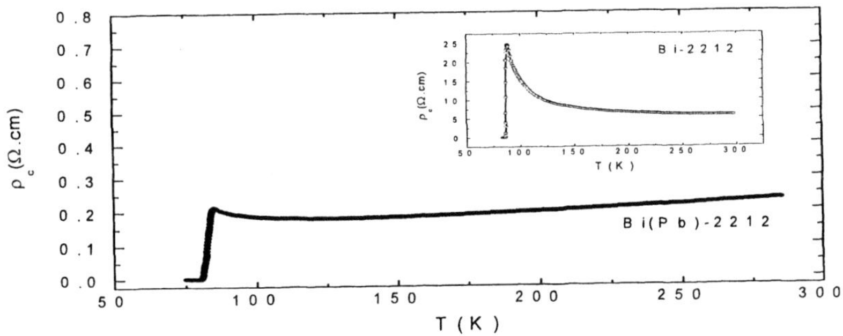

Fig.4. The temperature dependence of out-of-plane resistivity ρ_c of (Bi,Pb)-2212 (main panel) and Bi-2212 single crystals (inset) at zero magnetic field.

Cai et al. find that the anisotropy of Bi-2212 single crystals can be reduced by Pb doping [7]. To reduce the anisotropy of Bi-2212 superconductor, heavily Pb-doped (Bi,Pb)-2212 single crystals were grown. The lower anisotropy of (Bi,Pb)-2212 single crystal can be verified from its normal state resistivity along c-axis direction ρ_c value and temperature dependence of resistivity ρ_c (T) as shown in Fig.4. It shows that ρ_c value of 0.19 Ω cm at 100 K is about two orders of magnitude smaller than that of 15.44 Ω cm for Bi-2212 single crystal at the same temperature. Moreover, ρ_c (T) of (Bi,Pb)-2212 single crystal is metallic behavior in the normal state above 122 K, which is pronouncedly different from the semiconducting behavior in the whole temperature range of normal state for Bi-2212 single crystal. In addition, anisotropic parameter γ value can be calculated according to relation $\gamma^2 = \rho_c / \rho_{ab}$, here, ρ_{ab} is the resistivity along Cu-O plane, i.e, in-plane resistivity. At 100 K, $\gamma^2 = 21942$ and 1983 are obtained for Bi-2212 and (Bi,Pb)-2212 single crystals, respectively. Therefore, the anisotropy of (Bi,Pb)-2212 single crystal is reduced dramatically compared with Bi-2212 single crystal.

For (Bi,Pb)-2212 single crystal with $T_c = 69$ K and transition width $\Delta T_c < 1$ K, magnetic hysteresis loops M(H) at intermediate temperature of 28 K is shown in Fig.5. It shows that there is an anomalous magnetization peak at $H_{2p} = 1400$ G, which is almost two times higher than that of Bi-2212 single crystal at the same temperature (see Fig.2, 3).

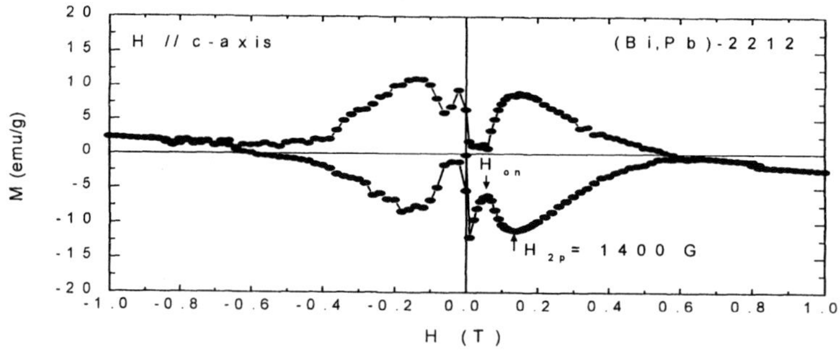

Fig.5. Magnetic hysteresis loops M (H) for $T_c = 69$ K (Bi,Pb)-2212 single crystal at T = 28 K.

Magnetic hysteresis loops M (H) at low temperature region from 13 to 18 K, intermediate temperature region from 33 to 48 K and high temperature region from 56 to 61 K are measured and plotted in Fig. 6 (a) (b) (c), respectively. It shows that the anomalous magnetization peak H_{2p} and the onset field H_{on} can be clearly observed from 13 K up to 61 K. At higher temperature of 62 K, a small H_{2p} of 160 G with H_{on} of 150 G can still be detected

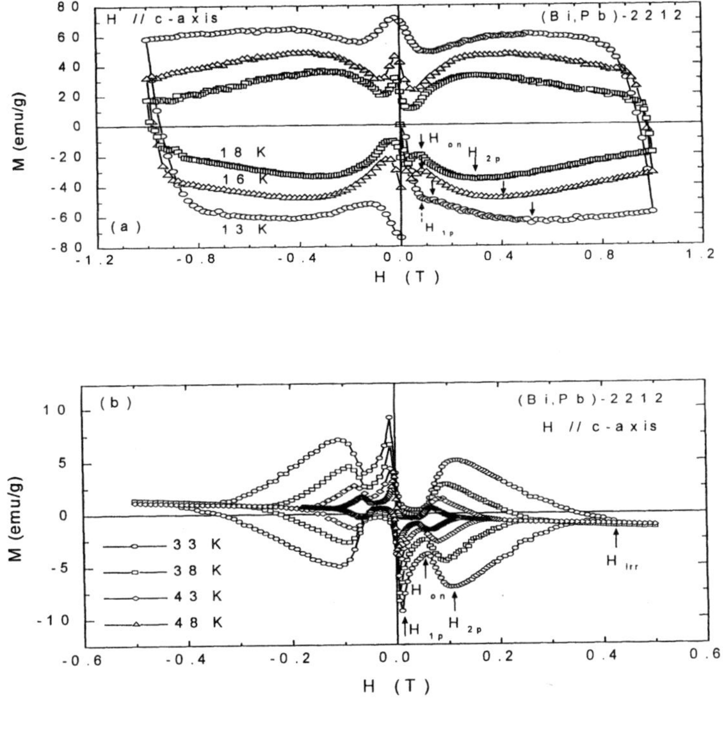

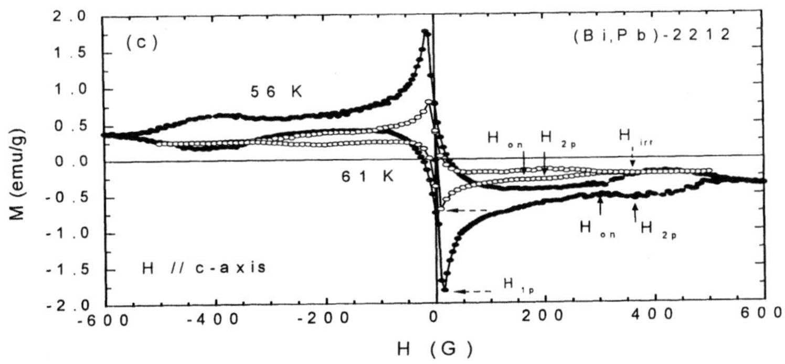

Fig.6. Magnetic hysteresis loops M (H) for $T_c = 69$ K (Bi,Pb)-2212 single crystal at T = 13, 16, 18 K (a); T= 33, 38, 43. 48 K (b); T = 56, 61 K (c).

(not shown here). For lower temperature of 11 K, H_{on} field of 2 kG can be detected even though H_{2p} is already difficult to resolve. H_{1p} and H_{irr} in Fig. 6 are initial-penetration first peak and the irreversibility field, respectively.

Fig.7 shows magnetic hysteresis loop M (H) of Bi-2212 single crystals at T = 40 K. It indicates that H_{2p} disappears completely at 40 K. Contrary to Bi-2212 single crystals with a narrow H_{2p} temperature window (0.2-0.4T_c) and weak temperature dependence (H_{2p} (T) < 1000 G) [4-6], H_{2p} and H_{on} in (Bi,Pb)-2212 crystals show a much wider temperature range of 11-62 K (0.16-0.90 T_c) with stronger temperature dependence of 160 G –5.2 kG for H_{2p}(T) and 150 G –2 kG for H_{on}(T). As shown in the low field H-T phase diagram in Fig.8, both H_{2p} (T) and H_{on}(T) vary continuously all the way up to about 62 K. The termination of H_{2p} and H_{on} at high temperatures seems to be due to the unavoidable crossing around 63 K with the irreversibility line H_{irr}(T).

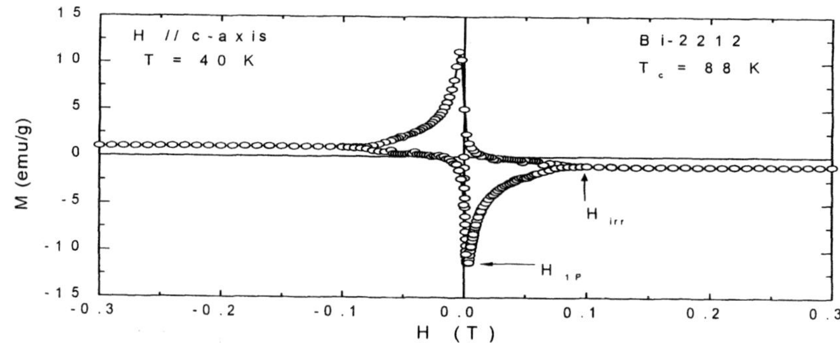

Fig.7. Magnetic hysteresis loops M (H) for T_c = 88 K Bi-2212 single crystal at T = 40 K.

As it is well known, the vortex phases diagram in the mixed state are the results of the competition between the vortex elastic energy, E_{el}, the thermal fluctuation energy, E_{th}, and the vortex pinning energy, E_{pin}, [26-29]. The competition between E_{el} and E_{th} determines the melting line H_m(T) while the competition between E_{th} and E_{pin} determines the irreversibility line H_{irr}(T). That is to say, H_{on}(T) is the intersection of E_{el} (B, T) and E_{pin}(B, T) surfaces. For overdoped (Bi,Pb)-2212 single crystal with lower T_c of 69 K, most relevant is the competition between vortex pinning energy and vortex elastic energy. They will share more weight in determining the behavior of the vortex structure. By equating the elastic energy and pinning energy and assuming that pinning mainly arises from local T_c fluctuation (i.e. ΔT_c disorder pinning), the following relation for $H_{on}(T)$ was predicted for a disorder-induced vortex lattice transition [10]:

$$H_{on}(T) = H_{on}(0)[1-(T/T_c)^4]^{3/2} \qquad (1)$$

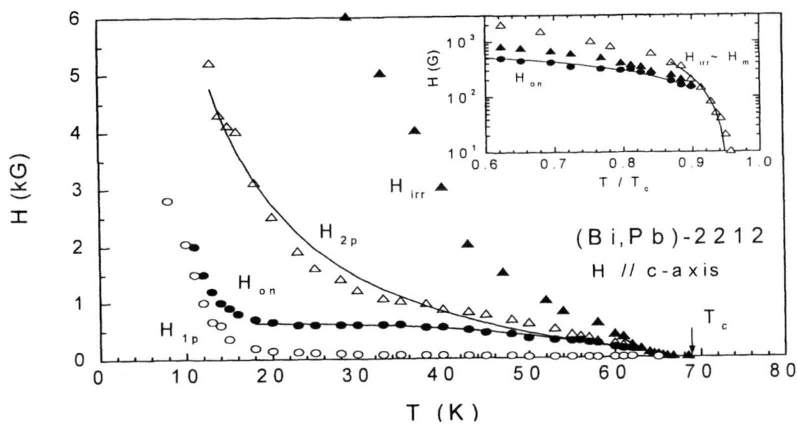

Fig.8 . Superconducting H-T phase diagram for (Bi,Pb)-2212 single crystal with initial field penetration first peak H_{1p} (T), anomalous magnetization peak H_{2p}(T), onset field H_{on}(T), and irreversibility field H_{irr}(T). The solid lines are fitting curves (see text). The inset shows semilogarithmic H (T/T_c) behavior near T_c

This is a transition from a relatively ordered, elastic vortex quasi-lattice (or Bragg glass) to an entangled, pinning-dominated vortex glass. That is to say, at low fields below H_{on}, the elastic interaction governs the structure of the vortex solid, forming a quasiordered lattice. However, as $H > H_{on}$, disorder dominates and vortex interaction with pinning centers giving rise to an entangled vortex solid. Eq. (1) fitted well for $T_c = 23$ K $(Nd_{1.85}Ce_{0.15})CuO_{4-\delta}$ crystals in the temperature range of 0.3-0.9 T_c [10,11], where local T_c fluctuation originated from inhomogeneous oxygen reducing. Applying Eq. (1) to (Bi,Pb)-2212 crystals with local T_c fluctuation from intrinsically disordered crystals due to the inhomogeneous distribution of Pb, which is evidenced by the appearance of Pb-rich domains with nanometer scale in the crystal through the observation of high-resolution electron microscopy (HREM) [30], it is found that $H_{on}(T)$ over a temperature range of 18-62 K (0.26-0.90 T_c) can be fitted well by Eq.(1) with fitting parameters $H_{on}(0) = 639$ G and $T_c = 69.4$ K as shown in Fig. 8.

As shown in the inset of Fig. 8, $H_{on}(T)$ seems to cross with the irreversibility line $H_{irr}(T)$ around 63 K. The inset of Fig. 8 indicates that the irreversibility line H_{irr} in the high temperature range of $T > 0.84\ T_c$ and at low magnetic fields can be well described by a power law $H_{irr}(T) \sim (1 - T/T_c)^{\alpha}$, with $\alpha = 1.75$ and $T_c = 67$ K. Since the deformation of vortex lattice induced by thermal fluctuation is dominated at high temperatures [31], the commonly accepted scenario is that $H_{on}(T)$ crosses the first order vortex lattice melting line at high temperatures. For the high temperature magnetization data, since no jump is observed in the reversible region of M (H), which indicates a first order melting transition, it is reasonable to assume that the irreversibility line $H_{irr}(T)$ is very close to the melting line $H_m(T)$ in the high-temperature and low-field region. Therefore, high temperature termination of $H_{on}(T)$ is indeed possible due to crossing the first order vortex lattice melting line near T_c. Since both H_{on} and H_{2p} fields are inherently enclosed by the initial penetration field first peak H_{1p} and H_{irr}, the extended stability region of (Bi,Pb)-2212 can be easily seen if one compares the irreversibility lines $H_{irr}(T/T_c)$ of overdoped 69 K (Bi,Pb)-2212 crystal with nearly-optimum-doped 88 K Bi-2212 crystal as shown in Fig.9. The shifting of irreversibility line H_{irr} to higher temperature T/T_c due to reducing anisotropy and increasing pinning in the Pb-doped crystal is directly responsible for the extension of the stability region up to 0.9 T_c for (Bi,Pb)-2212 single crystal.

At low temperature range of below 18 K (< 0.26 T_c), $H_{on}(T)$ enhancement and thus deviation from the fitting of Eq. (1) is observed. Similar phenomenon is also observed in (Nd,Ce)-214 crystals [11]. The exact nature of low temperature $H_{on}(T)$ behavior is not clear at present. It may also correlate with the boundary of a zero-dimensional pinning regime (individually pinned pancake vortices) at low temperatures [32, 33]. At low temperatures, point disorder is sufficiently strong and the phase coherent ordered vortex solid becomes metastable, which may result in a zero-dimensional pinning.

As to the anomalous magnetization peak $H_{2p}(T)$ of (Bi,Pb)-2212 single crystal, due to enhancing interlayer coupling and reducing anisotropy, the temperature window of H_{2p} is extended pronouncedly. It is obvious that the dimensional crossover model can't be regarded

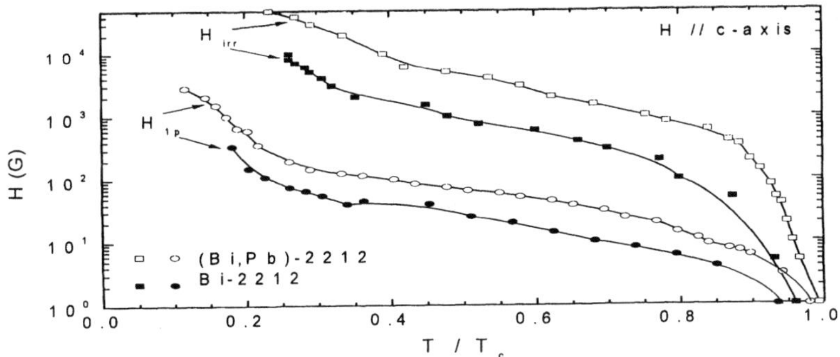

Fig.9. Irreversibility line $H_{irr}(T/T_c)$ and initial field penetration first peak H_{1p} (T/T_c) for 69 K (Bi,Pb)-2212 and 88 K Bi-2212 single crystals. The solid lines are guides to the eyes only.

as the origin of $H_{2p}(T)$ of (Bi,Pb)-2212 single crystal, because this model predicts that there is a temperature independent crossover field $H_{cr} = \phi_0/(\gamma s)^2$ [25]. Alternatively, $H_{2p}(T)$ can be accounted for according to the thermal-decoupling theory [34]. For moderate anisotropy parameter γ with $\xi_{ab}/s << \gamma << \lambda_{ab}/s$ in (Bi,Pb)-2212 crystals, where ξ_{ab} is in-plane coherence length, λ_{ab} is in-plane penetration depth and $s = c/2 = 1.539$ nm for CuO_2 interlayer distance, and assuming mean-field temperature dependence for $\lambda_{ab}^2(T) = \lambda_{ab}^2(0)/(1 - T/T_c)$, then the decoupling field $H_d = \Phi_o^3/[16\pi^3 e k_B \mu_o s \gamma^2 T \lambda_{ab}^2(T)]$, where Φ_o is flux quantum, e = 2.718..., k_B the Boltzmann constant and μ_0 the permeability of free space, can be used for $H_{2p}(T)$ [34]:

$$H_{2p}(T) = H_{2p}^0[(T_c/T) - 1)] \qquad (2)$$

This equation gives a moderate fitting for $H_{2p}(T)$ line from 13 to 62 K with $H_{2p}^0 = 1.06$ kG and $T_c = 71$ K as shown in Fig.8. From $H_{2p}^0 = \Phi_o^3/[16\pi^3 e k_B \mu_o s \gamma^2 T_c \lambda_{ab}^2(0)]$, using $\lambda_{ab}(0) = 260$ nm [35] and s = 1.539 nm, an anisotropy parameter $\gamma = 39$ is obtained, which is close to the value derived from single crystal resistivity measurement and is lower than that of 88 K Bi-2212 crystals. This result indicates that the $H_{2p}(T)$ in (Bi,Pb)-2212 may stem from the thermal-disorder-induced interlayer pancake vortex decoupling [36,37,38].

The $H_{on}(T)$ and $H_{2p}(T)$ are difficult to resolve below 11-13 K. Since the thermally induced quasi-lattice deformations are very weak at low temperatures, the quasi-lattice will be destroyed at some characteristic field through disorder-induced transition because increasing filed can also result in the increase of disorder, which is practically temperature-independent [31]. This argument is similar to zero-dimensional pinning limit at low temperatures [32,33]. However, the exact reason of low temperature termination is not very clear at present. Alternative explanation is that the quasi-lattice may be destroyed by two-dimensional (2D) melting of vortex solid at low temperatures [39,40].

In the limit of a thin-film type-II superconductor with thickness *s*, the spontaneous nucleation of edge dislocations in the flux line lattice is favorable for lowering the free energy of the system, which leads to dislocation-mediated two-dimensional melting of lattice at T_m^{2D} [39,40]. For layered high temperature superconductors, $T_m^{2D} = \Phi_o^2 s/[64\pi^2 k_B \mu_o \lambda_{ab}^2(0)]$ is predicted. Again using $\lambda_{ab}(0) = 260$ nm [35] and s = 1.539 nm, $T_m^{2D} = 13$ K is obtained. This value is close to the low temperature terminating point. The similar T_m^{2D} estimation values were also reported [41,42]. This indicates that order-disorder transition may cross a 2D melting line at low temperatures and make it difficult to observe H_{2p} and H_{on} below a critical temperature.

5. DYNAMIC MOMENT RELAXATION MESUREMENT IN THE VICINITY OF THE ANOMALOU MAGNETIZATION PEAK OF (Bi,Pb)-2212 SINGLE CRYSTALS

Recent experiment shows that both $H_{on}(T)$ and $H_{2p}(T)$ are frequency independent [33], which rules out the possibility that $H_{2p}(T)$ is a dynamic effect due to a change in pinning mechanism. However, due to finite isothermal field ramping rate (1-90 G/s), finite measurement internal (20-50 s) and metastability of the vortex structure, relaxation effect was always observed. To probe the vortex dynamics in the vicinity of H_{on} and H_{2p}, magnetic relaxation measurements M(t) have been carried out at different fields and temperatures below T < 62 K. Magnetic relaxation measurements are carried out within the time window 35 ≤ M(t) ≤ 10000 s. The magnetic relaxation behavior at T = 25 K in the vicinity of $H_{on} \sim 600$ G is shown in Fig.10 (a). It reveals that the slope of the M(t)/M(0) vs ln(t) curves exhibits a pronounced crossover at H = 600 G. This feature is clearly demonstrated from the curves at H = 700, 800 and 1000 G, whose slopes are smaller than those of curves for H < 600 G. At H =

2000 G, this kind of variation of slopes of M (t) curves is also observed as shown in Fig. 10(b). The magnetic relaxation results are analyzed according to the Anderson-Kim thermally activated flux creep model [43,44] and collective pinning theory [45]. It is found that M(t) can be described using the collective pinning formula $M(t) = M_0[1 + (\mu k_B T/U)\ln(t/t_0)]^{-1/\mu}$ [45], based on pinning potential power law $U(j) = U_0[(j_{c0}/j)^{\mu}-1]$ with attempt frequency $1/t_0$, supercurrent density j, critical current density j_{c0} and exponent μ depending on the dimensionality of the system, the field and current regime.

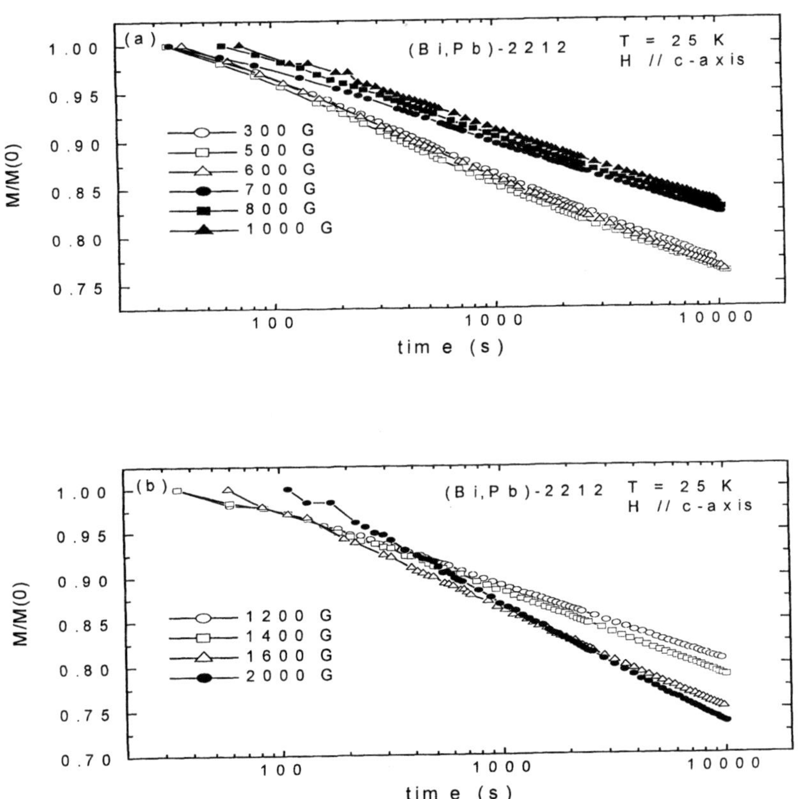

Fig. 10. Semilogarithmic plot of the normalized magnetic moment M(t)/M(0) vs time for (Bi, Pb)-2212 single crystal at T = 25 K, with H = 300 – 1000 G (a) and 1200 – 2000 G (b)

The normalized relaxation rate is defined by $S = -d\ln M(t)/d\ln t = k_BT/(U + \mu k_BT\ln(t/t_0))$. The magnetic field dependence of S is shown in Fig. 11. It indicates that S increases slightly with increasing field for H < H_{on}= 600 G, above which S decreases sharply with increasing field. The decrease of S indicates the improvement of the effectiveness of pinning. It can be explained in terms of the proliferation

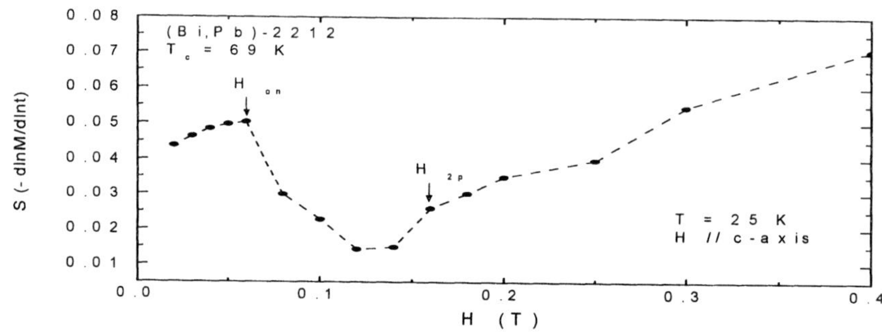

Fig.11 The normalized relaxation rate S as a function of H for (Bi,Pb)-2212 single crystal at T = 25 K

of topological defects in the ordered flux line lattice [46,47], which results in the entanglement of vortices and increases the effectiveness of pinning [26,27,28,29,33]. The minimum in S is about H = 1200 G, which is smaller than H_{2p} = 1600 G. For H > 1200 G, S increases with increasing field again. The field dependence of the exponent μ extrapolated by fitting the collective pinning theory is shown in Fig. 12. It indicates that μ ≈ 0.5 at low fields and it then increases to μ ≈ 2 at about 1200 G below H_{2p}. This shows that the magnetic relaxation below 1200 G can be described by the collective pinning theory. For the three-dimensional case, the collective creep theory predicts that μ = 1/7 for single-vortex creep at low temperatures and moderate magnetic fields, but relatively high currents. At intermediate current, μ = 3/2 for the creep of small vortex bundles ($a_0 < R_\perp$, $R_{//} \approx L_c < \lambda$) and the elasticity moduli exhibit large spatial dispersion. μ = 1 is for the creep of intermediate vortex bundles ($a_0 < R_\perp < \lambda < R_{//} \approx L_c$) and μ = 7/9 is for the creep of large vortex bundles ($\lambda < R_\perp$, $R_{//} \approx L_c$) at low current and high fields. Here, $R_\perp$ and $R_{//}$ denote the dimensions of the vortex bundles transverse and parallel to the flux motion; L_c is the longitudinal dimension of the vortex bundles along the field; a_0 is the flux line spacing and λ is penetration depth. For two-dimensional systems, $\mu = 9/8$ is for collective pinning at large currents and $\mu = 1/2$ for large bundle size and small currents. Fig.12 shows that at $H = H_{2p}$, $\mu \approx 1$ is closed to 9/8 implying that two dimensional creep occurs at H_{2p}, which may originate from thermal disorder induces pancake vortex decoupling transition mentioned above. As $H > H_{2p}$, μ drops to values below 0.2. This implies that there is a crossover to a single- vortex pinning corresponding to $\mu = 1/7$ which is expected only for low fields and high values of j [48]. Hence, the μ values below 0.2 above H_{2p} are inconsistent with the collective flux creep theory. This can be also further demonstrated by the field dependence of pining potential as shown in Fig.13. It shows that U decreases with increasing field for $H < H_{on}$. U(H) follows power law $U \propto H^{-1}$. As $H > H_{on}$, U increases with H up to a maximum value at H*~ 1200 G, below H_{2p}, then decreases upon further increasing field. As $H > H_{2p}$, U (H) can be fitted approximately by power law $U \propto H^{-0.7}$. Because U increases with increasing

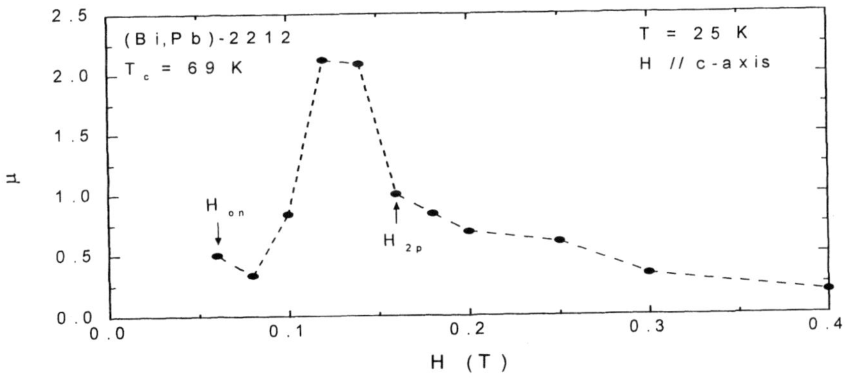

Fig.12 The field dependence of the critical exponent μ for (Bi,Pb)-2212 single crystal at T = 25 K

field H within the framework of collective flux creep theory, thus negative power law of U(H) observed is inconsistent with this theory [49,50,51]. Similar U(H) relation is also observed in $YBa_2Cu_3O_{7-x}$ single crystals by Abulafia et al.[52]. They propose that the negative power law of $U \propto H^{-0.7}$ can be well explained by a dislocation mediated plastic flux creep model based on the proliferation of dislocations in the vortex structure of the entangled phase similar to diffusion of dislocations in atomic solids. [53] This proliferation should influence the flux creep mechanism in the system. That is to say, the dislocation mediated plastic flux creep is expected at high fields. One of the main characteristics of the dislocation mediated plastic flux creep is the decrease of pinning potential U with the increase of H. [54] This is to be contrast with the collective (elastic) flux creep mechanism in which U increases with H. [42] The zero-current pinning potential is given by [54]

$$U_{pl} = \varepsilon\varepsilon_0 a_0 \propto H^{-0.5} \qquad (3)$$

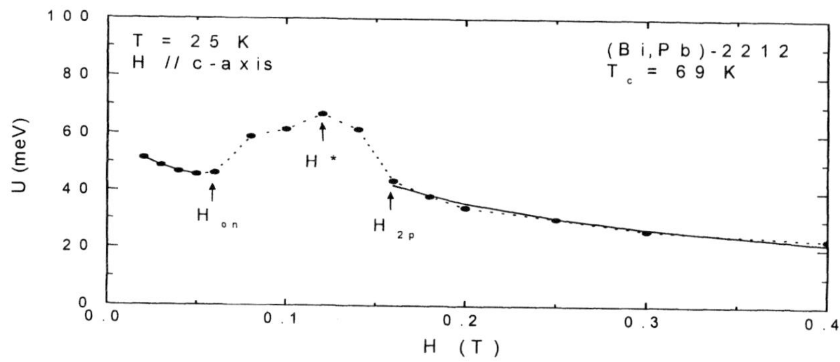

Fig.13 The pinning potential U as a function of field H for (Bi,Pb)-2212 single crystal at T = 25 K, The solid lines are fitting curves (see text)

where $\varepsilon_0 = (\Phi_0/4\pi\lambda)^2$ is the vortex line tension [48], $a_0 \approx (\Phi_0/H)^{1/2}$ is the mean intervortex distance, Φ_0 is the flux quantum and $\varepsilon = (m_{ab}/m_c)^{1/2}$ is the mass anisotropy parameter. Eq.(3) shows that the activation energy of plastic flux creep decreases with the increase of field. Therefore, our results indicates that the vortex motion mechanism in the high field of $H > H_{2p}$ is dominated by the dislocation mediated plastic flux creep. For $H_{on} < H < H_{2p}$, there is mixed mechanism including both the collective flux creep and the plastic flux creep simultaneously. The competition between them results in the appearance of U maximum at H* below H_{2p}. However, for $H_{on} < H < H^*$, the collective flux creep controls the flux dynamics though there also exists the plastic flux creep component, which gives rise to the increase of U with increasing field. For $H > H^*$, the plastic flux creep governs the flux dynamics though there also exists the collective flux creep component, which gives rise to the decrease of U with increasing field. That is to say, the flux creep mechanism between H_{on} and H_{2p} is complex and the flux creep process is governed by the smaller activation energy among the collective flux creep activation energy U_{el} and the plastic flux creep activation energy U_{pl}. In addition, our results also reveal that the plastic flux creep does not occur exactly at H_{2p} but does at H* prior to H_{2p}. Similar flux dynamic phenomena above are also observed at other different temperatures lying 18 K < T < 62 K. Fig. 14 is the plot of the pinning potential U as a function of field H for (Bi,Pb)-2212 single crystal at T = 33 K, which is obtained by a similar

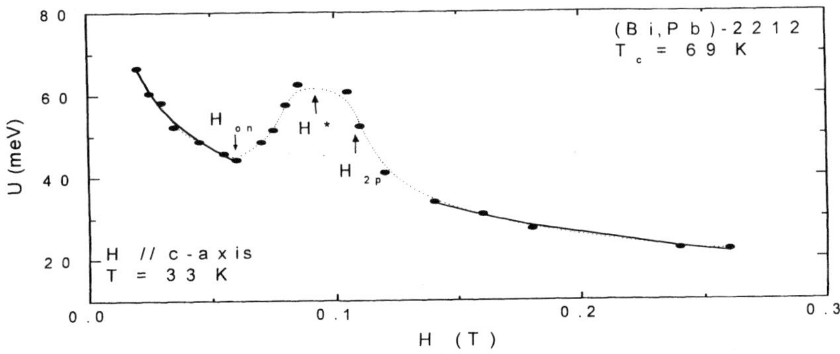

Fig.14 The pinning potential U as a function of field H for (Bi,Pb)-2212 single crystal at T = 33 K, The solid lines are fitting curves (see text)

analysis described above. It indicates that U decreases with increasing field for $H < H_{on}$. U(H) follows power law $U \propto H^{-0.4}$. As $H > H_{on}$, U increases with H up to a maximum value at H*~ 900 G below H_{2p}, then decreases upon further increasing field. As $H > H_{2p}$, U (H) can be fitted approximately by power law $U \propto H^{-0.7}$.

In the low temperature region of T < 18 K where $H_{on}(T)$ deviates from the fitting of Eq.(1), the magnetic relaxation in the vicinity of H_{on} and H_{2p} behaves as different behavior described above. Based on the magnetic relaxation measurement at T = 16 K, the pinning potential U as a function of field H is plotted in Fig.15. It indicates that the power law of U(H) for $H<H_{on}$ and $H>H_{2p}$ described above is absent and almost constant U(H) is observed as $H > H_{2p}$. In the collective vortex pinning and single vortex pinning regime U ~ constant is expected as $T < T_{dp}$, where T_{dp} is the depinning temperature. As $T > T_{dp}$, U grows with temperatures.[42]

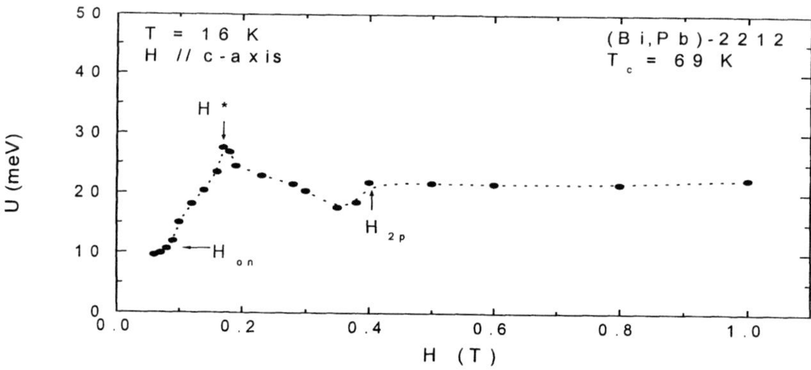

Fig.15 The pinning potential U as a function of field H for (Bi,Pb)-2212 single crystal at T = 16 K.

In short, the vortex dynamics variation in the vicinity of H_{on} (T) and H_{2p}(T) may also imply the underlying change of the vortex structure in essence.

6. CONCLUSION

In summary, the study of the anomalous magnetization peaks for optimum-doped Bi-2212 and overdoped 69 K (Bi,Pb)-2212 single crystals are reviewed. Experimental results show that the anomalous magnetization peak of Bi-2212 single crystals does not originate from the surface state and surface barrier of crystals but from intrinsic characteristic of crystals. In contrast to optimum-doped Bi-2212 crystals, both H_{on} and H_{2p} of (Bi,Pb)-2212 single crystals exhibit strong temperature dependence up to the vicinity of T_c. Analysis shows that the temperature dependence of H_{on} is consistent with a disorder-induced structure transition in the vortex solid. At the same time, the temperature dependence of H_{2p} seems to be related with the thermal disorder induced vortex pancake decoupling transition lying in different CuO_2 layers. The magnetic relaxation behavior in the vicinity of H_{on} and H_{2p} seems to support above scenarios.

ACKNOWLEDGEMENTS

I would like to thank my close collaborators J. J. Du, W. H. Song, B. Zhao for their contributions. My thanks also to Prof. H. C. Ku for his help in experiment. This work was supported by the National Key Basic Research under contract no. G19990646, National Center for R&D on Superconductivity under contract no. 863-CD010105, the National Natural Science Foundation under contract NSF 59872043 and the Fundamental Bureau, Chinese Academy of Sciences.

REFERENCES

[1] M.Daeumling et al, Nature (London) **346**, 332 (1990).
[2] U.Welp et al., Appl. Phys. Lett. **57, 84** (1990).
[3] L.Civale et al., Phys.Rev.B **43**, 13732 (1991).
[4].N.Chikumoto et al., Phys.Rev. Lett. **69**, 1260 (1992).
[5] G. Yang et al , Phys.Rev. B **48**, 4054 (1993).
[6] Y. Yeshurun et al., Phys. Rev. B **49**, 1548 (1994).
[7] X. Y. Cai et al., Phys. Rev. B **50**, 16774 (1994).
[8] T.Kobayashi et al., Appl. Phys.Lett. **62,** 1830 (1993).
[9] F.Zuo et al., Phys. Rev. B **49,** 12326 (1994).
[10] D. Giller et al., Phys. Rev. Lett. **79**, 2542 (1997).
[11] M. C. de Andrade et al., Phys. Rev. B **57**, 708 (1998).
[12] M. Xu et al., Phys. Rev. B **48,** 10630 (1993).
[13] V. N. Kopylov et al., Physica C **170** . 291(1990).
[14] V. Hardy et al., Physica C **232,** 347 (1994).
[15] F. Zuo et al., Phys.Rev.B **52,** 755 (1995).
[16] D. Pelloquin et al., Phys. Rev. B **54**, 16246(1995).
[17] T. Tamegai et al., Physica C **213,** 33 (1993).
[18] V. F. Correa et al., Phys. Rev. Lett. **87**, 57003(2001)
[19] X. L. Wang et al., Phys. Rev. B **55**, R3402(1997).
[20] C. P. Bean and J. D. Livingston, Phys. Rev. Lett. **12,** 14 (1964).
[21] R. Busch et al., Phys. Rev. Lett. **69,** 522 (1992).
[22] J. C. Martinez et al., Phys. Rev. Lett. **69**, 2276 (1992).
[23] R. Cubitt et al., Nature **365**, 407 (1993).
[24] S. L. Lee et al., Phys. Rev. Lett. **71**, 3862 (1993).
[25] G. Blatter et al., Phys. Rev. B **54**, 72 (1996).
[26] D. Ertas and D. R. Nelson, Physica C **272**, 79 (1996).
[27] T. Giamarchi and P. Le Doussal, Phys. Rev. B **55**, 6577 (1997).
[28] J. Kierfeld, T. Nattermann, and T. Hwa, Phys. Rev. B **55**, 626 (1997).
[29] D. S. Fisher, Phys. Rev. Lett. **78**, 1964 (1997).

[30] I. Chong et al., Science **276**, 770 (1997).

[31] B. Khaykovich et al., Phy. Rev. B **56**, 517 (1997).

[32] M. Niderӧst et al., Phys. Rev B **53**, 9286 (1996).

[33] M. F. Goffman et al., Phys. Rev. B **57**, 3663 (1998).

[34] L. L. Daemen et al., Phys. Rev. Lett. **70**, 1167 (1993) and Phys. Rev. B **47**, 11291 (1993); L. I. Glazman and A. E. Koshelev, ibid.**43**, 2835 (1991).

[35] T. Jacobs et al., Phys. Rev. Lett. **75**, 4516 (1995).

[36] Y. P. Sun et al., Phys. Rev. B **61**, 11301(2000).

[37] Y. P. Sun et al., Chin. J. of Phys. **38**, 237 (2000)

[38] Y. P. Sun et al., Appl. Phys. Lett. **76**, 3795(2000)

[39] B. A. Hubermann et al., Phys. Rev. Lett. **43**, 950 (1979).

[40] D. S. Fisher, Phys. Rev. B **22**, 1190 (1980).

[41] A. Schilling et al., Phys. Rev. Lett. **71**, 1899 (1993).

[42] G. Blatter et al., Rev. Mod. Phys. **66**, 1125 (1994)

[43] P. W. Anderson, Phys. Rev. Lett. **9**, 309(1962)

[44] P. W. Anderson and Y. B. Kim, Rev. Mod. Phys. **36**, 39(1994)

[45] M. V. Feigel'man et al., Phys. Rev. Lett. **63**, 2303 (1989).

[46] E. H. Brandt, Phys. Rev. B **34**, 6514 (1986).

[47] R. Wӧrdenweber et al., Phys. Rev. B **34**, 494 (1986).

[48] G. Blatter et al., Rev. Mod. Phys. **66**, 1125(1994).

[49] H. Küpfer et al., Phys. Rev. B **50**, 7016 (1994).

[50] A. A. Zhukov et al., Phys. Rev. B.**51**, 12704(1995).

[51] G. K. Perkins, et al., Phys. Rev. B **51**, 8513(1995).

[52] Y. Abulafia et al., Phys. Rev. Lett. **77**, 1596(1996)

[53] J. P. Hirth and J. Lothe, Theory of Dislocations (John Wiley& Sons, New York, 1982), Chap.15

[54] V. B. Geshkenbein et al., Physica C **162-164**, 239 (1989).

DYNAMICS OF TWO-DIMENSIONAL JOSEPHSON JUNCTION ARRAYS

Md. Ashrafuzzaman and Hans Beck
Institut de Physique
Université de Neuchâtel, Switzerland
e-mail: md.ashrafuzzaman@unine.ch
hans.beck@unine.ch

1 Introduction

1.1 Josephson junction arrays

Two-dimensional (2D) Josephson junction arrays (JJA) offer a unique opportunity for studying a variety of topics in 2D physics, such as phase transitions, non-linear dynamics, percolation, frustration and disorder, in relatively ≪clean≫ experimental realisation. Fabrication of arrays and their basic physical properties have been described in various articles [1], [2]. A JJA consists of islands, positioned in some periodic or irregular arrangement, that become superconducting below a given transition temperature T_c^o. Below this temperature each island l is characterized by its superconducting wave function

$$\psi = |\psi_l| e^{i\theta_l} \tag{1.1}$$

with its amplitude and its phase. For all matters and purposes one can assume that $|\psi_l|$ has the same value in each island, such that the phase is the only relevant variable. The islands are linked to each other by the Josephson coupling. The potential energy of the array is then given by

$$H = \sum_{\langle ll' \rangle} J_{ll'}[1 - \cos(\theta_l - \theta_{l'})] \tag{1.2}$$

The sum can usually be restricted to nearest neighbors in the array and the Josephson coupling J is related to the critical current I_c by

$$J = \frac{\hbar}{2e} I_c \tag{1.3}$$

If the array is placed in a perpendicular magnetic field the phase differences in (1.2) have to be replaced by the ≪ gauge invariant ≫ combination

$$\phi_{ll'} = \theta_l - \theta_{l'} - A_{ll'} \tag{1.4}$$

where $A_{ll'}$ is the line integral

$$A_{ll'} = \frac{2e}{\hbar} \int_l^{l'} d\mathbf{r} \cdot \mathbf{A}(\mathbf{r}) \tag{1.5}$$

of the applied vector potential. Most of our subsequent calculations will be based on the assumption of a large magnetic penetration depth. Then any vector potential appearing in our equations describes an applied external electromagnetic field, whereas internal field fluctuations are neglected.

When the phases of neighboring islands differ from each other a supercurrent

$$K_{ll'} = J_{ll'} \sin \phi_{ll'} \tag{1.6}$$

is flowing through the junction (l, l') and the two islands get charged. The electrostatic energy

$$E_{ll'} = \frac{1}{2} \sum_{ll'} Q_l (C^{-1})_{ll'} Q_{l'} \tag{1.7}$$

has to be added to (1.2). Here Q_l is the charge of island l, related to the excess number δN_l of Cooper pairs by $Q_l = 2e\delta N_l$, and $C_{ll'}$ is the capacitance matrix of the array. Its diagonal elements determine the charging energy with respect to ground and the nearest neighbor elements represent the junction capacitance. The excess number δN_l can be identified with a canonical momentum P_l conjugate to the phase [1,3,4]:

$$[P_l, \theta_{l'}] = \frac{\hbar}{i} \delta_{ll'} \tag{1.8}$$

The electrostatic energy (1.7) thus plays the role of a (non-local) kinetic energy that has to be added to the potential energy (1.2) yielding thus the complete Hamiltonian

$$H = \frac{1}{2} \sum_{ll'} (2e)^2 P_l (C^{-1})_{ll'} P_{l'} + \sum_{\langle ll' \rangle} J_{ll'} (1 - \cos \phi_{ll'}) \tag{1.9}$$

of the array on which our equations of motion will be based in section 2.

Much work has been done in order to elucidate the thermodynamic properties of (1.9) which is indeed a challenging problem of statistical mechanics. The simplest case taking into account only the potential energy (1.2) in zero applied field is the classical XY-model in 2 dimensions. In its ground state all the ≪spins≫ are parallel, respectively all the phases θ_l are equal. This system allows to study the critical behaviour of a 2D system with a complex order parameter. It gives rise to the well-known Berezinski-Kosterlitz-Thouless phase transition [5,6] bringing in the basic concept of vortex excitations. The basic features are summarized in the following subsection.

The full form (1.9) adds two complications:

a) A JJA in an applied magnetic field corresponds to the ≪frustrated≫ XY-model. Owing to the additional term $A_{ll'}$ in the coupling energy the phase configuration in the ground state is more complicated - corresponding to magnetic field induced vortices - and the energy per bond is less negative than $-2J$. The effect of $A_{ll'}$ is measured in terms of the flux threading through an elementary cell of the array, the ≪frustration parameter≫ f

$$f = \frac{1}{2\pi} \sum_{\langle ll' \rangle} A_{ll'} \tag{1.10}$$

the sum running over the bonds going around the cell. For simple - rational - values of f the ground state phase pattern is known (see [1] and the references given there). For certain f-values interesting infinite degeneracies have been found in triangular arrays [7] The behavior at finite temperatures, in particular the various kinds of phase transitions that a frustrated array is supposed to undergo is less well understood [1].

b) Owing to the commutation relations (1.8) the full Hamiltonian (1.9) describes a quantum system, sometimes called ≪quantum XY-model≫ although its operator algebra is, of course, different from the XY-model describing quantum spins. Quantum fluctuations, surviving even at $T = 0$, add to the thermal fluctuations of the phases already present in the classical XY-model. They tend to counteract the tendency to ≪topological≫ phase order which is characteristic of the 2D classical system. Taking into account only the charging energy with respect to ground the capacitance matrix has the form $C_{ll'} = C\delta_{ll'}$ and the influence of charging energy can be expressed in terms of the ratio

$$\alpha = \frac{(2e)^2}{2CJ} \tag{1.11}$$

The corresponding phase diagram (with and without dissipation) has been studied by various methods. Approximating the cosine phase interaction by a quadratic form in the phase differences corresponds to the ≪selfconsistent harmonic approximation≫ [4,8,9]. One finds that the critical temperature decreases with increasing α and finally vanishes at a critical threshold α_c on the order of 6. The same conclusion is drawn from Quantum Monte Carlo calculations [10]. The quantum model including dissipation is presumably also relevant for describing the superconductor-insulator transition occuring at the underdoped limit of cuprate superconductors [11]. It has been shown [12] that dissipation may lead to a new universality class characterized by an effective non-ohmic dissipation.

The present article mainly deals with the dynamical properties of $JJAs$. Section 2 presents various equations of motion for the phase variables which are based on the superconducting properties of the array including dissipation and charging effects. The essential excitations of the phase system are vortices (V) and antivortices (A) for which we derive appropriate equations of motion in section 3. The following section 4 deals with various linear response functions of an array, such as its electrical conductance and impedance, and relates such quantities to vortex variables. Various analytical and numerical theoretical techniques that are currently used in order to calculate dynamic response functions are summarized in section 5. Effects of ≪disorder≫, ie deviations from a periodic lattice structure of the array, are discussed in section 6. Section 7 prepares the interpretation of experimental data by giving an appropriate theoretical description of the relevant measurements. The main experimental observations of the dynamic behaviour of JJAs are summarized in section 8, whereas section 9 presents theoretical predictions found by analytical and numerical methods. Section 10 deals with the dynamic response of $JJAs$ to strong applied currents. The last section gives some hints to other aspects of JJA that could not be covered in this article. We would like to warn the reader at this point that our presentation has been made to be relatively concise, in order to keep the article at a decent size. However, we try to give pertinent references where more details can easily be found.

1.2 The Berezinski - Kosterlitz - Thouless scenario

Treating each phase θ_i as a classical variable ranging between 0 and 2π the Hamiltonian (1.2) describes the classical 2D XY-model, the phase angle giving the direction of a classical spin of unit length at site i. Its thermodynamic properties are governed by the Berezinski-Kosterlitz-Thouless (BKT) transition. The lowest energy configuration corresponds to all phases being equal, corresponding to the ≪ferromagnetic≫ ground state of the spins at $T = 0$. For any $T > 0$ this long range order is replaced by ≪topological≫ order due to thermal fluctuations [13]. Besides small amplitude variations, which give rise to ≪spin waves≫ in the dynamic generalization of (1.2), the main finite temperature configurations are given by thermally excited vortices (V) and antivortices (A). Their interaction varies essentially logarithmically with their distance. They thus behave like a 2D Coulomb gas. In the absence of frustration the latter is neutral (equal number of V and A). According to the BKT scenario, V and A are bound in pairs below the transition temperature T_{BKT}. When T grows above T_{BKT} more and more pairs with large separations dissociate yielding an increasing number of ≪ free ≫ V and A. The mean distance between the latter is given by the BKT correlation length ξ with its particular T-dependence

$$\xi(T) = \xi_o \, exp(\frac{b}{\sqrt{T/T_{BKT} - 1}}) \tag{1.12}$$

where ξ_o is a prefactor of the order of the lattice constant and b a parameter of order one. The topological order below T_{BKT} is characterized by a power-law spatial decay of phase (spin) correlations :

$$\langle e^{i(\theta_l - \theta_{l'})} \rangle = (\frac{|l - l'|}{a})^{-\frac{k_B T}{2\pi J(T)}} \tag{1.13}$$

with a coupling constant $J(T)$ renormalized by thermal fluctuations, and a finite helicity modulus Λ expressing the ≪phase stiffness≫ of the system. Λ is given by the variation of the free energy of the system with respect to a ≪phase twist≫ :

$$\Lambda = (\frac{\partial^2 F(\alpha)}{\partial \alpha^2})_{\alpha=0} \tag{1.14}$$

where the free energy

$$F(\alpha) = -k_B T \ln \int d[\theta] \quad e^{-\beta H[\theta,\alpha]} \tag{1.15}$$

is evaluated by imposing the boundary condition

$$\theta(la, Na) = \theta(la, 0) + \alpha \tag{1.16}$$

relating the phase $\theta(l, m)$ on a given column l of the array at the lower and the upper boundary, respectively, the lattice sites being parametrized in units of the lattice constant a by (la, ma). Λ has a particular temperature dependence. Starting from its $T = 0$ value, where it is given by the coupling constant J, it is a deacreasing function of temperature, since thermal fluctuations make the phase system softer. At T_{BKT} it jumps from a universal value

$$\Lambda_c = \frac{2}{\pi} \frac{k_B T_{BKT}}{J(T_{BKT})} \tag{1.17}$$

to zero. In section 4 Λ will be expressed by suitable phase correlation functions and it will be related to the electrical conductance of the array, as well as to the vortex dielectric function. This allows to observe the T-dependence of Λ by dynamic measurements, as discussed in sections 7 to 9.

Above the transition temperature the phase correlation function decays exponentially with the BKT correlation length given by (1.12).

In the frustrated case there are magnetic field induced V (or A, depending on the sign of the field) already in the ground state. The thermal V-A pairs coexist with the latter. The precise nature of the corresponding phase transition in the presence of frustration is still a matter of debate, see ch. 8 of ref. [1].

The precise link between the XY-Hamiltonian (1.2) expressed in terms of the phases and the 2-d Coulomb gas can be formalized by transforming it into a vortex representation with the help of a dual transformation [14, 15].. This procedure allows - at least approximately - to separate the spin wave contribution to the partition function and yields the vortex Hamiltonian

$$H_c = \frac{1}{2} \sum_{RR'} (m(R) - f) V_c(R - R')(m(R') - f) \tag{1.18}$$

The sum runs over the plaquettes of the array, $m(R)$ being the vorticity of plaquette R (in practice m = 0, 1 or -1). The interaction potential V_c depends logarithmically on the distance, at least for large values of the latter. The frustration f acts on the Coulomb gas of A and V as a ≪ background charge ≫. The precise critical behaviour of the BKT transition is obtained by introducing length scale dependent screening, which yields renormalization group equations [1,13,14] for the relevant parameters of the Coulomb system, namely its dielectric constant, related to the phase stiffness Λ (see section 4) and its fugacity.

We do not go into the renormalization group approach to the BKT transition [13] since it concerns mainly thermodynamics. Its generalization to dynamics - our present topic - as it was developed by Shenoy [16] will be presented briefly in section 5.5. A very extensive review of the BKT transition and the Coulomb gas model has been published by Minnhagen [17]. Other review articles, such as [18,19], are more directly concerned with dynamic properties of $JJAs$. In order to interpret correctly measurements on $JJAs$ in the light of the BKT scenario one should, however, bear in mind that the latter (as critical phenomena in general) is in principle valid for the thermodynamic limit of an infinitely large system. In section

7.1 we will mention finite size effects that can alter the picture to some extent, in particular through the presence of unbound vortices below the transition temperature.

2 The Dynamics of JJA

2.1 Equation of motion for the phases

In order to derive equations of motion for a JJA we start from our Hamiltonian (1.9) involving Josephson coupling and electrostatic energy. It is useful for later purposes to include a possible coupling of the array to external currents I_l flowing into (or out of) site l

$$H_1 = \sum_l I_l \theta_l \tag{2.1}$$

The resulting canonical equations read

$$\dot{\theta}_l = \frac{\partial H}{\partial P_l} = (2e)^2 \sum_{l'} (C^{-1})_{ll'} P_{l'} \tag{2.2}$$

$$\dot{P}_l = -\frac{\partial H}{\partial \theta_l} = -\sum_{l'} J_{ll'} \sin(\theta_l - \theta_{l'}) - I_l \tag{2.3}$$

Due to (1.6) and (1.8) the second equation is the continuity relation expressing charge conservation at each site of the array, the right hand side being the sum of supercurrents and external current flowing into site l.

Dissipation, due to normal current between islands and from islands to ground, is now easily added to the charge conservation expressed by (2.3) which then reads :

$$\dot{P}_l = -\sum_{l'} J_{ll'} \sin(\theta_l - \theta_{l'}) - \sum_{l'} (R^{-1})_{ll'} (\frac{\hbar}{4\pi e})^2 (\dot{\theta}_l - \dot{\theta}_{l'}) - R_o^{-1} (\frac{\hbar}{4\pi e})^2 \dot{\theta}_l - I_l \tag{2.4}$$

Here we have introduced the matrix $R_{ll'}$ of junction resistances and the resistance R_o to ground, and we have used the Josephson relation between the time derivative of the phase and a corresponding voltage. Equations (2.2) and (2.4) correspond to a description of the array in terms of an effective circuit involving capacitive, resistive and Josephson (≪ inductive ≫) elements. Neglecting the ≪local ≫ charging energy and normal current (ie between a given site and the ground) yields the so-called ≪ Resistively and capacitively shunted junction ($RCSJ$) model ≫.

Capacitive effects are important for ultrasmall superconducting islands. The interplay between charge and phase fluctuations lowers the superconducting transition temperature and can drive it completely to zero, leading thus to a quantum superconductor-insulator transition, as discussed in section 1.1. We will mostly concentrate on dynamic phenomena of ≪ classical arrays ≫ for which charging effects are unimportant. This corresponds to a motion without inertial acceleration and thus equation (2.4) with the left hand equal to zero will determine our phase dynamics. The two most frequently used equations of motion are then distinguished according to the type of phase damping involved, corresponding to the type of normal current dissipation :

a. Resistively shunted junction - RSJ - dynamics

Only the ≪ bond damping ≫ (the second term in (2.4)) is kept, describing dissipation by currents flowing through the junctions. This dynamics is therefore also called ≪ total current conserved ≫. With R_J being the junction resistance the RSJ equation for the phases then reads :

$$\dot{\theta}_l = -\sum_{l'} G_{ll'} (\frac{4\pi e}{\hbar})^2 \frac{1}{R_J} [\sum_{l''} J_{l'l''} \sin(\theta_{l'} - \theta_{l''}) + I_{l'}] \tag{2.5}$$

Here we have introduced the Green function of the lattice which allows to write the sum over bond differences of the phase θ as a sum over sites :

$$\sum_{l'} (\theta_l - \theta_{l'}) = \sum_{l'} (G^{-1})_{ll'} \theta_{l'} \tag{2.6}$$

b. Time dependent Landau-Ginzburg - $TDGL$ - dynamics

Taking into account only site damping due to losses to the ground (only R_o is considered in (2.4)) yields the simpler ≪relaxational≫ equation of motion

$$\dot{\theta}_l = -(\frac{4\pi e}{\hbar})^2 \frac{1}{R_o} [\sum_{l'} J_{ll'} \sin(\theta_l - \theta_{l'}) + I_l] \tag{2.7}$$

This corresponds to the usual time dependent Landau-Ginzburg equation representing ≪ nonconserved ≫ dynamics (charges are ≪ lost ≫ by currents going to the ground).

In the presence of an electromagnetic field the phase differences occurring in the various current expressions have to be replaced by the corresponding ≪ gauge invariant ≫ combinations with the applied vector potential, as indicated in (1.4).

Dynamic equations, such as (2.6) or (2.7) can now be used for various analytical considerations, such as the derivation of an equation of motion for the center of a vortex (section 3) or the calculation of dynamic linear response functions (section 4). They are also at the basis of numerical simulations of equilibrium dynamics (section 5.4) and of the response of the array to very strong driving currents (section 10).

2.2 Microscopic derivation of the coupling constants

The various coupling constants (resistances, capacities and Josephson couplings) of the $RCSJ$ model describe the physical properties of single Josephson junctions. They can be derived from a microscopic transfer Hamiltonian, describing the coupling between the two superconducting islands through a normal metal or an insulator by a tunneling interaction and including Coulomb repulsion between electrons. Besides numerous earlier publications on the physics of Josephson junctions useful reviews, focussing in particular on very small junctions which can now be fabricated, have been given by Schön and Zaikin [20] and by Simanek [4]. Using either an imaginary time or a real time representation the quantum mechanical partition function can be rewritten in terms of an action involving the phase difference between the two superconductors. Its dynamics is indeed governed by a capacitance and by generalized dissipation and Josephson coupling terms.

3 Vortex dynamics

3.1 Classical equations of motion

Vortex excitations are the main elements determining the physical properties of a JJA, in particular in connection with the Berezinski-Kosterlitz-Thouless scenario, as explained in section 1.2. Thus it is important to account for them explicitly in dynamic calculations. A classical equation of motion for a vortex configuration can be found, in analogy with the way the dynamics of localized excitations in magnetic models is treated [21]. We start from one of our general equations of motion for the phases of the array derived in section 2.1 which can be transformed into an equation for the bond supercurrent given by (1.6). In order to prepare a continuum description of the array it is useful to number each site (of a square array) by an integer I and to attach a horizontal and a vertical bond to each site, distinguished by the index $s = x$ or y, respectively(see figure 1). The bond current $K_{ll'}$ (1.6) is then replaced by $K_s(I)$ and the phase difference (1.4) by $\phi_s(I)$.

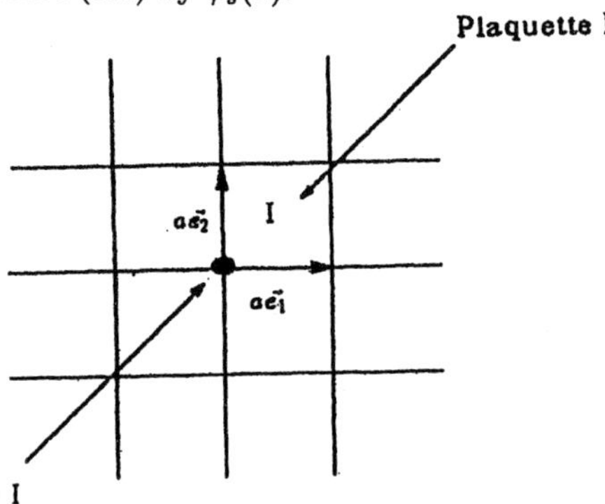

Figure 1 : Parametrization of a square lattice : each site is labelled by an integer I. Two nearest neighbor bonds attached to each site yield the two cartesian directions x and y, denotes by the corresponding unit vectors pointing along the plaquette I. They determine the x and y components of vector and tensor quantities in equations (3.1) to (3.5).

Our equations of motion for the phases, derived in section 2, can be used for determining the time evolution of K_s

$$C_o \frac{\partial^2 K_s(I)}{\partial t^2} = -E_s(I) \sum_{s'I'} \{J M_{ss'}(I-I') K_{s'}(I') + \eta_{ss'}(I-I')\dot{\phi}_s(I')\} \tag{3.1}$$

where $E_s(I) = \cos\phi_s(I)$ and the ≪dynamical matrix≫ M ensures the correct coupling between bonds.

In (3.1) we have specialized to the case of a local capacity, but we still admit both types of damping discussed in section 2.1 :

$$\eta_{ss'}(I-I') = \begin{cases} \delta_{ss'}\delta_{II'}\eta_L & \text{for local damping} \\ M_{ss'}(I-I')\eta_T & \text{for bond damping} \end{cases} \tag{3.2}$$

The friction constants are proportional to the inverse resistances showing up in (2.4), ie the ground (local damping), respectively the bond resistance (bond damping).

A moving vortex is now introduced by a trial solution[22]

$$\phi_s(I,t) = \phi_s^{(v)}(I,t) + \varphi_s(I,t) \tag{3.3}$$

where

$$\phi_s^{(v)}(I,t) = arc\sin K_s^{(v)}(I|\mathbf{R}(t)) \tag{3.4}$$

The first term represents the circular symmetric ≪hedge hog≫ phase pattern with a time dependent center $\mathbf{R}(t)$ and the second contribution describes deviations from this form. The form (3.2) is substituted into the phase equation of motion (3.1) yielding an equation for the bond currents $K_s^{(V)}$ forming the vortex pattern

$$C_o \frac{\partial^2 K_s^{(v)}(I)}{\partial t^2} = -E^s(I)\sum_{s'I'}\{JM_{ss'}(I-I')[E_{s'}(I')\varphi_{s'}(I')+K_{s'}^{(v)}(I')]+\eta_{ss'}(I-I')\frac{\partial}{\partial t}arc\sin K_{s'}^{(v)}(I')\} \tag{3.5}$$

We have neglected the small deformation field φ wherever it multiplies terms involving $\phi_s^{(v)}$. The first term on the right hand side of (3.5), involving the dynamical matrix M, represents the Peierls force acting on the moving vortex which originates in the discrete lattice structure of the JJA. Indeed, the potential energy of the phase pattern of the vortex increases when its geometrical center moves away from the middle of a plaquette. It reaches its maximum value when the vortex center crosses the bond separating two adjacent plaquettes. The height E_B of this energy barrier depends on the lattice structure : $E_B \approx 0.199J$ for the square and $E_B \approx 0.043J$ for the triangular array [23]. Going over from the discrete lattice to a continuum and integrating out the phase variables in a suitable way (see [21] and [22] for more details) allows to obtain a Newtonian equation for the center $\mathbf{R}$ of the vortex

$$M_v\ddot{\mathbf{R}} = -\Gamma\dot{\mathbf{R}} - 2\pi qaJ\hat{z}\times\vec{\varphi} \tag{3.6}$$

The parameter $q = \pm 1$ determines the sign of the excitation (V or A). The vortex mass M_v is expressed in terms of the grain capacity C_o and $a \ll$ core radius $\gg b$ of the vortex patterns (which can be taken of the order ot the lattice constant a) :

$$M_v = C_o\frac{20\pi}{9}(\frac{b}{a})^2 \tag{3.7}$$

Obviously, in the absence of an underlying discrete lattice structure there is no more energy barrier for the moving vortex. The two types (3.2) of dissipation, although yielding different equations for the phase field, lead to the same friction term in (3.6), the respective friction constants being only quantitatively different :

$$\Gamma = \begin{cases} \eta_L\frac{20\pi}{9}(\frac{b}{a})^2 & \text{for local damping} \\ \eta_L\frac{2\pi}{9}(\frac{b}{a})^2 & \text{for bond damping} \end{cases} \tag{3.8}$$

One should stress here that different calculations may yield different results for friction constant and mass: ≪simple≫ calculations like the above skteched trial function approach yield a ≪bare≫ mass, whereas taking into account the coupling of the vortex configuration with thermal fluctuations introduce a renormalization to mass and friction, which can even be (logarithmically) divergent due to the 2D ≪substrate≫, see, for example, [24]. The second term on the right hand side of (3.6) can be interpreted in different ways : it can represent the force acting on the vortex when a supercurrent, given by a phase gradient $\vec{\varphi} = \nabla\theta$, is flowing through the sample. It is usually called "Lorentz force" [1], [25], in analogy to the force a current-carrying wire experiences in a perpendicular magnetic field. In our later calculations we will insert into equation (3.5) the phase gradient $\nabla\theta$ resulting from the presence from another vortex located at $\mathbf{R}_1$ in order to introduce the effect of V-V interaction in the equation of motion :

$$M_v\ddot{\mathbf{R}} = -\Gamma\dot{\mathbf{R}} + q_1 q_2 a^2 J^2 \frac{\mathbf{R}-\mathbf{R}_1}{(\mathbf{R}-\mathbf{R}_1)^2} \tag{3.9}$$

When capacitive effects in the array can be neglected (for sufficiently large capacities or for low enough frequencies) the mass term disappears from (3.6).

3.2 Quantum dynamics

At sufficiently low temperatures the JJA, characterized by the Hamiltonian (1.9) and an appropriate dissipation function, should be treated by quantum mechanics. As sketched in section 1.1 this is necessary when the ≪mass term≫ in (1.9) becomes comparable to the Josephson coupling, ie when the ≪quanticity ratio≫ α given by (1.11) is large enough.

The corresponding dynamics of quantum vortices has been studied on the basis of a quantum action for the phases, which allows to treat dissipation on the same footing as the Hamiltonian terms. Vortex variables are introduced in two ways. Generalizing the Villain transformation, used for the same purpose for the classical XY-model [14] leads to dynamic plaquette vorticities [26]. The transformation is performed on the quantum form of the partition function involving an action including charging energy and dissipation. As a result one obtains a new action in which the (true) electric charges interact with V and A, acting like ≪topological charges≫. Integrating out the electrical charges yields an effective Lagrangian for the V-A system that attributes effective mass and friction to the latter. Alternatively, the action is expressed in terms of vortex positions and momenta which are again introduced by a trial form for the dynamic phase field of a vortex [27], as it was done above for the classical case.

Quantum vortices having a non-zero mass show interesting quantum effect[25], in particular when the lattice pinning potential is taken into account. While classical particles would oscillate in the minima of this potential and jump across barriers by thermal excitation, quantum vortices can tunnel through pinning barriers. Moreover, the quantum states of a vortex in the lattice potential will be a Bloch state with some given momentum. As in the case of band electrons Bloch oscillations can occur, as the particles move across the Brillouin zone, being driven by en effective electric field. Quantum interference and quantum localization in a random pinning potential (see also at the end of subsection 5.3) are another manifestation of the quantum nature of vortices.

We do not give more details here, since our subsequent analysis will mainly be focussed on ≪ classical ≫ arrays.

4 Linear response of a JJA

4.1 Kubo formalism for conductance and impedance

In this section we prepare the theoretical description of experiments that ≪ test ≫ the dynamic behaviour of a JJA. This can be done in different ways by applying suitable external disturbances to the system.

The frequency dependent conductance $G(\omega)$ is obtained from measuring the current that flows in the array as a response to a small time dependent external voltage. Alternatively, when a weak external current is fed into the array the impedance $Z(\omega)$ determines the resulting voltage drop across it. Explicit expressions for $G(\omega)$ and $Z(\omega)$ in terms of suitable time dependent correlation functions are obtained in this subsection in the framework of linear response theroy. In subsection 4.2 the response functions $G(\omega)$ and $Z(\omega)$ will be expressed in terms of quantities characterizing the vortex system. Other experimental approaches, such as the two-coil technique or the measurement of flux noise, will be described in section 7.

Since the JJA dynamics involves dissipation it is convenient to interpret the equations of motion (2.3, 2.5) as Langevin equations for the phase and the conjugate momentum by adding a random force. For calculating response functions it is useful to go over to the corresponding Fokker-Planck (FP) equation [38,39], governing the probability distribution $\rho(\theta_1, ...; P_1, ..)$ of the phases and the momenta :

$$\frac{\partial \rho}{\partial t} = L_{FP}\rho \tag{4.1}$$

The FP-operator $L_{FP} = L + L_{ex}$ has a contribution L containing convective and dissipative terms present for an isolated array and L_{ex} describing the coupling to external sources. Besides the coupling to external currents I_l in (2.1) an external electric field can be applied through a time dependent vector potential $A_{ll'}$ in the phase differences (1.4) yielding the following form of L_{ex}

$$L_{ex}\rho = -\sum_l I_l \frac{\partial \rho}{\partial P_l} + \sum_{ll'} J_{ll'} \cos(\phi_{ll'}) \frac{\partial \rho}{\partial P_{l'}} - \sum_{ll'} (\frac{\hbar}{4\pi e})^2 R_{ll'}^{-1} \dot{A}_{ll'} \frac{\partial \rho}{\partial P_{l'}} \tag{4.2}$$

Each term in (4.2) has the general structure $L_{ex} = X * F(t)$ where X is an observable of the array expressed in terms of the angles θ_l and the conjugate momenta P_l and $F(t)$ is a time dependent external force $F(t)$. In the FP formalism the linear response of an observable O in the presence of such an external perturbation is given by splitting the probability distribution

$$\rho = \rho_o + \rho_1 \tag{4.3}$$

into the canonical equilibrium part ρ_o and the deviation ρ_1 due to F. They are, respectively, given by

$$\begin{aligned} \rho_o &= \tfrac{1}{Z} exp(-\beta H) \\ \rho_1(t) &= \textstyle\int_{-\infty}^{t} d\tau \; e^{(t-\tau)L} L_{ex}(\tau)\rho_o \end{aligned} \tag{4.4}$$

assuming that the perturbation is adiabatically switched on at $t = -\infty$. The expectation value of O is then given by

$$\langle O(t)\rangle = \int_{-\infty}^{t} d\tau \; R(t-\tau)F(\tau) + O_o \tag{4.5}$$

where O_o is the equilibrium expectation value of O. The frequency Fourier transform of the response function R is given by the retarded time dependent correlation function of the variables O and X.

$$R(\omega) = \int_0^\infty dt\, e^{i\omega t} Tr(Oe^{Lt}X\rho_o) \tag{4.6}$$

the symbol $\ll$ Tr $\gg$ meaning integration over phase and momentum variables.

We now apply this general scheme to the two experimental configurations mentioned above :

a) For calculating the conductance $G(\omega)$ the observable O is the current (supercurrent plus normal current) flowing through some junction

$$O = J_{ll'} \sin\phi_{ll'} + (R^{-1})_{ll'}(\frac{\hbar}{4\pi e})^2 \dot{\phi}_{ll'} \tag{4.7}$$

The action of L_{ex} is given by the second and the third term of (4.2) which, according to (4.4), will act on ρ_o. The derivative with respect to P_l is easily evaluated using (4.4), (1.9) and (2.2) :

$$\frac{\partial \rho_o}{\partial P_l} = -\beta\rho_o \sum_{l'} (C^{-1})_{ll'} P_{l'} = -\beta\rho_o \dot{\theta}_l \tag{4.8}$$

We specialize to the situation where the applied electric field E_x is homogeneously spread across the N_x sites of the array, say in x-direction. Thus the time derivative ot the vector potential applied to a bond going from site l to site l' in the (positive) x-direction is given by

$$\dot{A}_{ll'} = -\frac{E_x}{N_x} = -\dot{A}_{l'l} \tag{4.9}$$

As for calculating the static helicity modulus it is important to impose suitable boundary conditions on the phases for the conductance calculation. Since we deal with the variables θ_l and P_l in the present context we use periodic boundary conditions (PBC) for the phases. Later on, in section 5.4 on numerical simulations, boundary conditions will be introduced which are more appropriate when the relevant variables are vortex positions. Taking (4.9) and the PBC into account sums like

$$S = \sum_{\langle ll'\rangle_x} (\dot{\theta}_l - \dot{\theta}_{l'}) \tag{4.10}$$

extending over all bonds in x-direction vanish. Thus the total normal current in (4.7) reduces to

$$I_x^{(n)} = -\eta \sum_{\langle ll'\rangle} \dot{A}_{ll'} = \eta E_x \tag{4.11}$$

which gives the ohmic contribution to the conductance. Moreover, for a regular array the parameters $J_{ll'}$ and $\eta_{ll'}$ are equal for all bonds $\langle ll'\rangle$ so that, due again to (4.9) and the PBC the third term in (4.2), representing the coupling of the applied vector potential in the normal current, also vanishes. We only have to evaluate the response of the supercurrent in (4.7) to the second term of the external perturbation in (4.2). It has two contributions :

$$\langle K_x(t)\rangle = J \sum_{\langle ll'\rangle_x} \langle \cos\phi_{ll'}\rangle A_{ll'}(t) - \beta\frac{J^2}{N_x}\int_{-\infty}^t d\tau \sum_{\langle ll'\rangle_x}\sum_{\langle nn'\rangle_x} \langle \sin\phi_{ll'}(t)\sin\phi_{ll'}(\tau)\dot{\phi}_{nn'}(\tau)\rangle A_{nn'}(\tau) \tag{4.12}$$

Partial integration of the second term and Fourier transforming with respect to time leads to the final result

$$\langle K_x(\omega)\rangle = G_s(\omega)E_x \tag{4.13}$$

where the superconducting part of the array conductance

$$G_s(\omega) = \frac{\Lambda}{i\omega} - \beta F_c(\omega) \tag{4.14}$$

is expressed in terms of the helicity modulus

$$\Lambda = J\sum_{\langle ll'\rangle_x}\langle\cos\phi_{ll'}\rangle - \beta\frac{J^2}{N_x}\sum_{\langle ll'\rangle_x}\sum_{\langle nn'\rangle_x}\langle\sin\phi_{ll'}\sin\phi_{nn'}\rangle \tag{4.15}$$

and a dynamic current-current correlation function

$$F_c(\omega) = \int_0^\infty dt\; e^{i\omega t}J^2\sum_{\langle ll'\rangle_x}\sum_{\langle nn'\rangle_x}\langle\sin\phi_{ll'}(t)\sin\phi_{nn'}(0)\rangle \tag{4.16}$$

Finally, the total conductance $G(\omega)$ is given by

$$G(\omega) = \frac{\Lambda}{i\omega} - \beta F_c(\omega) + \eta \tag{4.17}$$

The superconducting contribution to G has already been derived by Shenoy [28].

Thus the dynamic conductance G is expressed in terms of an effective circuit with parallel superconducting and normal channels. This picture can be made more explicit by writing G as

$$G(\omega) = \frac{1}{i\omega L(\omega)} + \frac{1}{R(\omega)} \tag{4.18}$$

with two real, frequency dependent response functions : an effective inductance L and an effective resistance R. Comparison of (4.17) with (4.18) shows that the helicity modulus which is one of the basic features of the BKT transition can be extracted from the low frequency behaviour of the dynamic conductance.

b) The impedance $Z(\omega)$ is obtained by calculating the expectation value of the voltage

$$V_{ll'} = \frac{\hbar}{2e}(\dot\theta_l - \dot\theta_{l'}) \tag{4.19}$$

across a given junction under the influence of an applied current acting through the first term of L_{ex} in (4.2). In order to allow the external current to enter the array open boundary conditions have to be used for this calculation. Again using (4.8) for the derivative of ρ_o in (4.3), the response function is immediately found to be

$$\langle V_{ll'}(t)\rangle = \int_{-\infty}^{t} d\tau\sum_n\langle(\dot\theta_l(t) - \theta_{l'}(t))\dot\theta_n(\tau)\rangle I_n(\tau) \tag{4.20}$$

The full impedance is found by summing the voltage drops over all bonds across the sample in x-direction and letting an equal current I flowing in on the left side through the sites $(0, ma)$, respectively out on the right sides (N_xa, ma) of the array. This yields the final expression for the impedance

$$Z(\omega) = \frac{\hbar}{2e}\sum_{\langle ll'\rangle_x}\int_0^\infty dt\; e^{i\omega t}\langle(\dot\theta_l(t) - \dot\theta_{l'}(t))(\dot\theta_{0,ma} - \dot\theta_{N_xa,ma})\rangle \tag{4.21}$$

The expressions for G and Z in terms of the corresponding dynamic correlation functions do not seem to have much in common. It is thus legitimate to ask the question whether the expected relation

$$G(\omega) = Z(\omega)^{-1} \tag{4.22}$$

holds. Expressing these correlation functions in the following subsection, for a classical array, by suitable vortex variables will indeed confirm (4.22). Alternatively, expressing the partition function of an array including capacitive effects and damping terms as an imaginary time functional integral, and going to a dual representation using the electrical charge on each site rather than the phase as variables, Panykov and Zaikin [29] have shown that the corresponding linear response expressions for G and Z indeed satisfy (4.22). However, relation (4.22), as plausible it may look, need not necessarily hold. Indeed, it has been shown [29,30] that for a quantum array the linear voltage response to an external current is not equivalent to the inverse of the linear current response to an applied voltage. The difference between the two arises because the voltage bias suppresses some important quantum fluctuations of the phase (instantons) due to the fact that the total phase difference across the sample is fixed by the applied voltage drop. Therefore the phases are hindered in their quantum fluctuations. The latter are not suppressed by an applied current since open boundary conditions are then used. As a result, the impedance Z sees the sample ≪ less superconducting ≫ than the conductance ! The same effect could, in principle, also occur with thermal fluctuations in the classical regime, but an explicit example seems not to have been constructed.

4.2 Dynamic correlation functions and dielectric function for the vortex system

As stated in section 1.2 our vortex-antivortex system is modelled by the 2D neutral Coulomb gas (2DCG). It is therefore useful to introduce some concepts that are currently used when treating the dynamics of the 2DCG (see, for example, the chapter ≪Charged fluids≫ of ref. [31]). For a hydrodynamic description, adapted to treat various ≪macroscopic≫ experiments treated in section 7, the basic variables are charge density $\rho(\mathbf{r})$ and current density $\mathbf{j}(\mathbf{r})$, the spatial Fourier transforms of which are related to the positions $\mathbf{R}_l$ and the sign $q_l = \pm 1$ of the charge of the N particles by

$$\begin{aligned} \rho(\mathbf{p}) &= \tfrac{1}{\sqrt{N}} \textstyle\sum_l q_l e^{i\mathbf{p}\cdot\mathbf{R}_l} \\ \mathbf{j}(\mathbf{p}) &= \tfrac{1}{\sqrt{N}} \textstyle\sum_l q_l \dot{\mathbf{R}}_l e^{i\mathbf{p}\cdot\mathbf{R}_l} \end{aligned} \tag{4.23}$$

In section 5.3 we will use Mori's formalism in order to calculate dynamic correlation functions of these basic observables, such as the ≪charge density correlator≫

$$\phi_{\rho\rho}(\mathbf{p}, z) = \int_0^\infty dt\ e^{-zt} \langle \rho(\mathbf{p}, t) \rho^*(\mathbf{p}, 0) \rangle \tag{4.24}$$

where the average is over a canonical ensemble and the time dependent particle positions have to be obtained by integrating suitable equations of motion. The equal time correlation function

$$S_c(\mathbf{p}) = \langle |\rho(\mathbf{p})|^2 \rangle \tag{4.25}$$

is the charge structure factor of the 2DCG. $\phi_{\rho\rho}$ and S_c are related to the dynamic charge susceptibility χ by [31]

$$\chi(\mathbf{p}, z) = -\frac{1}{k_B T}[S_c(\mathbf{p}) - z\phi_{\rho\rho}(\mathbf{p}, z)] \tag{4.26}$$

which, in turn, is linked with the dynamic dielectric function $\epsilon(\omega)$ by

$$\frac{1}{\epsilon(\omega)} = \lim_{p\to 0}[1 + \frac{2\pi q_o^2 n}{p^2}\chi(\mathbf{p}, -i\omega)] \tag{4.27}$$

where q_o is the charge of the particles and n their density. As usual in electrodynamics the dielectric function can be expressed in terms of a dynamic conductance $\sigma_v(\omega)$:

$$\epsilon(\omega) = 1 - 2\pi\frac{\sigma_v(\omega)}{i\omega} \tag{4.28}$$

Here we have used the letter σ and a subscript v in order to distinguish this ≪ vortex conductance ≫, related to the mobility of the vortices and thus to their response to an applied force, from the physical electric conductance G of the JJA.

For many purposes - in particular in order to illustrate the relation $G(\omega) = Z(\omega)^{-1}$- it is now useful to express our response functions directly by these dynamic vortex correlation functions. Various procedures have been proposed in order to transform the XY problem from its phase form (expression (1.2)) into a vortex representation. An approximate dual transformation allows to separate linear (spin wave) and vortex excitations, yielding the vortex Hamiltonian (1.18). This is particularly useful for dealing with equilibrium thermodynamics, but the same approach has also been applied to the action in a functional integral representation of the quantum version of the XY-model.

Another approach yielding insight in the dynamics of the system [33] consists in approximating the lattice phase Hamiltonian by a an energy functional for continuous phase and momentum fields

$$H = \frac{J}{2}\int d^2r(\nabla\theta)^2 + \frac{1}{2C_o}\int d^2r\mathbf{P}(r) \tag{4.29}$$

and separating the phase field θ into a ≪ regular ≫ (spin wave) part ϕ and a vortex part ψ, given by

$$\theta = \phi + \psi;\ \nabla\times\nabla\phi = 0,\ \nabla\cdot\nabla\psi = 0 \tag{4.30}$$

Côté and Griffin [33] transform the equations of motion for the phase and the conjugate momentum fields, without dissipative terms, into Maxwell equations in which the vortices act like effective charges being the source of effective electric and magnetic fields related to the phase field by

$$\mathbf{E} = \sqrt{2\pi}\nabla\theta\times\hat{z}$$
$$\mathbf{B} = \sqrt{\tfrac{2\pi}{C_o}}p\hat{z} \tag{4.31}$$

$\hat{z}$ being a unit vector perpendicular to the array. The charge of a vortex is related to the original coupling constant in (4.29) by

$$q_o^2 = 2\pi J \tag{4.32}$$

This concept can easily be generalized to include dissipation by adding a local damping term (normal currents flowing to the ground) to the equation of motion as in section 2. Faraday's equation is then generalized to

$$\frac{1}{c}\dot{\mathbf{B}} = -\nabla\times\mathbf{E} - \frac{1}{cC_oR_o}\mathbf{B} \tag{4.33}$$

with

$$c = \sqrt{\frac{J}{C_o}} \tag{4.34}$$

being the ≪ speed of light ≫ in this effective electrodynamics of the array. Expressing super and normal current densities by

$$\begin{aligned} \mathbf{K}_s &= J\nabla\theta \\ \mathbf{I}_n &= \tfrac{\hbar}{2e}\tfrac{1}{R_o}\nabla\dot{\theta} \equiv \eta\nabla\dot{\theta} \end{aligned} \tag{4.35}$$

allows now to express the response function found above for G and Z in terms of vortex quantities.

For the impedance Z we have to establish a linear relation between the total current, $\mathbf{I} = \mathbf{K}_s + \mathbf{I}_n$, and the voltage across the sample. We start from Maxwell's equation

$$\nabla \times \mathbf{B} = \frac{1}{c}\dot{\mathbf{E}} - \frac{2\pi}{c}\mathbf{j}_v \tag{4.36}$$

$\mathbf{j}_v$ being the vortex current density and rewrite it in terms of the time derivative of the phase gradient

$$\nabla\dot{\theta} = \frac{1}{\sqrt{2\pi c C_o}}(\hat{z} \times \dot{\mathbf{E}} - 2\pi\hat{z} \times \mathbf{j}_v) \tag{4.37}$$

Using (4.31) and (4.35) we replace the effective electric field by $K_s = I - K_n$. Then we relate, within our ≪effective electrodynamics≫, the vortex current density $\mathbf{j}_v$ to the driving electric field through the vortex conductance σ_v :

$$\mathbf{j}_v = \sigma_v\mathbf{E} \tag{4.38}$$

which yields

$$(1 + i\omega\frac{\eta}{J} - \eta\frac{2\pi\sigma_v}{J})\nabla\dot{\theta} = \mathbf{I}(\frac{i\omega}{J} - \frac{2\pi\sigma_v}{J}) \tag{4.39}$$

The dynamic vortex dielectric function is introduced by (4.28). Using the dynamic Josephson relation

$$V = \frac{\hbar}{2e}\nabla\dot{\theta} \tag{4.40}$$

we finally express the impedance $Z = \frac{V}{I}$ in terms of $\epsilon_v(\omega)$:

$$Z(\omega) = \frac{\hbar}{2e}\frac{1}{\eta + \frac{J}{i\omega\epsilon_v(\omega)}} \tag{4.41}$$

For the dynamic conductance, given by (4.17), we first note that the helicity modulus Λ is directly related to the static vortex dielectric function[15]:

$$\Lambda = \frac{J}{\epsilon_v(\omega = 0)} \tag{4.42}$$

In the continuum description the supercurrent density is given by (4.35). For our *PBC* the regular part of the phase gradient does not contribute to the dynamic current-current correlator in (4.16). According to (4.31) the singular part stands for the effective electric field related to the vortex charge density by Poisson's equation. Using also $q_o^2 = 2\pi J$ this allows to relate (4.16) to the dynamic vortex charge correlator (4.24) :

$$F_c(\omega) = \beta J^2 (2\pi)^2 n_v \lim_{p\to 0} \frac{\phi_{\rho\rho}(\mathbf{p}, -i\omega)}{p^2} \tag{4.43}$$

n_v being the vortex number density. The way towards the vortex dielectric function is found by using the relations (4.26) and (4.27), once for finite frequency ω and once for $\omega = 0$, which immediately yields

$$F_c(\omega) = \frac{J}{i\omega}[\frac{1}{\epsilon_v(\omega)} - \frac{1}{\epsilon_v(0)}] \tag{4.44}$$

Substituting this result and (4.17) in the expression (4.20) for G leads to

$$G(\omega) = \frac{J}{i\omega\epsilon_v(\omega)} + \eta \tag{4.45}$$

Manifestly it is the reciprocal of the impedance given by (4.41). Expression (4.45) is a particular case of the form (4.18), in which the original bond resistance becomes frequency dependent through the vortex dielectric function. Expressing the latter by the vortex conductivity through (4.45) one obtains

$$G(\omega) = \frac{J}{2\pi\sigma_v(\omega) + i\omega} + \eta \tag{4.46}$$

Thus, the superconducting contribution to G is directly related to the vortex mobility : $G(0)$ is inversely proportional to the latter. In particular, when the vortex system has a non vanishing mobility $G(0)$ is finite and superconductivity is destroyed.

5 Theoretical techniques for calculating dynamic vortex response function

5.1 Early approaches

Ambegaokar et al[34] calculated the dielectric function $\epsilon(\omega)$ of the vortex system

$$\epsilon(\omega) = 1 + 2\pi\chi(\omega) \tag{5.1}$$

for independent bound pairs by averaging the dynamic susceptibility χ of a dipole of length r over a Boltzmann probability distribution $\varphi(r,T) = e^{-\beta V(T,r)}$ of finding a V-A-pair of length r at a given temperature. A simple form for the effective interaction V (the Coulomb interaction devided by a dielectric constant) yields

$$\varphi(r) \propto (\frac{r}{a})^{-4\frac{T_{BKT}}{T}} \tag{5.2}$$

Alternatively one can use a length dependent static dielectric function [34 and 35]. The resulting dielectric function

$$\epsilon(\omega) = 1 + \frac{q_o^2}{4a}\int_a^\infty r dr \frac{\varphi(r)}{-i\omega\Gamma + \frac{k_B T}{r^2}} \tag{5.3}$$

has a power law dependence on frequency with exponents that depend on temperature.

For the free vortex system, above T_{BKT}, the simplest form for ϵ is given by Drude's form obtained by inserting a frequency independent vortex conductivity σ_o which is proportional to the inverse of the friction function in the equation of motion (3.6). This yields the well-known simple form for the inverse of the dielectric function :

$$Re(\tfrac{1}{\epsilon(\omega)}) = \tfrac{\omega^2}{\sigma_o^2+\omega^2}$$
$$Im(\tfrac{1}{\epsilon(\omega)}) = \tfrac{\omega\sigma_o}{\sigma_o^2+\omega^2} \qquad (5.4)$$

5.2 Minnhagen's phenomenology (MP)

This approach [17] starts from a Debye form of the wave number dependent static dielectric function $\frac{1}{\epsilon(k)} = \frac{k^2}{k^2+k_s^2}$ with a screening wave vector k_s and goes to $Re(1/\epsilon(\omega))$ by replacing the wave number k by frequency ω through

$$k^2 = \frac{\omega}{D} \qquad (5.5)$$

which corresponds to taking into account diffusive motion of the particles with diffusivity D. Performing a Kramers-Kronig transformation one obtains

$$Re(\tfrac{1}{\epsilon(\omega)}) = \tfrac{\omega}{\omega+k_s^2 D}$$
$$Im(\tfrac{1}{\epsilon(\omega)}) = \tfrac{2}{\pi}\tfrac{\omega D k_s^2}{\omega^2-(Dk_s^2)^2}\ln\tfrac{Dk_s^2}{\omega} \qquad (5.6)$$

Since $\epsilon(\omega) = 1 + \sigma(\omega)/(-i\omega)$ this amounts to using a vortex conductivity with a peculiar non-analytic frequency dependence for small ω

$$Re\sigma(\omega) = -\tfrac{\pi D k_s^2}{2}(\ln\tfrac{Dk_s^2}{\omega})^{-1}$$
$$Im\sigma(\omega) = -\tfrac{\pi^2 D k_s^2}{4}(\ln\tfrac{Dk_s^2}{\omega})^{-2} \qquad (5.7)$$

At first sight one would tend to reject this result, since - had one made the more correct replacement

$$k^2 = \frac{i\omega}{D}$$

as one should for a diffusive dynamics - one would come back to Drude's form rather than to (5.6) and (5.7). It turns out, however, that simulations and experimental data confirm the MP form of the dynamic response in an appropriate frequency domain, as we will show in sections 8 and 9.

A simple criterion allows to distinguish the MP form (5.6) from the more simple Drude expression : the ≪peak ratio≫

$$r = \frac{Im(\frac{1}{\epsilon(\omega_o)})}{Re(\frac{1}{\epsilon(\omega_o)})} \qquad (5.8)$$

between imaginary and real parts of the inverse dielectric function, evaluated at the frequency ω_o where $Im(1/\epsilon)$ has its maximum, is given by $r = 1$ for Drude's form and by $r = \pi/2$ for MP [17].

5.3 Equations of motion (Mori, Fokker-Planck)

The dynamic charge density correlator $\phi_{\rho\rho}(\mathbf{p}, z)$ defined in (4.24) is a central quantity for treating vortex dynamics. Indeed, it is directly related to experimentally accessible quantities such as the dynamic dielectric function (see section 4.2) and the flux noise spectrum that will be introduced in section 7.3. It is therefore important to rely on systematic methods for calculating such classical time dependent correlation functions.

Mori's approach [36] for evaluating dynamic correlation functions of systems for which relevant static correlation functions are known is particularly useful for our 2D Coulomb gas. It can be based on the two hydrodynamic variables ρ and $\mathbf{j}$ introduced in section 4.2. The time evolution of these variables is easily found by using the equations of motion for the center of a vortex (see section 3) including force and friction terms. Mori's technique yields a formally exact expression for $\phi_{\rho\rho}(\mathbf{p}, z)$

$$\phi_{\rho\rho}(\mathbf{p}, z) = \frac{S_c(p)}{z + \frac{k_B T p^2}{S_c(p)\gamma(p,z)}} \tag{5.9}$$

where $S_c(p)$ is the charge structure factor (4.25) of the vortex system and $\gamma(p, z)$ - the ≪ memory function ≫ given by the dynamic correlator of the ≪random force≫ - summarizes the complicated effects of the Coulomb interaction on charge dynamics [35]. Various approximations can be used in order to obtain manageable expressions for $\gamma(p, z)$. We first focus on the region above the BKT temperature treating the vortex system in its ≪metallic≫ phase, ie by neglecting possible binding in pairs. We assume that the zero wave vector limit $\gamma(\mathbf{0}, z) = \gamma(z)$ is sufficient to describe hydrodynamic phenomena. The general form of this quantity is given by [35] :

$$\gamma(z) \approx \Gamma + \frac{q_o^2}{2k_B T n_v \Omega^3} \sum_{\mathbf{kk}'} \mathbf{k}\cdot\mathbf{k}' V_c(\mathbf{k}) V_c^*(\mathbf{k}') \int_0^\infty dt\; e^{-zt} \langle n^*(\mathbf{k}, t)\rho(\mathbf{k}, t) n(\mathbf{k}', 0)\rho(\mathbf{k}', 0)\rangle \tag{5.10}$$

Here Ω is the total area of the system, $V_c(p) = \frac{2\pi q_o^2}{p^2}$ is the Fourier transform of the Coulomb potential and $n(\mathbf{p})$ is the particle density

$$n(\mathbf{p}) = \frac{1}{\sqrt{N}} \sum_l e^{i\mathbf{p}\cdot\mathbf{R}_l}$$

corresponding to the charge density in (4.8). In the above statement we have - as it is usually done - neglected the action of the ≪Mori projector≫ [36] in the time evolution of the higher order correlation function on the right hand side of (5.10). In the absence of interparticle interaction $\gamma(z)$ simply reduces to the bare friction Γ. Thus $\gamma(z)$ is a generalized friction function and its inverse is proportional to the ≪charge mobility≫ of the system. This can be seen by inserting the specific form (5.8) of $\phi_{\rho\rho}(\mathbf{p}, z)$ into the relations (4.26) and (4.27) determining the dielectric function. One finds the form (4.28) with

$$\sigma_v(\omega) = \frac{2\pi q_o^2 n_v}{\gamma(-i\omega)} \tag{5.11}$$

This expressions illustrates the role of the inverse of the friction function as a ≪charge mobility≫.

As usual in treating interacting systems a higher order correlator, involving both charge and number densities, shows up in (5.10). In the spirit of the ≪ mode coupling ≫ approximation, currently used in liquid dynamics [37], this higher order correlator is factorized into a product of $\phi_{\rho\rho}(\mathbf{p}, z)$ and of the corresponding number density correlator

$$\phi_{nn}(\mathbf{p}, z) = \int_0^\infty dt\, e^{-zt} \langle n(\mathbf{p}t) n^*(\mathbf{p}0) \rangle \tag{5.12}$$

However, in order that such a factorization be a useful approximation, the long range Coulomb potential - which induces long range correlations between particles - should be replaced by the screened interaction, as it is customary in treating the dynamic properties of Coulomb systems [40,41]

$$V_{sc}(p) = \frac{2\pi q_o^2}{p^2 + p_f^2} \tag{5.13}$$

which, in this simple form, involves the Debye screening wave number

$$p_f^2 = \frac{k_B T}{2\pi q_o^2 n_f} \tag{5.14}$$

determined by the charge q_o and the density n_f of free vortices. For the static structure factor $S_c(p)$ we use the simple form

$$S_c(p) = \frac{p^2}{p^2 + p_o^2} \tag{5.15}$$

with

$$p_o^2 = \frac{k_B T}{2\pi q_o^2 n_v}$$

which satisfies the sum rule [31] for the metallic phase of a Coulomb gas. Note, that p_o, the relevant wave number for $S_c(p)$, involves the total particle density. For the number density correlator (5.11) we use a simple diffusion form

$$\phi_{nn}(p, z) = \frac{n_v}{z + \frac{k_B T p^2}{\Gamma}} \tag{5.16}$$

Equation (5.9) is still an implicit expression for $\gamma(z)$ since the right hand side involves the same quantity again through $\phi_{\rho\rho}(\mathbf{p}, z)$. To lowest order in the interaction $\gamma(z)$ can be evaluated by replacing it by Γ under the integral on the right hand side of (5.9). Results for $\gamma(z)$ and thus for the flux noise spectrum and the dielectric function, obtained by using this procedure will be mentioned in section 9.

Mori's approach also allows to take into account pinning (Peierls) forces acting on the vortices, as they result from equation (3.5) (see the discussion following that equation). One then obtains another contribution to the friction function $\gamma(z)$, having a similar structure as (5.10), with the Coulomb force being replaced by the pinning force. This approach has been applied [42] to take into account the effect of random pinning in disordered lattices (described in section 6.). A mean-field like description of disorder induced localization of vortices is found in this way [42].

The Fokker-Planck equation, presented in section 4.1 in connection with linear response calculation using phases as variables, can also be used for determining the vortex distribution function, providing thus another systematic approach for evaluating dynamic correlation functions. It has been used by Timm [43] for calculating the flux noise spectrum in an independent pair approximation. The corresponding Fokker-Planck operator takes into account the free diffusion of pairs, the internal interaction between the partners and a possible annihilation of the V and A constituting the pair. The results of this approach will be mentioned in section 9.

5.4 Numerical simulations

The dynamic properties of JJAs have been studied by solving numerically different types of equations of motion:

a) for the phases of the XY-model using RSJ dynamics (see our equation (2.5)) : References 44-54.

b) for the phases of the XY-model using TDGL dynamics (see our equation (2.7)) : References 44-46, 50, 53, 55-60.

c) for the positions of the charges in the 2D V-A-Coulomg gas (see our equation (3.9)) : References 59, 61-63.

d) for the 2D lattice Coulomb gas using Monte Carlo dynamics : References 64, 65.

Details about the simulation technique can be found in these references. The correct choice of boundary conditions is particularly important. Indeed, various quantities that are relevant for the BKT transition, such as the phase stiffness, describe the response of the system to forces applied to its boundaries. For simulations of the XY-model (a and b) periodic boundary conditions (PBC) imposed in the phase angle θ_l are most commonly used when the total current in the sample is zero (see our discussion of boundary conditions in the context of linear response calculations in section 4.1). However, as shown by Olsson [66], PBC for the phase angles lead to non-periodic boundary conditions for the vortex interaction. On the other hand, Coulomb gas simulations (c and d) prefer PBC for the motion of the point particles representing the vortices. This requires imposing ≪fluctuating twist boundary conditions≫ ($FTBC$) on the phase angles [66]. In this case a variable $D = (D_x, D_y)$ is added to the phase difference (1.4) between neighboring sites l and l' :

$$\phi_{ll'} \rightarrow \phi_{ll'} - \mathbf{e}_{ll'} \cdot \mathbf{D} \tag{5.17}$$

the vector $\mathbf{D}$ being projected onto the unit vector $\mathbf{e}$ of the bond (ll'). For the phase angles themselves PBC are used. $\mathbf{D}(t)$ is itself a dynamic quantity which is related to vortex motion : whenever a vortex has crossed the sample of length L from one side to the other $\mathbf{D}$ changes by $2\pi/L$. Its equation of motion is established by considering [45] current conservation across the sample

$$\frac{V_x}{R} + J \sum_{\langle ll' \rangle_x} \sin(\phi_{ll'} - D_x) + \eta_x = 0 \tag{5.18}$$

Here, the first term expresses the normal current in x-direction in terms of the voltage drop V_x and the normal resistance R, and the last term represents the thermal noise current used in the usual Langevin equation in order to impose a given temperature T on the system. By the dynamic Josephson relation the voltage is proportional to the time derivative of the total phase change across the sample. Since PBC are imposed on θ_l it is directly related to the time derivative of D_x :

$$V_x = -\frac{\hbar}{2e} L \dot{D}_x \tag{5.19}$$

Combining these relations leads to an equation of motion for $\mathbf{D}$ [66]. By considering the Fokker-Planck equation for the probability distribution of the phase angles and the additional variables (D_x, D_y) one obtains the standard white noise correlations for the noise current :

$$\langle \eta_x(t) \eta_x(0) \rangle = \frac{2T}{L^2} \delta(t) \tag{5.20}$$

A current passing through the sample can either be incorporated in the current conservation condition that determines $\mathbf{D}(t)$, or - alternatively - open boundary conditions for the phase angles can be used (see section 4.1). The former approach has the advantage that one can continue to use periodic boundary conditions, which is more convenient for numerical calculations.

5.5 Dynamic scaling

The BKT scenario governs the phase transition in a 2D model involving a complex order parameter with an amplitude and a phase. It has the particularity that the correlation length, diverging as usual when T goes to its critical value from above, stays infinite for all $T < T_{KT}$. Thus the usual dynamic scaling relation

$$\begin{aligned} F(k, \omega) &= \hat{F}(k\xi, \omega\xi^z) \\ F(r, t) &= \hat{F}(r/\xi, t\xi^{-z}) \end{aligned} \tag{5.21}$$

for functions depending on space and time variables, is valid a priori only above T_{BKT}. The dynamic critical exponent z summarizes the essential information about vortex dynamics. Numerous calculations, mainly numerical, have been aimed at obtaining z. Apriori, one expects $z = 2$ [50], according to the usual classification of dynamic critical phenomena [67], which is indeed bourne out by various calculations, and it is also reflected by the flux noise measurements of ref [68]. It corresponds to the fact that motion of vortices is of diffusive nature.

Below T_{BKT}, where the model remains critical, the situation is more complex. Since $\xi = \infty$ the relevant scaling now involves other typical lengths, such as the sample size or the size of the detection device (such as the coil used for flux noise measurement, see section 7). Morevoer, z depends on the dynamic phenomenon to which it is applied and thus on the boundray conditions used. Jenssen et al [50] give a rather complete overview over previous results and over their own calculations. They have used RSJ, as well as LG dynamics for the phases (see section 2), applying, for both cases, either periodic or fluctuating twist boundary conditions and they have analysed resistance, supercurrent scaling and the short time relaxation of a non-equilibrium configuration.

An interesting observation has recently been made for the three dimensional(3D) XY-model : there is evidence [59] for the existence of two dynamic critical exponents signalling the dependence of dynamic correlation functions on wave number k and on the correlation length ξ:

$$F(k,t) = \hat{F}(k\xi, t\xi^{-z_o}, tk^{-z}) \tag{5.22}$$

with $z_o = 1.5$ and $z = 1$, respectively $z = 2$ for the 3D lattice Coulomb gas, respectively the 3D XY-model.

From an experimental point of view a particularly interesting application relates the value of z below T_{BKT} to the non-linear current-voltage characteristics $V \sim I^{a(T)}$ (see section 7.1). Ambegaokar et al [34] found

$$a(T) = \frac{\pi J(T)}{k_B T} + 1 \tag{5.23}$$

where $J(T) = J/\epsilon(T)$ is the effective XY-coupling constant, renormalized by spin wave and VA-pair fluctuations represented by the static dielectric constant $\epsilon(T)$. The non-linear I-V relation relies on the fact that VA pairs that are separated by an applied current (and thus contribute to a finite voltage like ≪ free ≫ vortices) can recombine, see section 7.1. Expression (5.22) is found be setting the pair breaking and the recombination rates equal to each other [34]. At the critical temperature one obtains $a = 3$, which together with $z = 2$ satisfies the dynamical scaling relation [69] $a = z + 1$. More recently Minnhagen et al [49] have found a different value

$$\hat{a}(T) = \frac{2\pi J(T)}{k_B T} - 1 \tag{5.24}$$

by combining static and dynamic scaling arguments. They then link this expression to a temperature dependent dynamical exponent

$$\hat{z}(T) = \hat{a}(T) - 1 \tag{5.25}$$

Interestingly enough, both expressions yield the same value $a = 3$ at T_{BKT}, since (see section 1) :

$$J(T_{BKT}) = \frac{2k_B T_{BKT}}{\pi} \tag{5.26}$$

Bormann [70] has established a link between (5.23) and a more involved recombination scenario - a collective ≪partner transfer≫ between bound pairs - which is supposed to take place when the vortex density is not too small. A recent analysis of I-V data obtained for various 2D superconductors [71] has produced a much larger exponent on the order of 5 or greater. Such a value might be justified [35] by introducing a ≪thermal escape time≫ for a VA pair out of their logarithmic interaction potential into the generalized friction function determining the dynamic vortex correlator (see section 5.3). On the other hand it has been pointed out [51,94] that finite size effects play in important role in determining $a(T)$.

Shenoy [16] has developed a dynamic renormalization group approach for $JJAs$. The evaluation of the dynamic current-current correlation function $F_c(\omega)$, defined in (4.16), is based on the Fokker-Planck equation (4.1) for the probability distribution of the phases. Transforming the phase action (Hamiltonian (1.2) plus friction term) to vortex variables allows borrowing results from static BKT-scaling and imposing it onto dynamics, by introducing an extra frequency scaling. The dynamic conductivity $\sigma(\omega)$ of the array is then expressed as an integral over length scales l of a scale dependent conductance $\sigma(\omega, l)$. This approach was somewhat refined later on [107]. While the aproximations for $\sigma(\omega, l)$ in the first calculation [16] essentially yielded a Drude-like frequency dependence, a more refined treatment of allowed to reproduce the MP form (5.6) in some intermediate frequency range, and thus to understand this anomalous behaviour from a more microscopic point of view.

6 Disorder effects

6.1 Linear dynamics of disordered arrays

In the considerations of section 5 the energy barrier that a moving vortex has to cross when moving from one plaquette of the array to a neighboring one (see section 3.1) has been neglected. This approximation must be abandoned when the array deviates from perfect translational invariance. Different types of disorder can be considered.

a. For **bond disorder** the parameters characterizing the individual junctions, namely the Josphson couplings $J_{ll'}$, the junction resistances $R_{ll'}$ and the mutual capacities $C_{ll'}$, are random. This type of disorder is realized when the positions of the individual superconducting islands deviate in a random way from their ideal lattice positions or if they have random size, since $J_{ll'}$, $R_{ll'}$ and $C_{ll'}$ depend on the distance between the adjacent superconductors and on their geometry.

b. For **site disorder** the properties of the superconducting islands, ie their resistance and capacitance to ground, are random. More simply, in a ≪ dilute ≫ array certain islands are totally missing.

Dynamic measurements have in particular been done on dilute, or ≪ percolative ≫, arrays in which only a fraction p of superconducting islands is present [95]. The experiments were done for p-values somewhat larger than the percolation threshold $p_c = 0.5$ where the disorder has a marked influence on the properties of the array. Indeed for $p < p_c$ the average array has no more connected path joining opposite boundaries and superconductivity is completely suppressed. At temperatures that are low compared to the T_{BKT} (which in a dilute array is lower than for the regular counterpart, due to the absence of certain bonds, see, for example, section 17 of reference [1]) the equations of motion can be linearized, taking into account only small amplitude ≪spin wave≫ excitations. The problem at hand is then the same as determining the vibrational modes of a disordered harmonic solid, which is a well studied field [72]. In reference [73] the impedance $Z(\omega)$ of such a dilute JJA has been determined by evaluating the dynamic voltage correlation function (4.20) by different methods. The coherent potential approximation (CPA) has proven to be a useful approach for treating dynamic disorder problems. On the other hand Mori's method, applied in section 5.3 to vortex dynamics, is also useful. The inverse of the impedance is expressed by an effective coupling function $K(\omega, p)$ depending on frequency and on the dilution fraction p :

$$Z(\omega) = (\eta + \frac{K(\omega,p)}{i\omega})^{-1} \tag{6.1}$$

corresponding to the form (4.41) or to the effective inductance L and resistance R of (4.18).

As a main result these calculations reveal the existence of three different frequency regimes:

a) For very low frequency inductance and resistance are frequency independent. Such low frequencies effectively show the long wavelength response of the dilute array. The details of the disordered structure are averaged out and the ≪macroscopic≫ behaviour which is seen corresponds to a homogeneous superconductor. However, $1/L$ and R no more correspond to single junction parameters. They are reduced and go to zero when the percolation threshold p_c is reached:

$$\frac{1}{L(0,p)} \propto (p - p_c)^t$$
$$R(0,p) \propto (p - p_c)^t \tag{6.2}$$

where t is the critical conductivity exponent introduced in percolation theory [74]. For the 2D lattice at hand $t = 1.3$, whereas the more simple ≪ mean field ≫ treatments like CPA yield $t = 1$[73].

b) In an intermediate frequency range the dynamic conductance varies like a power of frequency :

$$G(\omega) \sim \omega^{-u} \tag{6.3}$$

reflecting dynamic scaling that results from the self-similar structure of the array near the percolation threshold. The precise value of u depends on the type of disorder and on the method (such as CPA or Mori) used for the calculation. CPA applied to bond disorder and bond dissipation yields $u = 0.5$, which corresponds to the exact result found in percolation theory [75] by taking into account the self-duality of the random network complex impedances by which the dilute JJA can be modelled. On the other hand site dissipation yields frequency exponents for $1/L$ and R which differ from each other and are slightly smaller, respectively larger, than 0.5, their sum being, however, equal to 1. The cross-over frequency where regime a goes over into regime b is itself a function of the dilution parameter p :

$$\omega_c(p) \propto (p - p_c)^{2t} \tag{6.4}$$

This regime corrsponds to excitations which have a wave length smaller than the percolation correlation length and are therefore a direct witness of the disorder in the array. In the field of disordered harmonic solids such modes are called ≪ fractons ≫.

c) At even higher frequencies $1/L$ and R are again frequency independent and take the original values of a single junction multiplied by p. The dynamic response is now governed by excitations of single bonds and the relevant parameters correspond to averages over the array.

Experimental data obtained on dilute arrays will be presented in section 8.

6.2 Vortex dynamics of disordered arrays

Deviations from a regular lattice structure will have an influence on the vortices of an array through the Peierls force or pinning potential in equation (3.5), which has to be reintroduced in the equation of motion (3.6) or (3.9) for the vortex centers. The equilibrium arrangement of the (thermal or field induced) vortices and antivortices will correspond to a minimum of the free energy in the given random pinning potential landscape. It is generally accepted that weak disorder results in some sort of glassy vortex state [76,77]. At sufficiently low temperatures such a vortex configuration would then be trapped by the underlying potential, and the system should have a vanishing linear resistance, ie it would be truly superconducting. This should be valid for dilute JJAs, as they have been described in the preceding subsection, as well as for ultrathin superconducting films [78] that are intrinsically disordered. There is, however, an important difference between 2D and 3D systems : in two dimensions the V-system has thermally excited defects, in particular dislocation pairs, which allow for local flux flow (and thus ohmic resistance), even though any motion of the V-A-structure as a whole is quenched by the pinning potential. Such a motion can, however, be very slow, so that on the time scale of a given experiment the V-system still appears to be frozen. Recent measurements showing such a ≪dynamic freezing≫ will be presented in section 8 [78].

In the percolative arrays described in the preceding subsection the sign of disorder has also been detected in the vortex dynamics. Above the BKT transition the vortex contribution to the dynamic impedance $Z(\omega)$ shows power law behaviour with a temperature dependent exponent. These data [98] will be interpreted at the end of subsection 8.2.

7 Theoretical description of experiments probing equilibrium dynamics

Three major experimental techniques have been developed in order to get insight into the dynamic behaviour of $JJA's$ at or near equilibrium:

a) I-V characteristics (V-response to static driving fields)

b) Flux noise (spontaneous V fluctuations)

c) Two-coil induction (V-response to weak dynamic driving fields)

Before reporting on experimental findings we briefly explain the link between the measured quantities and the theoretical techniques exposed in section 5.

7.1 Current-voltage characteristics

Below the BKT transition there are no free vortices and the dc resistance is zero. However, an applied current will act on the vortices according to equation (3.6) which leads to a finite probability for a pair to unbind and thus to dissipation. Experimentally this phenomenon can be observed through the non-linear current-voltage characteristics

$$V \propto I^{a(T)} \tag{7.1}$$

The form of the temperature dependent exponent $a(T)$ for $T < T_{BKT}$ can be found setting the escape rate over the energy barrier for a pair equal to the probability for recombination. Details can be found in the original work of Ambegaokar et al [34] or in the review by Newrock et al [1]. The result was given in (5.23). Thus, for $T < T_{BKT}$, the I-V relation is non-ohmic, while above the unbinding temperature the presence of free vortices leads to ohmic dissipation.

As mentioned in section 5.4 dynamic scaling leads to result (5.24) which is different except at $T = T_{BKT}$. Bormann [70] has refined and extended the picture of this balance between pair breaking and recombination. At not too low densities of vortices charge transport is supposed to occur predominantly through direct particle transfer from one bound pair to a neighboring one. Taking into account such ≪ partner transfer ≫ Bormann indeed obtains the I-V exponent of [49] thus providing a more physical insight than simply dynamic scaling. However, in contrast to reference [49] Bormann's result for the exponent $a(T)$ goes over into the ≪ classial ≫ form proposed in [34] for very low currents. Thus the apparent contradiction between the two forms of [34] and [49] for $a(T)$ is resolved.

Theoretical values for $a(T)$ resulting from numerical simulations will be presented in section 8 and experimental findings for the I-V characteristics of $JJAs$ will be shown in section 9. For an adequate comparison between theory and experiment one has to bear in mind that finite size effects can influence the BKT scenario in particular through the presence of some free vortices which are not the result of BKT unbinding, but rather of the finiteness of the system. Such finite size effects have been discussed by different authors, see [79], [51], [71]. An observable effect of this phenomenon will be discussed in section 8.1.

7.2 Two-coil induction

These measurements use a drive coil of N_D turns of radius R_D and a pair of astatically wound receive coils of N_R turns of a smaller radius R_R mounted at some small distance d from the plane of the array (see *figure* 2). In order to study the dynamic response of the array an ac magnetic field is created by the drive coil, which excites the vortex system. The resulting motion of the vortices creates a voltage signal δV in the receive coil. This technique was pioneered by Hebard and Fiory [80] for superconducting films and has proven to be very successful for studying the dynamic aspects of the BKT transition [1,81-85].

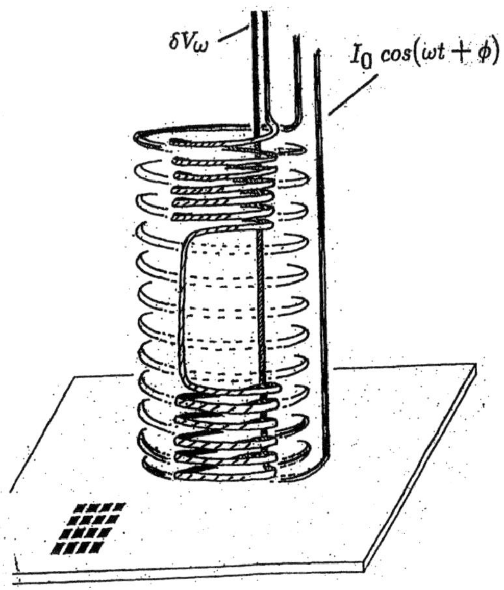

Figure 2 :Two-coil set-up used in the measurement of the dynamic conductance of the array, described in section 7.2.

The measured signal δV can be related to the dynamic conductance of the array by a straightforward calculation. Since the latter can be found in various references [81,83], we only sketch the main steps without giving an explicit derivation :
- An ac current $I_D(\omega)$ of frequency ω sent through the drive coil produces a time dependent vector potential $\mathbf{A}(\omega)$
- The time derivative of $\mathbf{A}$, in the plane of the array, is an electric field $\mathbf{E}(\omega)$ acting on the latter
- An electric current density $\mathbf{j}(\omega) = G(\omega)\mathbf{E}(\omega)$ is induced in the array that acts on the vortices
- The resulting time dependent flux threads through the receive coil and can be detected by the induced voltage $\delta V(\omega)$.

Thus the signal $\delta V(\omega)$ is linearly related to the drive current I_D:

$$\delta V(\omega) = i\omega I_D(\omega) \int_0^\infty dx \frac{M(x)}{1 + \frac{2x}{\mu_o h} \frac{1}{i\omega G(\omega)}} \tag{7.2}$$

Here h is the sum of the closest distances of the drive and the receive coils, respectively, from the array plane and the function $M(x)$ takes into account the geometry of the experimental set-up

$$M(x) = \pi\mu_o \frac{R_D R_R}{h^2} J_1(\frac{R_D}{h}x) J_1(\frac{R_R}{h}x) e^{-x} \frac{1 - e^{-N_D \delta h_D x}}{1 - e^{-\delta h_D x}} \tag{7.3}$$

δh_D, δh_R being, respectively, the distance between the turns of the drive and the receive coils, in units of h and J_1 is the first order Bessel function. In order to bring in vortex dynamics into the expression (7.2) one then has to use a link between the electric conductance G of the array and the vortex dielectric function $\epsilon(\omega)$, such as (4.45). More details concerning the derivation of (7.2) and (7.3) can be found, for example, in [83].

Even though finite frequencies are involved this technique allows to gain interesting experimental insight into the BKT-transition. Moreover it has been used to detect the frequency dependence of $\epsilon(\omega)$. Results will be presented in section 9.

7.3 Flux noise

Currents flowing in the array produce a magnetic field that extends into the space above and below the plane of the array. This field can be detected by measuring the corresponding magnetic flux through a given surface. In the usual experiment a circular wire of radius R delimiting such a closed surface is fixed at a distance d from the JJA. If the field, and thus the flux, varies with time a voltage is induced in the wire according to Faraday's law. In the absence of an external excitation (which was used in the two-coil experiment) the time variation of the flux is due to thermal fluctuations in the array. Such a ≪flux noise≫ experiment measures $S_\phi(\omega)$, the frequency Fourier transform of $S_\phi(t)$, which is the dynamic correlation function $\langle\phi(t)\phi(0)\rangle$ of the time dependent flux $\phi(t)$ through the area surrounded by the wire.

The usual theoretical analysis of flux noise attributes the fluctuating currents in the array to moving vortices. The flux ϕ is expressed as an integral over the dynamic charge density correlator of the vortex system in the plane by means of the following steps (we give some more details here since the relevant references do not give a complete derivation):

- ϕ is the line integral of the vector potential $\mathbf{A}$ along the wire

$$\phi = \oint d\mathbf{s} \cdot \mathbf{A} \tag{7.4}$$

- The vector potential $\mathbf{A}$ which is produced by the electric current density $\mathbf{j}$ flowing in the array is found by solving Maxwell's equation

$$\mathbf{A}(\mathbf{r}) = \mu_o \int d^2k_{||} e^{i\mathbf{k}_{||}\cdot\mathbf{r}_{||}} \frac{e^{-k_{||}r_{||}}}{2k_{||}} \mathbf{K}_{in}(\mathbf{k}_{||}) \tag{7.5}$$

Here the position vector $\mathbf{r} = (\mathbf{r}_{||}, \mathbf{r}_{\perp})$ is decomposed into its components in the plane of the array and perpendicular to the latter and $\mathbf{K}_{in}$ is the 2-dimensional Fourier transform of the areal current density in the array.

- For a single vortex the current $\mathbf{K}$, circulating around its center, is given by

$$\mathbf{K}(\mathbf{r}_{||}) = \frac{2e}{\hbar} J(\nabla\theta(\mathbf{r}_{||}) - \frac{2e}{\hbar}\mathbf{A}(\mathbf{r}_{||})) \tag{7.6}$$

Equations (7.5) and (7.6) have to be taken into account simultaneously which leads to magnetic screening in the final expression for the vector potential:

$$\mathbf{A}(\mathbf{r}) = \mu_o \int d^2k_{||} e^{i\mathbf{k}_{||}\cdot\mathbf{r}_{||}} \frac{e^{-k_{||}r_{||}}}{2k_{||}} \frac{\mathbf{k}_{in}(\mathbf{k}_{||})}{1 + \frac{1}{\lambda_{||}k_{||}}} \tag{7.7}$$

The effective (in-plane) screening length $\lambda_{||}$ is given by

$$\lambda_{||} = \frac{\hbar^2}{2e^2 J \mu_o}$$

- Assuming that the relevant contributions to the phase gradient are due to moving vortices the current density $\mathbf{K}$ is expressed by the Fourier transform of the density of the latter :

$$\mathbf{K}_{in}(\mathbf{k}_{||}) = -2\pi i (\frac{2e}{\hbar}) J \frac{\mathbf{e}_{\perp} \times \mathbf{k}_{||}}{k_{||}^2} \rho_v(\mathbf{k}_{||}) \tag{7.8}$$

- Introducing expressions (7.5) to (7.8) into (7.4) leads to an explicit expression of the flux ϕ through the plane of the wire in terms of the vortex density in the plane of the array :

$$\phi = \pi R \mu_o (-2\pi i) \frac{2e}{\hbar} J \int_0^{\infty} dk_{||} k_{||} J_1(k_{||}R) e^{-k_{||}d} \frac{\lambda_{||}k_{||}}{1 + \lambda_{||}k_{||}} \frac{\rho_v(k_{||})}{k_{||}^2} \tag{7.9}$$

where J_1 is the first order Bessel function bringing in the geometry of the pick up wire.

- The flux noise correlation is given by

$$S_{\phi}(\omega) = \int dt\ e^{i\omega t} \langle \phi(t)\phi(0) \rangle \tag{7.10}$$

- The statistical average concerns the vortex density showing up in statement (7.9). It leads to the dynamic density correlator of the V-A system introduced in equation (4.24). The latter is diagonal in the wave vector k (as it is defined in (4.24)). Thus the $S_{\phi}(\omega)$ is expressed by the Fourier transform of the dynamic vortex correlation function which, in turn, is related to the Laplace transform (4.24) by

$$\int dt\ e^{i\omega t}\langle\rho_v(k,t)\rho_v^*(k,0)\rangle = 2Re\phi_{\rho\rho}(k,-i\omega) \tag{7.11}$$

- The final result for the flux noise spectrum reads

$$S_\phi(\omega) = (2\pi^2 R\mu_o)^2(\frac{2e}{\hbar})^2(J\lambda_{||})^2\int_0^\infty dk\frac{J_1(kR)^2e^{-2kd}}{k(1+\lambda_{||}k)^2}2Re\phi_{\rho\rho}(k,-i\omega) \tag{7.12}$$

In case several concentric wires are set up on top of each other (7.12) has to be replaced by a sum of terms of the same structure, but each with the corresponding distance d_n from the array, showing up in the exponent. Results obtained through expression (7.10) by using different forms for the dynamic vortex correlation function $\phi_{\rho\rho}(p,z)$ will be shown in section 8. They will be compared with experimental data in section 9.

In order to build a bridge between flux noise and electrical conductance one can use the links between charge density correlator, dynamic susceptibility and dielectric function shown in (4.26,27) in order to rewrite (7.12) as

$$S_\phi(\omega) = (2\pi^2 R\mu_o)^2(\frac{2e}{\hbar})^2(J\lambda_{||})^2\int_0^\infty dk\frac{J_1(kR)^2e^{-2kd}}{k(1+\lambda_{||}k)^2}\frac{2}{\omega}\frac{k_BTk^2}{2\pi q_o^2 n}\frac{1}{\epsilon_v(\omega)} \tag{7.13}$$

When the pick-up coil is very close to the plane of the array (d small) only small k-values contribute to (7.11) which can be approximated by

$$S_\phi(\omega) = C\frac{1}{\omega}Im(\frac{1}{\epsilon_v(\omega)}) = C'ReG_s(\omega) \tag{7.14}$$

relating thus - through (4.45) - flux noise approximately to the long wave length conductivity, or - more precisely - to its superconducting part [48].

The above flux noise calculations have been based on taking into account only the supercurrent in the array (see equation (7.6)) and by relating the phase gradient in (7.6) directly to vortex configurations. The link between flux noise and conductance can be made more general [86] by expressing first the flux correlator (7.10) by the correlator of the total array current and by relating the latter to the array conductance via the considerations exposed in section 4 on linear response.

8 Experimental data

We present experimental data for the three types of ≪standard≫ experiments for which we have related the measured observables to theoretically accessible quantities in the preceding section.

8.1 Current-voltage characteristics

Various measurements of the relationship between an applied current and the resulting voltage drop, yielding the exponent $a(T)$ in (7.1), are cited in the review by Newrock et al [1]. There (in section 11) the interplay between different relevant length scales implied in such an experiment, where currents of different intensity are used, is also explained in detail. The jump from the non-ohmic behaviour given by a value of $a(T)$ that differs from 1 to the high temperature linear I-V relation is clearly bourne out by experiment.

As explained in sections 4 and 6 a new T-dependence [49] of the coefficient a, differing from the ≪ classical ≫ form has recently been proposed. Weber et al [64] have reanalysed I-V data (see *Figure* 3) obtained on Hg-Xe alloy films [88], $Bi_2Sr_2CaCu_2O_x$[89] and $Bi_{1.6}Pb_{0.4}Sr_2Ca_2Cu_3O_x$ [90] single-crystal films, finding indeed good agreement with the new form of $a(T)$ of Ref. [49] and [70]. Herbert et al [79] have come to the same conclusion for proximity-coupled and high-T_c weak-link arrays. They also find evidence for the finite size effects mentioned at the end of section 7.1. Indeed, the I-V curves for very weak currents bend over into a linear form even below the transition temperature, since the free vortices present in a finite array, lead to ohmic behaviour.

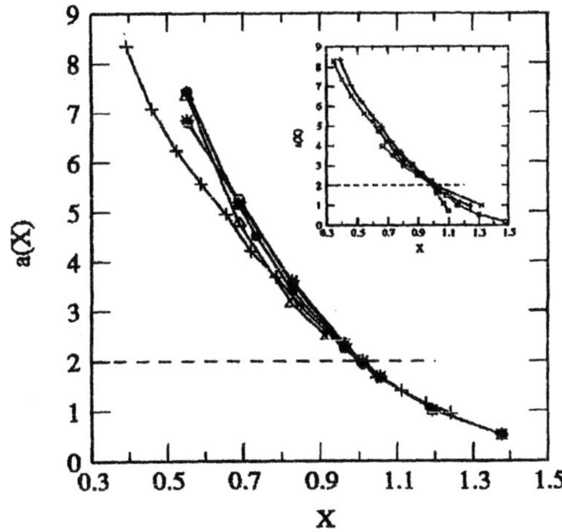

Figure 3 : Exponent $a(T)$ of the I-V characteristics (see equation (7.1)), according to Ref. 87, plotted as a function of $X = T/T_c$. The different curves have the following meaning :

- Plusses : Reference 88
- Other symbols : Monte Carlo results for an array of length of 16 (stars), 24 (open circles), 32 (filled circles) and 48 (triangles) unit cells.

The dashed line corresponds to the universal value $a = 2$, at the critical temperature.
The inset shows experimental data for $Hg-Xe$ alloy films, Ref. 88 (plusses), $Bi_2Sr_2CaCu_2O_x$ single-crystal films, Ref. 89 (filled squares) and $Bi_{1.6}Pb_{0.4}Sr_2Ca_2Cu_3O_x$ single-crystal films, ref. 90 (open squares).

8.2 Two-coil induction

The two-coil mutual induction technique, from which the dynamic conductance $G(\omega)$ of the array can be extracted (see section 7), has extensively been used in order to study the BKT

transition. One of the basic features is the temperature dependence of the phase stiffness or the helicity modulus Λ, reaching a universal value at T_{BKT} before jumping to zero. As explained in section 4 the low frequency limit of $G(\omega)$ is directly related to Λ. Ref. [81] presents data obtained on a proximity-effect JJA consisting of Pb-islands on a Cu substrate. The voltage induced in the pick-up coil is analysed according to section 7.2 and the corresponding dynamic dielectric function $\epsilon(\omega)$ is extracted for different temperatures in the vicinity of T_{BKT}. The fall-off of $Re(\epsilon^{-1})$ near T_{BKT} is characteristic of the ≪insulator-conductor transition≫ of the vortex system and $Im(\epsilon^{-1})$ shows that this transition is accompanied by dissipation. Going to higher frequency shifts the transition region to higher T, since larger ω tests shorter length scales for which the unbinding of V-A pairs happens at higher temperature T_ω, which can be interpreted as an ≪ effective frequency dependent critical temperature ≫. In order to extract the helicity modulus Λ which is a thermodynamic response coefficient one has to extrapolate to zero frequency. This can be done by identifying $Re(\epsilon(\omega, T))$ for a given frequency ω with the static dielectric function $\epsilon(l_\omega, T)$ for a length scale l_ω corresponding to ω. The T-dependence of Λ obtained in this way is in good agreement with a renormalization group calculation integrating the scaling equations up to a length of the order of $\xi(T_\omega)$, the BKT correlation length at $T = T_\omega$ where pairs of length $\xi(T_\omega)$ begin to dissociate. The extrapolation to infinite length scale agrees very well with the universal jump of Λ at T_{BKT} prredicted by theory. More details concerning this scaling procedure used for getting at Λ can be found in [81] and [91].

It is important to mention that whenever experiment and theory is compared in this way one has to take into account the T-dependence of the Josephson coupling given, according to (1.3), by the one of the critical current

$$I_c(T) = I_o(1 - \frac{T}{T_{BCS}})^2 exp(-\frac{L}{\xi_n(T)}) \tag{8.1}$$

see [92], where T_{BCS} is the BCS transition temperature of the individual islands of the array, separated by the distance L and ξ_n is the effective coherence length of the normal metal forming the junction between two neighboring islands. One can then compare experimental data with XY-model calculations by using a dimensionless temperature parameter

$$\hat{T} = \frac{k_B T}{J(T)} = \frac{2ek_B T}{\hbar I_c(T)} \tag{8.2}$$

Moreover, spin wave excitations produce an additional renormalization of the Josephson coupling. The latter can be taken into account by using an ≪ effective Coulomb gas temperature ≫ [93]

$$\tau = \frac{\hat{T}}{1 - \hat{T}/4} \tag{8.3}$$

The two-coil technique has also allowed to shed light on the dynamics of 2D Coulomb systems. In section 5 two phenomenological approaches to $\epsilon(\omega)$ have been presented : Drude's model and Minnhagen's phenomenology (MP). One might expect that above T_{BKT}, where free vortices are supposed to dominate the response to a driving force, Drude's model would be appropriate. It thus came as a surprise when measurements on weakly frustrated triangular $JJAs$ [82] revealed a frequency dependence of $\epsilon(\omega)$, see *figure* 4, which is well described by MP and support the idea of a non-analytic dynamic mobility, as explained in section 5.2. Triangular arrays are better suited for studying the intrinsic dynamics of interacting vortices since the barriers due to lattice pinning (section 3.1) are very low [23]. One has to bear in mind that the system of field induced vortices in a frustrated array should actually be described as a ≪ one-component plasma in a neutralizing background ≫ (OCP) rather than the ≪neutral two-component plasma≫ (TCP) of the thermally excited V-A system in an unfrustrated array. Thus, theoretical approaches developed for the TCP, as in section 5, are not immediately applicable. However, numerical simulations of frustrated arrays (section 9) have shown that the two types of plasmas behave the same way.

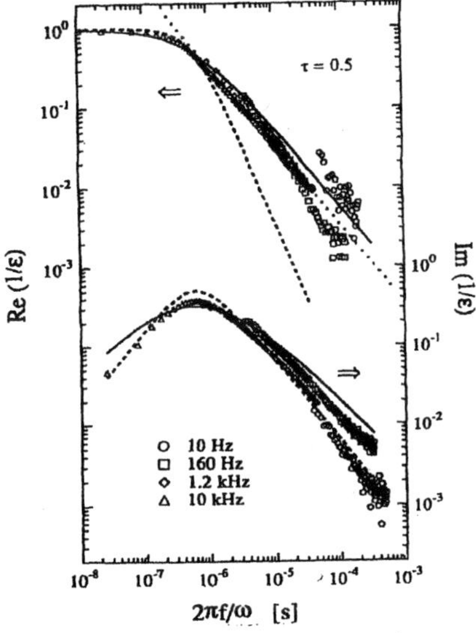

Figure 4 : Real and imaginary parts of the inverse of the vortex dielectric function measured in reference 82 on a JJA in a small magnetic field. The two functions are plotted as a function of the reciprocal frequency. The full lines are fits to the data using Minnhagen?s expressions (5.6), whereas the Drude form is given by the dashed lines.

It is clear that such a vortex mobility of the form (5.6) cannot hold down to arbitrarily low frequency above the vortex unbinding transition. Indeed, the electrical conductance $G(\omega)$ is related to the vortex mobility according to (4.46). The vanishing vortex mobility (5.6) for $\omega \to 0$ of MP would therefore imply that superconductivity is maintained even above T_{BKT}.

Another interesting application of the two-coil technique has shed light on the dynamics of percolative arrays [95] discussed theoretically in section 6, see *figure* 5. The data confirm that the transition termperature goes to a lower value when the concentration of missing sites increases, ie when the percolation threshold is approached. Moreover, both the inverse inductance and the resistance show very clearly two of the three frequency regimes discussed in section 6 : for low frequencies both quantities have constant values given by (6.2), whereas above some cross-over frequency the fractal behaviour is observed. The critical exponent $t \approx 1.4$ is extracted from the data which is in good agreement with the prediction $t \approx 1.3$ for percolation in two dimensions [96]. The frequency dependence in the fractal regime is compatible with $u \approx 0.55$, see equation (6.3) which is somewhat larger than what is expected from duality arguments for bond percolation [75] and for a mean field approximation. On the other hand, different types of approximations used in the theoretical description of disordered

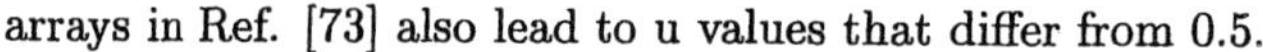
arrays in Ref. [73] also lead to u values that differ from 0.5.

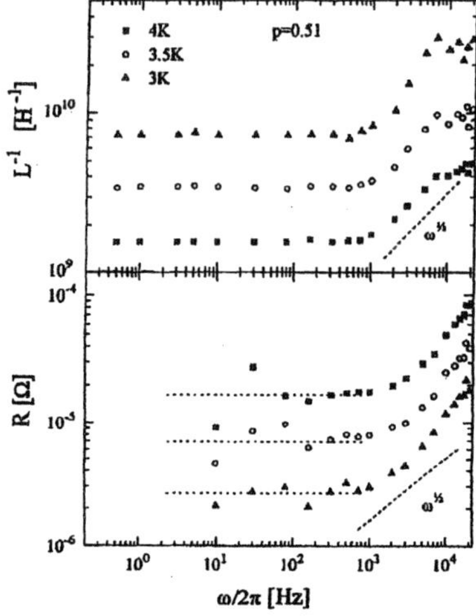

Figure 5 : Log-log plot of the frequency dependence of the inverse kinetic inductance $L-1$ and resistance R (according to equation (4.18)) of a dilute array close to the percolative limit ($p = 0.51$) from reference 95. The dashed lines indicate $\omega^{1/2}$ power laws.

Dynamic response measurements on the dilute arrays near their perconlation threshold have also been extended to temperatures near and above the expected BKT temperature [98], where vortex motion is expected to dominate over the linear spin wave dynamics discussed before. The contribution $Z_v(\omega)$ of the thermally excited V-A-system to the impedance $Z(\omega)$ is extracted from the measured values by representing the impedance of the array, in the spirit of the RSJ model, by two parallel channels - an inductive one for ≪ superfluid ≫ conduction and a dissipative Ohmic path - and by assuming that the vortex response renormalizes the superfluid part. A similar decomposition as (4.18) fo the dynamic conductance is used for Z_v

$$Z_v(\omega) = R_v(\omega) + i\omega L_v(\omega) \tag{8.4}$$

(whereby one has to bear in mind that resistances and inductances introduced, either by (4.18) or by (8.4) are not identical). Above T_{BKT} both, R_v and L_v, show a power law behaviour with frequency

$$R_v(\omega) \propto \omega^{u(T)} \ , \ L_v(\omega) \propto \omega^{u(T)-1} \tag{8.5}$$

with an temperature exponent. The latter is close to 1 at T_{BKT} and decreases slowly when T increases. Such a $T-$dependent exponent can be obtained by assuming that the relevant contribution to the dynamic response comes from pairs and by calculating their polarisability as indicated in subsection 5.1. However, a simple calculation using statement (5.3) yields an exponent $u(T)$ that decreases much faster with rising T than the ones observed [98]. Thus, one should probably revise the ideas of subsection 5.1 for disordered arrays. The response at temperature much higher than T_{BKT} again points to the motion of free vortices, the latter being activated across the pinning barriers.

Glassy behaviour of the vortex system, mentioned in section 6.2, has been seen in measurements of the dynamic impedance $Z(\omega)$ in ultrathin $YBa_2Cu_3O_7(YBCO)$ films in a perpendicular magnetic field [78]. At high enough temperatures the dissipative component of Z has an activated T-dependence resulting from activated motion of single (field induced) vortices out of local pinning wells. With decreasing temperature Z crosses over to a more or less T independent value. The cross-over temperature T^*, that depends weakly on ω, turns out to be much higher than a reasonable estimate for the melting temperature of a 2D V-lattice, due to unbinding of dislocation pairs. This points to the effect of pinning on the V-system. The frequency dependent response of such a pinned elastic manifold has been studied theoretically by various authors [99]. The corresponding predictions agree well with the measurements on $YBCO$ films. It would, of course, be interesting to see in more refined experiments whether $T^*(\omega)$ decreases without limit when the freq.uency ω becomes lower and lower. This would be a true sign of the fact that the 2D ≪ glassyness ≫ is really a dynamic phenomenon.

8.3 Flux noise

The flux noise spectrum $S_\phi(\omega)$ has been measured in overdamped $JJAs$ by Shaw et al [68] and more recently by Candia et al [100], as well as in high-$T_c(YBCO)$ films by Festin et al [101]. Shaw's data cover the range from $T = 1.825K$ to $T = 2.379K$ for a Nb on Cu array with $T_{BKT} \approx 1.63K$, see *figure* 6. At each T, $S_\phi(\omega)$ is white for $\omega < \omega_c$ and varies like $1/\omega$ for $\omega \geq \omega_c$. The cross-over frequency ω_c is proportional to the inverse square of the Kosterlitz-Thouless correlation length with the T dependence as given by (1.12). This result is consistent with dynamic scaling (see section 5.5), for a dynamic exponent $z = 2$ and with $b \approx 2$ in expression (1.12). The white noise level increases for decreasing T as $\xi_{BKT}(T)^2$. The authors conclude that, even at the lowest temperature shown (ie., for T as close to T_{BKT} as possible), ξ_{BKT} is on the order of the size of the pick-up coil and, as a consequence, the prefactor ξ_o in (1.12) is actually smaller than the lattice constant of the array. The fact that scaling is obeyed down to such a microscopic length scale is somewhat surprising.

The data of ref. [101] confirm previous observations [102] showing that high-T_c films also show typical 2D vortex fluctuations. Data are shown for 500 A thick $YBCO$ films from $T = 84.84K$ to $T = 84.89K$, just slightly above T_c. At low frequency $S_\phi(\omega)$ is again white, near some frequency ω_o the data have a common tangent corresponding to

$$S_\phi(\omega) \sim \omega^{-1.05} \tag{8.6}$$

before crossing over to $1/\omega^{3/2}$. However, contraray to what has been seen by Shaw et al [68] the noise spectrum does not have a true $1/\omega$ dependence over any appreciable frequency range. Thus, the two materials (JJA, respectively high T_c film) have a different characteristic behaviour : while the JJA shows an impressive $1/\omega$ noise, the film data only have a $1/\omega$ tangent whereas the high frequency behavior is $1/\omega^{3/2}$. The same is found in [102] for $Bi_2Sr_2CaCu_2O_{8+x}$ films.

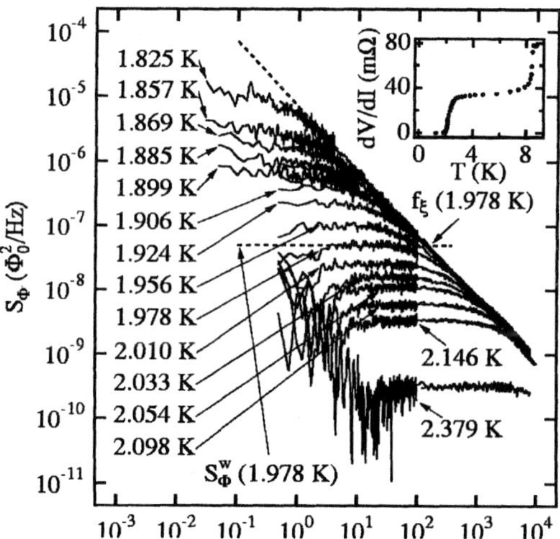

Figure 6 : Flux noise spectrum versus frequency for different temperatures of a JJA consisting of Nb islands on Cu (ref. 68).The dashed lines have slope -1 and 0, respectively, signalling white noise and $1/\omega$ noise in different frequency regimes. The inset shows the derivative of the current voltage characteristics, dV/dI, as a function of temperature. The BKT temperature of the array is at $1.63K$.

The most recent measurements on a proximity effect $Pb/Cu/Pb$ triangular JJA [100] confirm the $1/\omega$-dependence of S_ϕ above T_{BKT} (see *figure* 7), but the analysis of the low frequency white noise level has been interpreted in terms of an exponentially activated (total) number $n(T)$ of vortex excitations. This would show that the vortex mobility determining (7.13) through (4.28), (see also (9.5)), which is supposed to go to zero when the critical temperature is approached, does not really influence the white noise level. This may be a finite size effect (see section 9.3 where results of theoretical flux noise calculations will be presented). Candia et al [100] have also extended their measurements to temperatures close to, but below T_{BKT}. In the frequency range covered by the measurements the noise keeps to behave like $1/\omega$ but, for a given frequency, the level decreases when T goes down.

The ubiquity of the $1/\omega$-dependence is somewhat surprising. As shown in section 9 it can be reproduced when pair dynamics is taken into account, but with an exponent in the probability distribution for the relevant pair distances in statement (5.2) which is frozen at $T = T_{BKT}$. Thus it seems that the array $\ll$ stays critical $\gg$ even when the temperature differes from its critical value, as it has been remarked by Shaw et al [68]. In the context of disordered arrays [98], the corresponding exponent has also been extracted from dynamic data and shown to be much less temperature dependent than statement (5.2) would imply. As a matter of fact, the $1/\omega$ dependence could quite generally be a disorder effect. Indeed, even small variations of the distance d between neighboring superconducting islands implies a relatively strong $\ll$ disorder $\gg$ in the Josephson coupling constants, since the latter depend exponentially on d. In such a disordered array the free vortices that can be present at any temperature will have to move on a $\ll$ random substrate $\gg$. The corresponding dynamic response would typically behave like $1/\omega$.

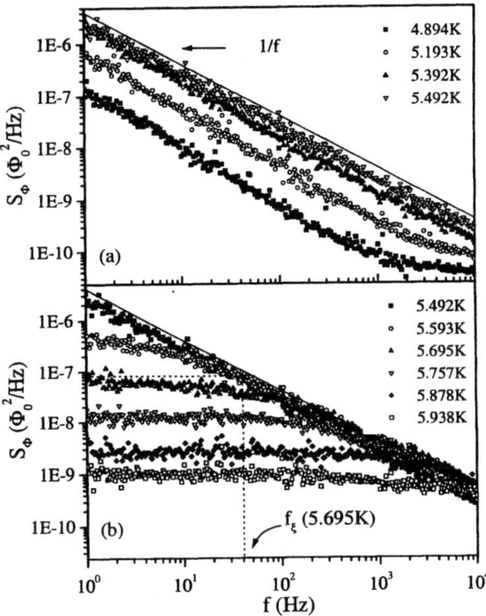

Figure 7 : Flux noise spectrum from reference 100, measured on a Pb/Cu array as a function of frequency f at temperatures slightly below (a) and slightly above (b) the BKT transition.

9 Theoretical results

9.1 Dynamic critical exponents

The dynamic exponent z, introduced and discussed in section 5.5, has been determined through various numerical simulations. Tiesinga et al [46] solved the equations of motion for the phase field using alternatively $TDGL$ and RSJ dynamics, as explained in section 5.4. The corresponding value of z is obtained by studying the scaling form (5.21) of their correlation functions, ie by identifying the link between the relevant relaxation frequency and the BKT correlation length. Interestingly, they find that z is different for the two types of dynamics : $z(TDGL) \approx 2$ corresponds to what is expected, whereas $z(RSJ) \approx 0.9$ points to a different type of critical dynamics(see the insets in figure 10 that will be discussed in connection with flux noise). This would be an example of a system with a given static critical behaivour (BKT in this case) which goes over into different dynamic universality classes, depending on the precise equations of motion used to describe time evolution.

The scaling analysis of the $TDGL$ and RSJ simulations of Kim et al [53], who used periodic and fluctuating twist boundary conditions ($FTBC$, see section 5.4), yields $z = 2$

at the transition temperature for both types of dynamics for $FTBC$. This value is found by finite size scaling (extrapolation of calculated parameter values to the thermodynamic limit) of the array resistance. However, they observe that finite size scaling with PBC gave different results, or - more precisely - that PBC are actually inadequate for this procedure, since the relaxation time which reveals the exponent z is proportional to the macroscopic resistance that is actually zero for periodic boundary conditions. This shows once again that correct boundary conditions are crucial whenever transport coefficients are considered, which describe the response of the system to an external influence that couples to the boundary. Below T_{BKT} the value of z is slightly larger than 2, which is in agreement with the finding for a 2-d lattice Coulomb gas with Monte Carlo dynamics [64].

9.2 Vortex dielectric function

The fact that the vortex dielectric function, measured by the two-coil technique described in the previous section, has a frequency dependence resembling more the ≪Minnhagen phenomenology≫ (MP) than the classical Drude behaviour, has initiated both analytical and numerical work aiming at understanding this ≪anomalous≫ dynamic response of $JJAs$. Such an effect is likely to be related to vortex-vortex interaction, to pinning by the underlying lattice or to the fact that vortices are not just point particles, but singular excitations of the phase field which change shape when they move and when they interact with each other.

Coupling to spin waves [103,104], which are the ≪Bosons≫ mediating the Coulomb interaction between vortices in the framework of the effective electrodynamics described in section 4.2, has been shown to lead to a vortex dielectric function of the form given in section 5.2. The non-analytic low frequency dependence, which is characteristic of MP, is an ≪ infrared divergence ≫ that can be understood by counting powers of wave number in the relevant integrals. It is the combined effect of the spin wave propagator and the dimensionality of the array. Other authors [105,106] have also found non-analytic effects in vortex dynamics, for instance in the vortex mass of arrays with capacitive effects, caused by coupling to spin waves.

Capezzali et al [107] have treated the problem of interacting vortices, both by a refined dynamic scaling (a generalization of the earlier approach in ref. [16] which did not yield any MP features) and by a Mori calculation of the vortex density correlator (see section 5). They find three different regimes corresponding to different ranges of the scaling variable

$$Y = \frac{1}{\omega \xi^z} \tag{9.1}$$

z being the dynmic exponent discussed in section 5.5 and ξ the correlation length. The experimental data of ref. [82] fall in the intermediate regime where MP is valid. The high frequency regime corresponds to ≪ critical dynamics ≫ as it had already been exhibited in [16]. Contrary to the ≪ pure MP ≫ form there is a low frequency regime where normal Drude behaviour should be restituted. This is an important complement to MP, since taking the MP expressions (5.5) to (5.7) straight down to $\omega = 0$, and inserting them into expression (4.45) for the electrical conductance of the array would yield superconducting low frequency behaviour even above the transition temperature.

The calculation of the dynamic charge correlator $\phi_{\rho\rho}$(see section 4.2) by Mori's technique has more recently been generalized [108] by replacing the Coulomb potential by its screened form, as it is usually done in calculations for quantum Coulomb systems [40,41]. Since screening is due to free vortices the relevant screening length (5.13) is the BKT correlation length (1.12) which is directly related to the inverse density of free vortices. This allows to impose static critical behaviour on dynamics. The resulting friction function, determining $\phi_{\rho\rho}$ and the dielectric function has indeed a logarithmic frequency dependence, leading to MP behaviour, over a frequency interval that is increasing in length when the transition temperature is reached at which ξ_{BKT} diverges. This result shows that dynamics is ≪ normal ≫ as long as the long range Coulomb interaction is well screened by the unpaired particles and it gets more and more MP-like when screening becomes less efficient (see *figure* 8). This manifests itself in a friction function that increases more and more for low frequency, making the motion more ≪ sluggish ≫. At the transition temperature this ≪ critical slowing down ≫ is completed : the zero frequency friction function diverges.

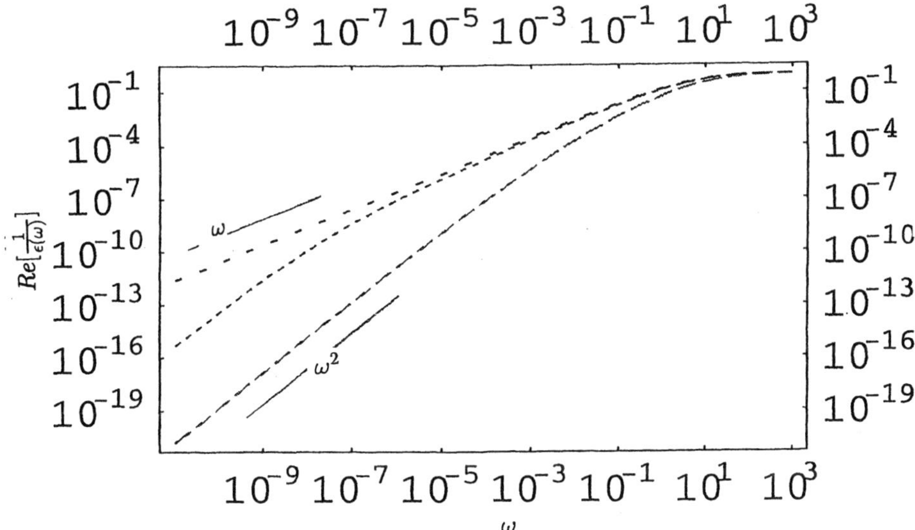

Figure 8 : Real part of the inverse of the vortex dielectric function, as a function of frequency, calculated in the framework of Mori's approach taking into account the motion of unbound vortices moving in a screened Coulomb potential, for different temperatures above $T_{BKT}(T/T_{BKT} = 1.02$, 1.09 and 1.37 from top to bottom respectively). The frequencies are given in units of a scale frequency $\omega_a = k_B T/\Gamma a^2$, involving the lattice constant a and the bare damping Γ (see equation (5.10)). The slopes of the two straight lines correspond to $1/\epsilon(\omega) \sim \omega$ (Minnhagen phenomenology) and $1/\epsilon(\omega) \sim \omega^2$ (Drude behaviour).

While these analytic calculations treat the motion of ≪ free ≫ vortices in the force field of the Coulomb gas, the calculation of the vortex dielectric function based on the pair relaxation picture [34] has also been refined [35] in order to study possible deviations from a simple Drude behaviour. Using a length scale dependent static dielectric constant, obtained by numerical integration of Minnhagen's version [17] of the BKT scaling relations below and above T_{BKT}, yields an $\epsilon(\omega)$ that is indeed closer to MP than to Drude in some intermediate frequency range.

Minnhagen's group has studied JJA dynamics in great detail [55,57,58,60,63] using dynamic simulations some aspects of which have been discussed in section 5.4. These authors have used the different description (a) to (d) of $JJAs$ quoted at the beginning of section 5.4, and they have treated both unfrustrated and frustrated arrays, in which field induced vortices are also present. They also replaced the simple cosine interaction of the phase Hamiltonian (1.2) by the form

$$H = \sum_{\langle ll' \rangle} J[1 - \cos^{2p^2}(\frac{\phi_{ll'}}{2})] \tag{9.2}$$

This coupling reduces to (1.2) for $p = 1$, but for larger value of p it approaches a ≪ potential well ≫ favoring the creation of vortices [109]. The dielectric function obtained by calculating the relevant phase correlation function indeed follows closely the MP prediction (see *figure* 9): contrary to the simple Drude form $Re(1/\epsilon)$ varies linearly with ω, $Im(1/\epsilon)$ has a long ≪ flat ≫ region, and the peak ratio (5.8) is close to $\pi/2$. Deviations from MP for small ω, leading back to Drude, appear for $p > 1$. Frustrated systems [54,57] show similar behaviour : when the magnetic field (and with it the effective array resistance) is increased at a fixed temperature the response crosses over from normal to anomalous. In ref. [57] it was observed that increasing the magnetic field - ie the frustration of the array -not only leads to an obvious higher density of field induced vortices, but it also increases the number of V-A pairs. This seems to corroborate the idea that V-A correlations are responsible for the anomlous dynamics observed in simulations and in experiments. The screened Coulomb potential used in the Mori calculations [108] may indeed be one of the manifestations of the presence of a sufficient number of thermal vortex excitations.

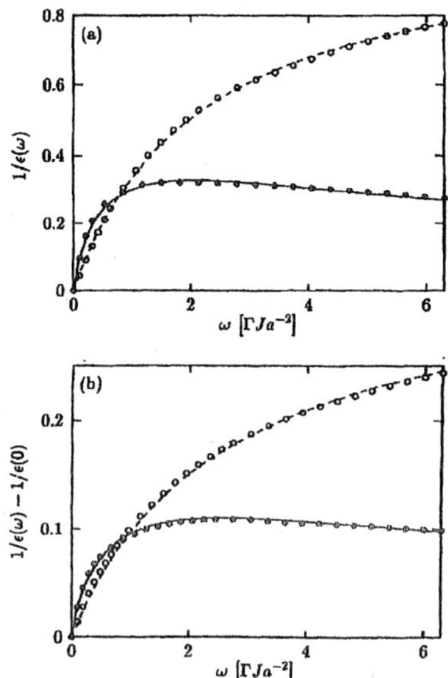

Figure 9 : Real and imaginary parts of the inverse vortex dielectric function of an unfrustrated array using an interaction potential with $p = 2$ in equation (9.2), The open circles ($Re(1/\epsilon)$) and the filled circles ($Im(1/\epsilon)$) are the results of dynamic simulations using $TDGL$

dynamics (method b of section 5.4) of Ref. 55. The dashed and full curves are fits to the numerical data using the MP relations (5.6).

The effect of (regular or random) pinning on the motion of field induced vortices of weakly frustrated JJAs has been investigated by Monte Carlo simulation [65] of the 2D one-component lattice Coulomb gas ($OCCG$). Anomalous MP behaviour of the dynamic dielectric function (see subsection 5.2) is found, and it seems to be related to pinning. It is well developed when sufficiently strong random pinning is present. In the absence of random pinning it is restricted to temperatures close to the depinning of the vortex system from the underlying lattice structure. On the other hand, the continuum $OCCG$, which experiences no pinning at all, shows no anomalous response. In contrast to this, the neutral 2D two-component Coulomb gas [62], which respresents the VA-system in an unfrustrated array shows MP behaviour below the BKT transition. One may therefore wonder whether in the latter case vortices of one sign of charge act as pinning centers for antivortices, and vice-versa [62].

9.3 Flux noise

Dynamic simulations have also been used for evaluating the flux noise spectrum either by evaluating the full integral (7.12) or by using approximate forms of the latter, such as (7.14). At low frequency the expected white noise is found with a level that goes up when T approaches T_{BKT} from above. The behaviour for higher frequencies varies between the different authors. Some calculations [56], [110], simulating $TDGL$ dynamics for a 2D XY model with a phase coupling of the type (9.2) yield

$$S_\phi(\omega) \sim \omega^{-3/2} \tag{9.3}$$

for some intermediate frequency range and $S_\phi(\omega) \sim \omega^{-2}$ for high frequencies. The broken power of (9.3) can be understood as the sign of conventional random walk of vortices across the boundary of the pick-up coil [110,111]. When the curves bend over from white noise to the power law (9.3) a common tangent can be drawn to the curves which has a slope close to $1/\omega$. However, there is no extended regime showing the measured $1/\omega$ noise (see section 8.3).

Other dynamic simulations [46] find an intermediate frequency range, above the white noise regime, where the spectrum follows in power law : $(1/\omega)^\alpha$ (see $figure!$ 10). The exponent α is T-dependent and differs slightly between different types of dynamics used (see section 5.4) :

$$\begin{aligned} 0.85 \leq \alpha \leq 0.95 \qquad & for\ TDGL\ dynamics \\ 1.17 \leq \alpha \leq 1.27 \qquad & for\ RSJ\ dynamics \end{aligned} \tag{9.4}$$

which, again points to different 'critical dynamics' where different equations of motion are used. Both cases are relatively close to the measured $1/\omega$ dependence, but - as the authors state - the $TDGL$ dynamics is closer to the measured $1/\omega$ dependence. This is somewhat surprising, since the ≪usual≫ picture one has of a Josephson juncion and of an array is precisely the RSJ effective circuit model. It is possible that the arrays of Shaw et al [68], consisting of large Nb islands deposited on a Cu substrate have very small resistances to ground, such that the local dissipation term in the $TDGL$ equation dominates the bond term in RSJ. Other RSJ-simulations [112] also yield the behaviour of (9.3), both for $f = 0$ and for frustrated arrays. The same holds for the solution of the Fokker-Planck equation for pairs [43], discussed at the end of section 5.3. On the other hand a calculation [113] of the flux noise spectrum starting from a Euclidean action for the XY-model with local damping also finds a cross-over from white to $1/\omega$ noise.

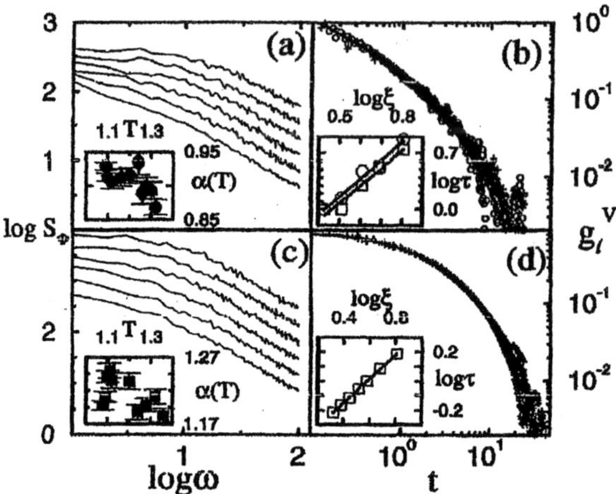

Figure 10 : Flux noise spectra versus frequency obtained in Ref. 112 for $TDGL$ (a) and RSJ (c) dynamics for different temperatures above T_{BKT}. The insets of the panels (a) and (c) show the exponent α (see our expresssion (9.4)) as a function of temperature. Panels (b) and (d) show a scaling plot of the noise spectrum as a function of time. The insets present the dependence of the characteristic time t on the correlation length x, yielding the values $z = 2$, respectively 0.9 of the dynamic exponent introduced in subsection 9.1.

The results obtained by the Mori approach to the dielectric function (section 9.2) have also been used for evaluating flux noise spectra. Preliminary calculations [108] show that the friction function, obtained by taking into account motion of unbound particles in a screened Coulomb potential, although reproducing nicely the MP behaviour in $\epsilon(\omega)$, does not lead to $1/\omega$ noise. These calculations are thus in good agreement with the above mentioned numerical simulations [56], [110]. Another approach starts from the averaged pair susceptibility presented in section 5.1, integrating the single pair susceptibility χ over a pair distribution of the form (5.2) from the lattice constant (the smallest possible pair distance) up to the BKT correlation length ξ and going from χ to the density correlator $\phi_{\rho\rho}$ with the help of (4.303) yields a flux noise spectrum with a white noise level and a subsequent power law with a temperature dependent exponent [97], similarly as in [46]. In fact, the $1/\omega$ dependence is found for $T = T_{BKT}$. The white noise level and the cross-over frequency between the latter and the power-law regime varies like ξ^2. Taking the limit of zero frequency in the integral (7.10) indeed yields the following approximate statement for the white noise level

$$S_\phi(\omega = 0) = C\frac{\gamma(0)k_BT}{n_v(T)} \tag{9.5}$$

where the constant is the result of the wave number integral involved in (7.10), $n_v(T)$ is the total number of vortices and $\gamma(0)$ the zero frequency limit of the friction function. The data of reference [100] for the white noise level can be understood by limiting the integral over the pair distribution by the sample size L, whenever $\xi > L$. This explains why no more critical behaviour is seen in the white noise data when the critical temperature is approached : the relevant length is fixed by L and the only major T-dependence left in (9.5) comes from $n_v(T)$, which is indeed thermally activated as seen in the data of ref. [100].

10 Driven arrays : dynamics far from equilibrium

$JJAs$ are interesting examples of systems with non-linear equations of motion (EM). The relevant EM for the phases, depending on the type of dynamics, have been found in section 2.1, whereas section 3 ends up with an equation of motion for the vortex positions. While, in sections 5 to 7, we have mainly been interested in equilibrium dynamics and in response to weak driving fields we now turn to work dealing with the dynamic behaviour of $JJAs$ under moderate and strong driving currents.

The interplay between vortex pinning due to the discrete underlying lattice structure (the ≪ egg cartoon potential ≫) and V-V interaction can be tested by changing the applied perpendicular magnetic field, i e the frustration. In the ground state the field induced vortices tend to form a regular arrangement minimizing their interaction as well as the pinning energy. The critical current needed to drive this vortex lattice across the sample is the higher the better the ≪ matching ≫ between array and vortex structure, which is the case for $f = p/q$ with small integers p and q. Marconi and Dominguez have studied the depinning and melting of driven moving vortex lattices for small frustration [114], $f = 1/25$, and for ≪ full ≫ frustration [115], $f = 1/2$, by solving the RSJ Langevin equations of motion for the phases in presence of the applied vector potential and a random force which maintains a fixed finite temperature. The results can be summarized in a non-equlibrium phase diagram in the plane of temperature T_{dep} and drive current I. For very low current it reveals the equilibrium depinning transition temperature above which the vortex lattice floates freely on the arrays, thanks to its thermal energy. Stronger currents shift this depinning transition to lower T and produce phases of different properties above T_{dep}. At low T the moving lattice is still ≪transversely pinned≫ : it has anisotropic Bragg peaks, quasi-long range order and maintains superconducting coherence (and thus a finite critical current) in the direction perpendicular to the direction of the strong drive current (one has to bear in mind that a vortex moves in the direction perpendicular to the drive current, see section 3.1). At higher T the vortex lattice is simply ≪ floating ≫ .

The presence of random pinning adds new complexity. As mentioned in section 6.2, an irregular underlying lattice structure forces the vortices to ≪adapt≫ their equilibrium structure in order to minimize their free energy. In order to drive the vortex arrangement across the ≪pinning landscape≫ the applied current I, which acts as a transverse force on the vortices, has to exceed some threshold value I_c. Once the vortices are depinned, which is seen by the appearance of a finite voltage and thus a finite resistance of the array in the current direction, their motion will, of course, be influenced by the random pinning potential. Dominguez [116] has observed, in his numerical simulations, a dynamic transition occuring for a current $I_p(> I_c)$. For currents with a strength between I_c and I_p the moving ≪vortex liquid≫ has an isotropic structure and the turbulent flow regime is plastic. Above I_p there is a regime with homogeneous ≪laminar≫ flow in the direction of motion and with a structure with anisotropic short range order of smectic type characterized by the appearance of corresponding Bragg peaks.

Various other work has been published about the possiblity of such a dynamic regime in superconducting films (for references, see for example [117]). Two dynamical phase transitions for thin films are found theoretically in [118] : for a given strength of the exciting force the system goes again over from plastic flow, where the vortices move in an intricate network of channels, to smectic flow. This transition is characterized by a maximum in the differential resistance (the derivative of the averaged vortex velocity with respect to the drive current strength). At a higher driving force there is a sharp transition to a frozen transverse solid. Here the motion of the vortices transverse to the flow direction, which consisted of jumps between different channels in the smectic regime, is strongly reduced. More details about the temporal behaviour of the system (noise spectrum, diffusion exponents etc.) can be found in [118]. Moving anisotropic vortex structures have also been observed experimentally [117] in small high quality single crystals of $NbSe_3$.

Non-linear dynamic systems, such as our $JJAs$, can also show various phenomena related to the occurrence of chaos. Indeed, already a single, capacitively and resistively shunted junction with an applied voltage or current corresponds to a damped driven pendulum, which is known to show very rich structures in its dynamic phase diagram. Regimes of periodic and quasiperiodic motion alternate with chaotic regions when parameters, such as the drive force, are varied. Josephson junctions coupled in series have a spatial variable in addition to time. They thus allow for spatio-temporal chaos. Various such systems have been studied [9] numerically. They show coherent, ordered, partially ordered, fully turbulent and quasiperiodic phases. When time is discretized in the coupled non-linear equations of motion for the JJA the latter are of the same structure as the equations of ≪globally coupled maps≫ [9]. Their mathematical properties (transition from regular to chaotic motion, structure of attractors, etc.) have widely been studied. Chaotic motion in $JJAs$ should in principle be observable experimentally through the characteristic features of the current-voltage characteristics of strongly driven arrays, but not much systematic analysis seems to have been done up to now.

11 Miscellaneous

Josephson junction arrays are an interesting example of articifial systems. They allow us to profit from the microscopic phenomenon of superconductivity in order to observe challenging collective effects. This article is mainly devoted to the dynamic manifestation of the latter. The interested reader should find elsewhere more information about the static and thermodynamic behaviour of $JJAs$ in order to fully appreciate the rich spectrum of time dependent phenomena discussed here. For obvious reasons we had to make a choice concerning the topics to be presented (and this choice is certainly biased by our own experience in the field !). In fact, quite a variety of other aspects of JJA dynamics have not been covered. Let us just mention a few of them:

Hall effect

In addition to the ≪Lorentz force≫, mentioned in section 3, a vortex is also subject to a ≪Magnus force≫ [25] when an external charge introduces a potential between the array and the ground. This force is perpendicular to the vortex velocity. The motion resulting from the combination of Lorentz force (showing up in equation (3.6)) and Magnus force leads to a ≪vortex Hall effect≫ that has been observed for classical and quantum arrays [25]. Since the Hall angle is inversely proportional to the damping constant Γ showing up in the vortex equation of motion, Hall measurements give information about the type of dissipation prevailing in the array. As a function of frustration, ie of the applied magnetic field, the Hall resistance shows a rich structure [120]. Reversal of the sign of the Hall constant, which is a sign that the relevant type of charge carriers changes, appears at several frustrations.

Ballistic vortex motion

When the mass term in the equation of motion (3.6) dominates over the friction vortices can in principle move almost freely across the sample. Such ≪ballistic≫ vortex motion has been observed, both in one- and two-dimensional arrays [1,25], in a frequency window, bounded from below by the presence of pinning forces and from above by various (non-linear) damping mechanisms setting in at high velocities. It is interesting to note that, from a more elaborate theoretical point of view, ≪ideal≫ ballistic motion is not really likely to be observable [1]. Simulations on finite systems have to cope with boundary effects (a vortex being attracted by its image when it approaches the boundary). Moreover, non-linear viscosity can lead to localized oscillations rather than propagation of the vortex. Thus ballistic vortex motion definitely needs to be studied in more detail [1].

Shapiro steps

In section 4.1 we have introduced a weak applied external electric field into the equations of motion and we have calculated the linear response of the array. For an applied voltage of finite strength one can use the Josephson relation (4.19) between a voltage drop across a junction and the corresponding time dependence of the phase difference. When the applied voltage is of the form

$$V(t) = V_o + V_1 \cos \omega_1 t \tag{11.1}$$

the corresponding phase difference is easily found by integrating (4.19). Substituting it into (1.6) yields an statement for the resulting Josephson current for a single junction. Expressing the sin of a trigonometric function by a series of Bessel functions [1] shows that one should observe a dc super current whenever

$$V_o = V_n = n\frac{\hbar\omega_1}{2e} \tag{11.2}$$

These voltage steps are called ≪Shapiro steps≫ [121] of a single junction. Including the normal current flowing through the resistive channel yields a current voltage characteristics showing steps at the voltage values given by (11.2). Current driven arrays are somewhat more difficult to treat theoretically, because one really has to solve the equation of motion (2.4) for a given drive current. Shapiro steps again occur at the voltage values of (11.2).

Under ≪ideal≫ conditions, ie in zero temperature and external field, an $M \times N$ array of such junctions should act as a single junction with an applied current I/M. The total voltage drop across the array should then be N times the one across each junction and ≪giant≫

Shapiro steps should occur at voltages

$$\hat{V}_n = NV_n \tag{11.3}$$

Reference 1 gives many an excellent overview over various more elaborate aspects of Shapiro steps in JJA, such as the occurrence of fractional giant steps and of suharmonic steps.

Application for microwave radiation

A Josephson junction is a voltage controlled oscillator capable of generating electromagnetic radiation [122]. The dynamics of the supreconducting phase can be formulated in the framework of the capacitively and resistively shunted junction model presented in section 2.1. A dc bias voltage which excites the damped non-linear oscillator has to be added to the equations of motion given there. Quantitative estimates show, however, that the linewidth of a single junction is rather large [122]. One obvious way to overcome this drawback is now precisely to use one- or two-dimensional arrays of junctions and to have them oscillate in phase. In the 2D case treated in this article the rf current flows, for example, along the rows of the array, whereby care has to be taken that the phases of the current in each row are the same. The line width is then reduced, with respect to a single junction, by the number of the array sites. Details about the realisation of such voltage controlled microwave oscillators in the frequency range between 100 GHz to 1 THz can be found in reference [122].

Many of the concepts developed for arrrays can, of course, also be applied to the description of thin superconducting films which are, in some sense, a continuum version of the discrete lattice. Thus we have mentioned results found on such films in sections 6 and 8. Moreover, high temperature superconductors have more or less strongly anisotropic structures of rather weakly coupled lattice planes. Thus, for many purposes the phenomenon of superconductivity can be restricted to lattice planes, so that many aspects of two-dimensional suprerconductivity described in the present article are again applicable.

Acknowledgements

We gratefully acknowledge continuous support from the Swiss National Science Foundation and the Swiss Commission fédérale des bourses for our own research activity in this field and we thank D.Bormann, M.Capezzali, D.Dominguez, R.Fazio, S.Korshunov, M.Kosterlitz, P.Martinoli, P.Minnhagen, M.Mombelli, S.Shenoy, A.Varlamov, A.Zaikin for many interesting discussions.

On the other hand we express our apologies to all the many workers in the field whose results have not been adequately appreciated and quoted in this article.

References

[1] R.S.Newrock, C.J.Lobb,U.Geigenmüller, M.Octavio ; Solid State Physics 54, 263 (2000)

[2] P.Martinoli. C.Leemann ; Journal of Low Temperature Physics 118, 699 (2000)

[3] P.W.Anderson ; in ≪ Lectures on the Many-Body-Problem ≫, Vol 2, ed. E.R.Caianiello, Academic Press, New York (1964)

[4] E. Simanek ; ≪ Inhomogeneous superconductors ≫ , Oxford University Press, 1994, Ch. 4 and 5

[5] V.L.Berezinskii ; Sov Phys JETP 32, 493 (1971)

[6] J.M.Kosterlitz, D.J.Thouless ; J Phys C6, 1181 (1973)

[7] S.E.Korshunov, A.Vallat and H.Beck ; Phys Rev B51, 3071 (1995)

[8] E. Simanek ; Sol. State Comm. 31, 419 (1979)

[9] D. Ariosa, H.Beck, Phys Rev B45, 819 (1992)

[10] L.Jacobs, J.V.Jose, M.A. Novotny, A.M.Goldman, Phys Rev B38, 4562 (1988) and references therein

[11] M. Capezzali, D. Ariosa, H.,Beck; Physica B230-232, 962 (1997)

[12] K.H.Wagenblast, A.v Otterlo, G.Schn, G.T.Zimanyi ; Phys Rev Lett 78, 1779 (1997)

[13] J.M.Kosterlitz ; J.Phys C7, 1046 (1974)

[14] J.V. Jose et al ; Phys Rev B 16, 1217 (1977)

[15] A.Vallat, H.Beck ; Phys Rev B 50, 4015 (1994)

[16] S.R.Shenoy ; J Phys C, Solid State Physics 18, 5163 (1985)

[17] P.Minnhagen ; Rev Mod Phys 59, 1001 (1987)

[18] D.Dominguez, J.V.Jose ; Int J Mod Phys B8, 3749 (1994)

[19] J.C.Ciria, C.Giovanella ; J Phys Cond Mat 10, 1453 (1998)

[20] G.Schön, A.D.Zaikin ; Physics Reports 198, 237 (1990)

[21] V.L.Pokrovsky, M.V.Feigel'man, A.M.Tsvelick ; ≪ Spin waves and magnetic excita tions ≫, ed. A.S.Borovik-Romanov and S.K.Sinha ; Elsevier Science Publishers B.V. 1988, p. 67

[22] H.Beck, D.Ariosa ; Solid State Communications 80, 657 (1991)

[23] C.J.Lobb, D.Abraham, M.Tinkham ; Phys Rev B27, 150 (1983)

[24] U.Eckern, A.Schmid ; Phys Rev B 39, 6441 (1981)

[25] R.Fazio, H.van der Zaant ; Physics reports 355, 235 (2001)

[26] R Fazio, U Geigenmüller, G Schön ; in ≪ Quantum fluctuations in Mesoscopic and Macroscopic Systems ≫, eds. H.A.Caldeira et al, World scientific, Singapore, 1991, p. 214

[27] R. Fazio, G. Schön ; Phys Rev B 43, 5307 (1991), U Eckern, E B Sonin, Phys Rev B 47, 505 (1993)

[28] S.R.Shenoy ; J Phys C (Solid State Physics) 18, 5163 (1985)

[29] S.V.Panyukov, A.D.Zaikin ; Physics Letters A 156, 119 (1991)

[30] P.Bobbert, R.Fazio, G.Schön, A.D.Zaikin ; Phys Rev B 45, 2294 (1992)

[31] N.H.March, M.P.Tosi ; ≪ Atomic Dynamics in Liquids ≫, John Wiley and Sons, New York, 1976

[32] V.Ambegaokar, S.Teitel ; Phys Rev B 19, 1667 (1979)

[33] R.Côté, A.Griffin ; Phys Rev B 34, 6240 (1986)

[34] V.Ambegaokar, B.I.Halperin, D.R.Nelson, E.D.Siggia; Phys Rev Lett 40, 783 (1978) and Phys Rev B21, 1806 (1980)

[35] D.Bormann, H.Beck, O.Gallus, M.Capezzali ; J Phys IV France 10, Pr-447 (2000)

[36] H.Mori ; Prog.Theor.Phys. (Kyoto) 34, 423 (1965)

[37] J.-P. Hansen, I.R. McDonald ; Theory of Simple Liquids (2nd edition), Academic Press Inc., San Diego, CY, 1986

[38] D.Forster, ≪ Hydrodynamic Fluctuations, Broken Symmetry and Correlation Functions ≫, Benjamin/Cummings,s London, 1975

[39] N.G. van Kampen ; ≪ Stochastic Processes in Physics and Chemistry ≫, North-Holland, Amsterdam, 1981

[40] R.Zimmermann, H.Stolz ; Phys Stat Sol (b) 131, 151 (1985)

[41] H.E.DeWitt, M.Schlanges, A.Y.Sakakura, W.D.Kraeft ; Phys Lett A197, 326 (1995)

[42] M.Capezzali, M.Mombelli, P.Béran, H.Beck ; in ≪Macroscopic Quantum Phenomena and Coherence in Superconducting Networks ≫,ed. C.Giovannella and M.Tinkham, World Scientific, Sikngapore, 1995, p. 270

[43] C.Timm ; Phys Rev B 55, 3241 (1997)

[44] B.J. Kim, P Minnhagen, P. Olson ; cond-mat/9806231, Phys Rev B 59, 11'506 (1999)

[45] L.M. Jensen, B.J. Kim, P. Minnhagen ; Europhys Lett 49, 644 (2000) ; Phys Rev B 61, 15'412 (2000)

[46] P.H.E.Tiesinga, T.J.Hagenaars, J.E.van Himbergen, J.V.Jose ; Phys Rev Lett 78, 519 (1997)

[47] I-J.Hwang, S.Ryu, D.Stroud ; cond-mat/9704108

[48] B.J.Kim, P.Minnhagen ; cond-mat/9902148 ; Phys Rev B 60, 6834 (1999)

[49] P.Minnhagen, O.Westman, A.Jonsson, P.Olsson ; Phys Rev Lett 74, 3672 (1995)

[50] L.M.Jensen, B.J.Kim, P. Minnhagen ; cond-mat/0003447 ; Phys Rev B 61, 15412 (2000)

[51] M.V.Simkin, J.M.Kosterlitz ; Phys Rev B 55, 11'646 (1997)

[52] I-J Hwang, D.Stroud ; Phys Rev B 57, 6036 (1998)

[53] B.J.Kim, P.Minnhagen, P.Olsson ; cond-mat/9806231 ; Phys Rev B 59, 11506 (1999)

[54] B.J.Kim, P.Minnhagen ; Phys Rev B 60, R15'043 (1999)

[55] A. Jonsson, P. Minnhagen ; Phys Rev B 55, 9035 (1997)

[56] J.Houlrik, A.Jonsson, P.Minnhagen ; Phys Rev B 50, 3953 (1994)

[57] A.Jonsson, P.Minnhagen ; Physica C 277, 161 (1997)

[58] A.Jonsson, P.Minnhagen ; Phys Rev Lett 73, 3576 (1994)

[59] P.Minnhagen, B.J.Kim, H.Weber ; Phys Rev Lett 87, 037002 (2001)

[60] P. Minnhagen, O.Westman ; Physica C 220, 347 (1994)

[61] K.Holmlund, P.Minnhagen ; Physica C 292, 255 (1997)

[62] K.Holmlund, P.Minnhagen ; Phys Rev B 54, 523 (1996)

[63] P.Minnhagen ; Physica B 222, 309 (1996)

[64] H.Weber, M.Wallin, H.J.Jensen ; Phys Rev B 53, 8566 (1996)

[65] B.J.Kim, P.Minnhagen ; cond-mat/0011297 (Nov 2000)

[66] P.Olsson ; Phys Rev B 46, 14'598 (1992), Phys Rev B 52, 4511 (1995), Phys Rev B, 4526 (1995)

[67] P.C.Hohenberg, B.I.Halperin ; Rev Mod Phys 49, 435 (1977)

[68] T.J.Shaw, M.J.Ferrari,L.L.Sohn, D.H.Lee, M.Tinkham, J.Clarke ; Phs Rev Lett 76, 2551 (1996)

[69] D.S.Fisher, M.P.A.Fisher, D.A.Huse ; Phys Rev B 43, 130 (1991) ; A.Dorsey, Phys Rev B 42, 7575 (1991)

[70] D.Bormann ; Phys Rev Lett 78, 4324 (1997)

[71] S.M.Ammirata, M.Friesen, St.W.Pierson, LeRoy A.Gorham, J.C.Hunnicutt, M.L.Trawick, C.D.Keener ; Physica C313, 225 (1999)

[72] T.Nakayama, K.Yakubo, R.L.Orbach ; Rev Mod Phys 66, 381 (1994)

[73] M.Mombelli, H.Beck ; Phys Rev B57, 14'397 (1998)

[74] A.L.Efros, B.I.Shklovskii ; Phys Status Solidi B76, 475 (1976)

[75] A.M.Dykhne, Zh.Eksp.Teor.Fiz. 59, 110 (1970) [Sov.Phys.JETP 32, 63 (1971)], J.P.Straley, Phys Rev B 15, 5733 (1977)

[76] M.P.A.Fisher, Phys Rev Lett 62, 1415 (1989)

[77] M.V.Feigel'man, V.B.Geshkenbein, A.I.Larkin, V.M.Vinokur; Phys Rev Lett 63, 2303 (1989)

[78] M.Calame, S.E.Korshunov, Ch.Leemann, P.Martinoli ; Phys Rev Lett 86, 3630 (2001)

[79] S.T.Herbert, Y.Jun, R.S.Newrock, C.J.Lobb, K.Ravindran, H.K.Shin, D.B.Mast, S.Elhamri ; Phys Rev B 57, 1154 (1998)

[80] A.F.Hebard, A.T.Fiory ; Phys Rev Lett 44, 291 (1980)

[81] P.Martinoli, Ph.Lerch, Ch.Leemann, H.Beck ; Japanese Journal of Applied Physics 26-3, 1999 (1987)

[82] R.Théron, J.B.Simond, C.Leemann, H.Beck, P.Martinoli, P.Minnhagen ; Phys Rev Lett 71, 1246 (1993)

[83] B.Jeanneret, J.L.Gavilano, G.A.Racine, Ch.Leemann, P.Martinoli ; Appl Phys Lett 55, 2336 (1989)

[84] Ch.Leemann, Ph.Lerch, G.A.Racine, P.Martinoli ; Phys Rev Lett 56, 1291 (1986)

[85] B.Jeanneret, Ph.Fluckiger, Ch.Leemann, P. Martinoli ; Jpn J Appl Phys 26-3, 1417 (1987)

[86] S.Korshunov ; unpublished

[87] H.Weber, M. Wallin, H.J.Jensen ; Phys Rev B53, 8566 (1996)

[88] A.M.Kadin, K.Epstein, A.M.Goldman, Phys Rev B27, 6691 (1983), P.Minnhagen, Phys Rev B 28, 2463 (1983)

[89] I.G.Gorlova, Yu.I.Latyshev, Pis'ma Zh Eksp Teor Fiz 51, 197 [(1990) JETP Lett 51, 224 (1990)]

[90] A.K.Pradhan, S.J.Hazell, J.W.Hodby, C.Chen, Y.Hu, B.M.Wanklyn ; Phys Rev B47, 11'374 (1993)

[91] J.E.Mooij ; in ≪ Percolation, Localization and Superconductivity ≫, ed. A.M.Goldman and S.A.Wolf (Plenum Press, New York, 1984) NATO ASI SERIES Vol. 8109, p. 325

[92] P.G. DeGennes ; Rev Mod Phys 36, 225 (1964)

[93] P.Minnhagen ; Phys Rev B 32, 7548 (1985)

[94] J.Holzer, R.S.Newrock, C.J.Lobb, T.Aouaroun, S.T.Herbert ; preprint (2001)

[95] A.L.Eichenberger, J.Affolter, M.Willemin, M.Mombelli, H.Beck, P. Martinoli, S.E.Korshunov; Phys Rev Lett 77, 3905 (1996)

[96] Fractal and Disordered Systems, ed. A.Bunde and S.Havlin (Springer-Verlag, Berlin, 1991)

[97] Md.Ashrafuzzaman, H.Beck ; unpublished

[98] J. Affolter ; Ph-Thesis, University of Neuchâtel (2001)

[99] S.E.Korshunov ; cond-mat/0007387, and references therein

[100] S.Candia, Ch.Leemann, S.Mouaziz, P.Martinoli ; cond-mat/0111212, to appear in Physica C

[101] O.Festin, P.Svendlindh, B.J.Kim, P.Minnhagen, R.Chakalov, Z.Ivanov ; Phys Rev Lett 83, 5567 (1999)

[102] C.T.Rogers, K.E.Myers, J.N.Eckstein, I.Bozovic ; Phys Rev Lett 69, 160 (1992)

[103] H.Beck ; Phys Rev B 49, 6153 (1994)

[104] S.E.Korshunov ; Phys Rev B 50, 13616 (1994)

[105] U.Eckern, A.Schmid ; Phys Rev B 39, 6441 (1989)

[106] U.Geigenmüller, C.J.Lobb, C.B.Whan ; Phys Rev B 47, 348 (1993)

[107] M.Capezzali, H.Beck, S.R.Shenoy ; Phys Rev Lett 78, 523 (1997)

[108] M.Capezzali, H.Beck ; unpublished

[109] E.Domani, M.Schick, R.Swendson ; Phys Rev Lett 52, 1535 (1984)

[110] B.J.Kim, P.Minnhagen ; Phys Rev B 60, 6834 (1999)

[111] R.F.Voss, J.Clark ; Phys Rev B 13, 556 (1976)

[112] I.J.Hwang, D.Stroud ; Phys Rev B 57, 6036 (1998)

[113] K.H.Wagenblast, R.Fazio ; cond-mat/9611177 ; JETP Lett (USA)68, 312 (1998), english translation of Pis'ma Zh Eksp Teor Fiz (Russia) 68, 291 (1998)

[114] V. I. Marconi and D. Dominguez, Phys. Rev. B 63, 174509 (2001). This article also hspace0.6cmcontains a rather exhaustive list of other references

[115] V. I. Marconi and D.Dominguez, Phys. Rev. Lett. 87, 017004 (2001).

[116] D.Dominguez ; Phys Rev Lett 82, 181 (1999)

[117] F.Pardo, F.de la Cruz, P.L.Gammel, E.Bucher, D.J.Bishop; Nature (London) 396, 348 (1998)

[118] A.B.Kolton, D.Dominguez, N.Gronbech-Jensen ; Phys Rev Lett 83, 3061 (1999)

[119] D.Dominguez, H.A.Cardeira ; Phys Lett A200, 43 (1995) ; Phys Rev B 52, 513 (1995) ; and other references quoted therein

[120] C.D.Chen, P.Delsing, D.B.Haviland, T.Claeson ; in ≪ Macroscopic Quantum Phenomena and Coherence in Superconducting Networks ≫, World Scientific, Singapore (1995), p. 121

[121] S.Shapiro ; Phys Rev Lett 11, 80 (1963)

[122] J.Mygind, N.F.Pedersen ; in ≪ Macroscopic Quantum Phenomena and Coherence in Superconducting Networks ≫, World Scientific, Singapore (1995), p. 339.This book also contains other references dealing with microwave radiation from JJAs.

SMALL-CAPACITANCE JOSEPHSON JUNCTIONS: ONE-DIMENSIONAL ARRAYS AND SINGLE JUNCTIONS

MICHIO WATANABE

Semiconductors Laboratory, RIKEN (The Institute of Physical and Chemical Research), 2-1 Hirosawa, Wako-shi, Saitama 351-0198, Japan

DAVID B. HAVILAND

Nanostructure Physics, The Royal Institute of Technology (KTH), SCFAB, Roslagstullsbacken 21, 106 91 Stockholm, Sweden

I INTRODUCTION

Small-capacitance superconducting tunnel junctions provide an ideal system for studying the interplay between the Josephson phase and the charge on the junction electrode, which are quantum mechanically conjugate to each other. This interplay can be probed through the superconductor-insulator (SI) transition [1], which is a quantum phase transition occurring at $T = 0$. The SI transition has been extensively studied in two-dimensional (2D) systems. Experiments have been carried out on granular [2–5] and homogeneous films [6, 7]. Theoretical studies have modeled the films as 2D arrays of small-capacitance Josephson junctions (JJs) [8–12], and experiments with such arrays have also been reported [13, 14]. In 1D, however, experimental study of the SI transition in JJ arrays has been less extensive, while the theoretical investigation has been done [1, 8, 15–20]. Experimental data on long and narrow films are available [21, 22]. In contrast to films, JJ arrays can be fabricated with a high degree of uniformity, and the parameters of interest in the theory can be measured. Furthermore, one can design a JJ array in such a way that one of the important parameter in the theory, the Josephson coupling energy E_J between adjacent islands can be tuned *in situ* [23, 24]. The other important parameters are the charging energy associated with the junction capacitance, $E_C \equiv e^2/2C$, and the stray capacitance of each island to the ground, $E_{C_0} \equiv e^2/2C_0$. Depending on the values of these parameters either superconducting or insulating behavior is expected for an array with infinite length.

In long arrays, it is possible to observe a well developed Coulomb blockade [25] for Cooper pairs in the current-voltage (I-V) characteristics, even when the Josephson energy is dominant, $E_J \geq E_C$. Such a Coulomb blockade is extremely interesting because our usual notions about phase coherence in the sense of the Josephson effect do not apply. The phase of the superconducting state is washed out by strong quantum fluctuations, and the number of Cooper pairs on the island becomes well defined. Nevertheless, the large Josephson coupling causes the potential associated with one excess Cooper pair to spread out, and in this sense the single excess Cooper pair becomes delocalized. In the Coulomb-blockade state, the single excess Cooper pair can be described as a charge soliton [26, 27], which is dual to the Josephson fluxon of 1D parallel arrays. The charge-soliton length, $\lambda_s = 2e/2\pi C_0 V_c$, gives the length scale over which the potential is screened. Here V_c is the critical voltage for Cooper-pair tunneling in a single junction, which is a function of the ratio E_J/E_C [28]. For $E_J \geq E_C$, V_c is reduced exponentially to zero as E_J increases. This weakening of the Coulomb blockade causes $\lambda_s \to \infty$, and we expect the insulating state of the array to eventually give way to superconductivity when $E_J \gg E_C$. In Sec. III, we describe experimental data which display this evolution of the insulating state as E_J is tuned *in situ.* We interpret the data qualitatively in terms of a theoretical model for a $T = 0$ quantum phase transition [1].

We have also used the 1D JJ arrays to bias a single Josephson junction in order to control the electromagnetic environment for the single junction [29]. In single junctions, experimental observation of Coulomb blockade has been considered to be extremely difficult because a high-impedance environment is necessary, and special care should be taken with the measurement leads [25]. For this reason, thin-film resistors [30] and tunnel-junction arrays [31, 32] were employed for the leads, and an increase of differential resistance around $V = 0$ was reported. In contrast to the earlier works [30–32], our leads are tunable, and we can therefore study the *same* single junction in different environments. We show that the I-V curve of the single junction is indeed sensitive to the state of the environment. Furthermore, we can induce a transition to a Coulomb blockade in the single junction when the zero-bias resistance of the JJ arrays is much higher than the quantum resistance $R_K \equiv h/e^2 \approx 26$ kΩ. In addition to Coulomb blockade, we have clearly observed a region of negative differential resistance in the I-V curve. The negative

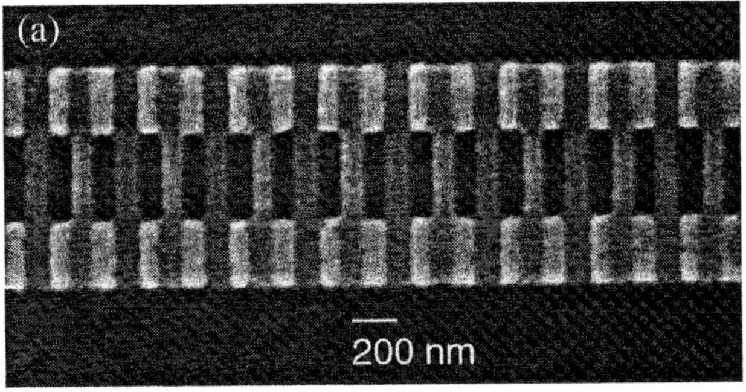

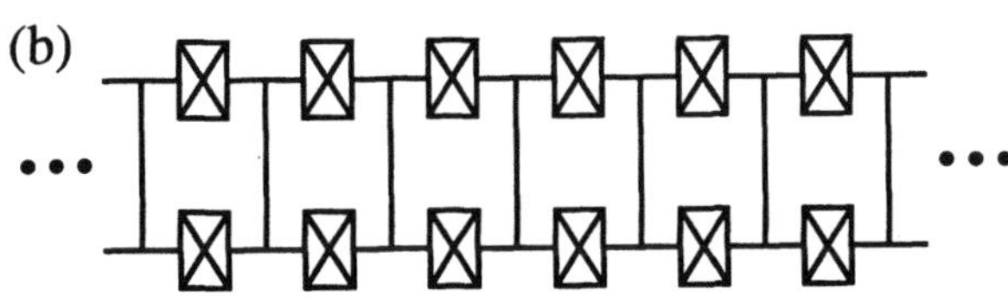

FIG. 1: One-dimensional array of small-capacitance dc SQUIDs. (a) Scanning electron micrograph. (b) Schematic diagram.

differential resistance appears as a result of coherent tunneling of single Cooper pairs according to the theory of current-biased single Josephson junctions [25,33]. Based on the theory, we have calculated the I-V curves numerically. The measured I-V is consistent with the numerical calculation.

II EXPERIMENT

A Sample fabrication and characterization

The 1D JJ arrays were fabricated on a SiO_2 substrate with electron-beam lithography and a double-angle-evaporation technique [34]. The arrays are made of Al with an Al_2O_3 tunnel barrier, and each of the Al electrodes in the array is connected to its neighbors by two junctions in parallel, thus forming a superconducting quantum interference device (SQUID) between nearest neighbors. Figure 1 shows a scanning electron micrograph of a section of an array and the schematic diagram. The advantage of the SQUID geometry is that we can control the effective E_J by applying an external magnetic field, B, perpendicular to the substrate,

$$E_J = E_{J0} \left| \cos\left(\pi \frac{BA_{\text{loop}}}{\Phi_0}\right) \right|, \tag{1}$$

where A_{loop} is the effective area of the SQUID loop and $\Phi_0 \equiv h/2e = 2 \times 10^{-15}$ Wb is the superconducting flux quantum. Because the geometrical inductance of the SQUID loop, $L_0 \ll \Phi_0/2\pi I_{c0}$, an external magnetic field creates a phase shift so that the critical current between nearest neighbors is modulated in a periodic way. Here,

$$E_{J0} \equiv \left(\frac{\Phi_0}{2\pi}\right) I_{c0} \tag{2}$$

and

$$I_{c0} \equiv \frac{\pi\Delta_0}{2eR_n} \tag{3}$$

is the Ambegaokar-Baratoff critical current [35], which is calculated from the superconducting energy gap, Δ_0 (= 0.2 meV for Al), and normal-state tunnel resistance of the junction, R_n. Henceforth, we will refer to the lumped SQUID as an effective junction with a tunable E_J, and a fixed charging energy $E_C \equiv e^2/2C$, where C is the sum (parallel combination) of two junction capacitances.

There are a couple of ways to obtain R_n. The resistance of the array divided by the number of junctions, N, measured above the superconducting transition temperature, T_c (= 1.2 K for Al), or in a magnetic field strong enough to completely suppress the superconductivity (> 0.1 T for Al), is R_n by definition. It is also possible to find R_n by taking the slope of the I-V curve at high bias, $V > N(2\Delta_0/e)$. The capacitance $C = c_s A$ is estimated from the junction area A, where the specific capacitance c_s is on the order of 10^2 fF/μm^2 as we will see later in Sec. IV.D. Another important parameter of the array is the capacitance of each electrode to the ground, $C_0 \sim 10$ aF [36], which depends on N logarithmically. In Sec. III, we will discuss three arrays with nominally identical junction parameters [$R_n = 4.9$ kΩ, $A = (0.4 \times 0.1\ \mu\text{m}^2) \times 2$, and $A_{\text{loop}} = 0.7 \times 0.2\ \mu\text{m}^2$], but having a different N: 255, 127, and 63.

The samples for Sec. IV.C and D are single Josephson junctions biased with the SQUID arrays. (See Fig. 2.) The single junction, which is in the center of Fig. 2a, has an area of $0.1 \times 0.1\ \mu\text{m}^2$. On each side of the single junction there are two leads enabling four-point measurements of the single junction. A part of each lead close to the single junction consists of the SQUID array with $A = (0.3 \times 0.1\ \mu\text{m}^2) \times 2$, and $A_{\text{loop}} = 0.7 \times 0.2\ \mu\text{m}^2$. The samples are characterized by R_n, N, and the normal-state tunnel resistance of the single junction, r_n.

B Low-temperature measurements

The I-V curves and the zero-temperature resistance were measured in a ^{3}He-^{4}He dilution refrigerator at 0.02 − 1 K. The temperature was determined by measuring the resistance of a ruthenium-oxide thermometer [37] fixed at the mixing chamber. Special care was taken to filter the sample from high-frequency electromagnetic radiation [34]. The preamplifier stage of our measurement scheme was specially designed for the high resistances associated with the Coulomb blockade.

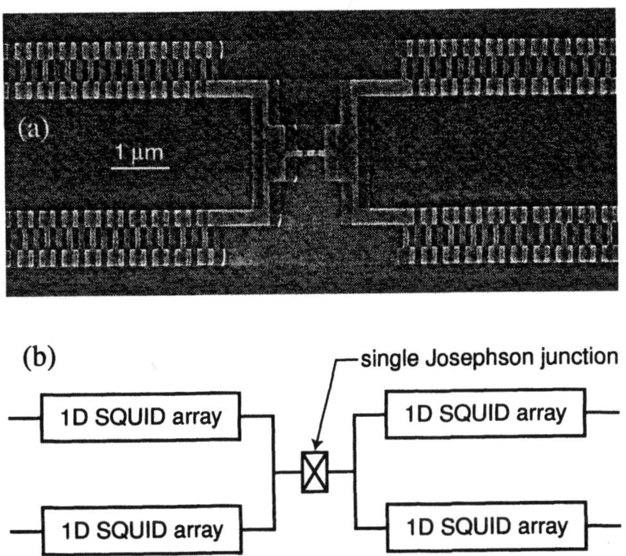

FIG. 2: single Josephson junction biased with arrays of small-capacitance dc SQUIDs. (a) Scanning electron micrograph. (b) Schematic diagram.

The temperature dependence of the zero-bias resistance (Fig. 4) was determined with lock-in technique at 13 Hz [23]. The power dissipation was kept at 10^{-16} W, which was just large enough to yield a detectable signal, and at the same time, small enough to probe the "linear" response.

The *I*-*V* curve of the single junction (Fig. 9) were measured in a four-point configuration, where the potential difference was measured through one pair of SQUID-array leads with a high-input-impedance instrumentation amplifier, and through the other pair of SQUID-array leads, the bias was applied and the current was measured with a current preamplifier [29]. When the voltage drop at the SQUID arrays was much larger than that at the single junction, the single junction was practically current biased. The SQUID arrays could be measured in a two-point configuration (same current and voltage leads) on the same side of the single junction. Note that the two arrays are connected in series and that current does not flow through the single junction.

III SUPERCONDUCTOR-INSULATOR TRANSITION IN ONE-DIMENSIONAL ARRAYS

A Current-voltage characteristics and the zero-bias resistance

Figure 3a shows the *I*-*V* curve of the three arrays at zero magnetic field. The arrays

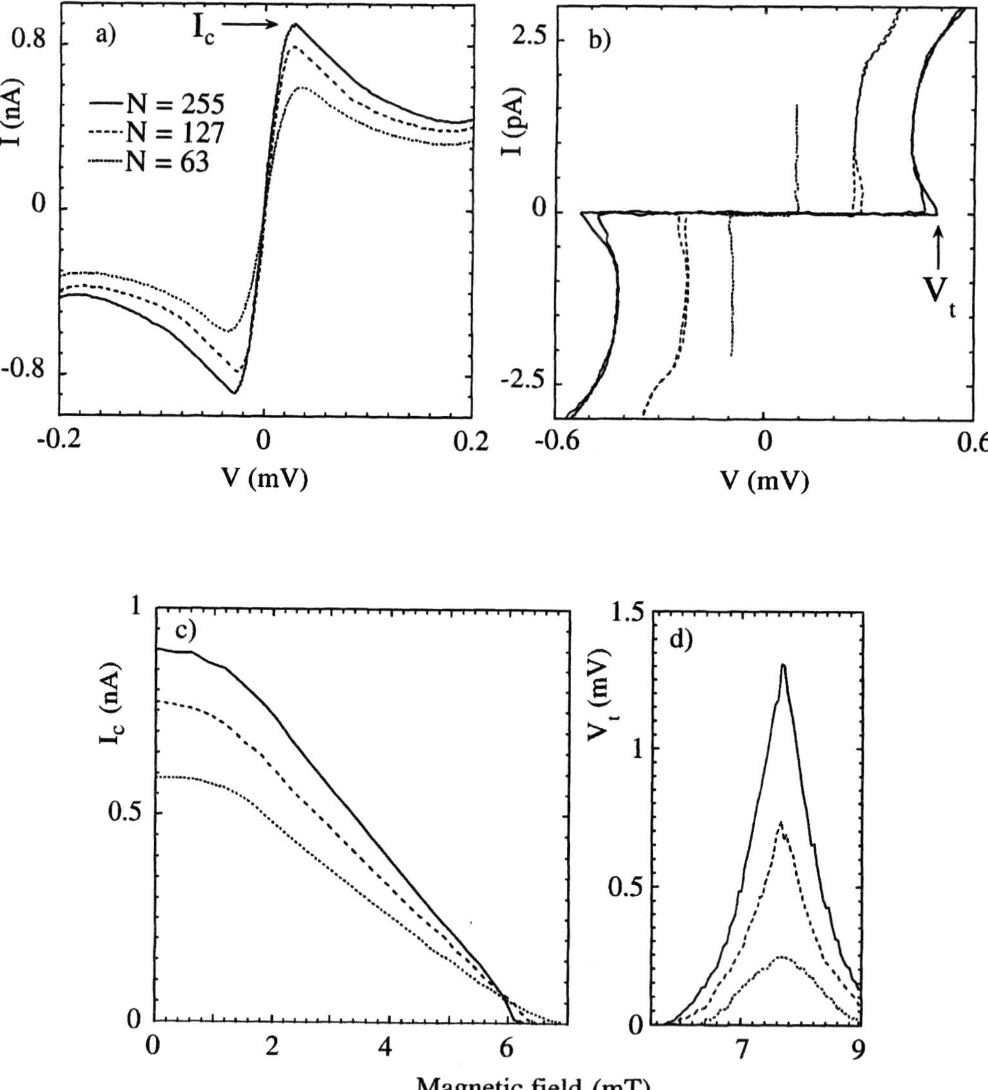

FIG. 3: Dependence of the I-V curves on the array length N, at $T = 0.05$ K. (a) The I-V curves at $B = 0$ showing Josephson-like behavior and the critical current I_c. (b) The I-V curves at $B = 7.1$ mT showing the Coulomb blockade of Cooper-pair tunneling and the threshold voltage V_t. (c) The magnetic field dependence of I_c. (d) The magnetic field dependence of V_t.

were made on the same chip, in the same vacuum cycle, using masks written to the same dimensions. Thus, all junctions in each array should be identical. The arrays are not truly superconducting, and there is actually a slope on the "zero-voltage" branch of the I-V curve, which gives a finite resistance. Furthermore, the observed "critical currents," i.e., the first local current maximum at ≈ 0.03 mV, are only 1% of the classical Ambegaokar-Baratoff value. This critical current shows a clear dependence on the array length. The longer the array, the lager the critical current, indicating that superconducting behavior is favored in the longer array. As E_J is suppressed below E_{J0} with an externally applied magnetic field, the measured critical current of each array is reduced, and the resistance on the "zero-voltage" branch increases. Figure 3c shows the magnetic-field dependence of the critical current. In the neighborhood of $B_c = 5.8$ mT, the curves in Fig. 3c cross one another, so that for $B > B_c$, the longer the array, the smaller the critical current.

Figure 3b shows the I-V curve of the three arrays at $B = 7.1$ mT ($>B_c$). Here we see a new type of behavior which is dual to the $B < B_c$ behavior. The I-V curve is characterized by a zero-current state for voltage below a threshold voltage, where the array switches to a finite current state. The magnetic-field dependence of the threshold voltage for $B > B_c$ is shown in Fig. 3d for the three arrays. We see that the longer the array, the larger the threshold voltage, indicating that insulating behavior is favored in the longer array.

Figure 4 shows the temperature dependence of the zero-bias resistance, $R_0(T)$, taken at the same magnetic fields (same E_J) for two arrays of different length: $N = 255$ (solid) and 63 (dashed). Each set of the curves shows qualitatively similar behavior. At zero magnetic field, as the temperature is lowered R_0 decreases to a value which is temperature independent. As the magnetic field is increased, the resistance of this "flat tail" increases, until it reaches a critical value, where $R_0(T)$ curves make a sharp turn to increasing resistance as $T \to 0$. Further increasing of the magnetic field drives the array into the insulating state, where R_0 increases rapidly as $T \to 0$.

If we examine the bottom two curves in each set of Fig. 4, we can see that at high temperatures, the 63-junction array has a smaller resistance than the 255-junction array, as expected for a classical resistor. However, at low temperatures, the resistance of the 63-junction array becomes *larger* than the 255-junction array. This increasing of the resistance for shorter arrays is a clear sign that quantum fluctuations are responsible for the measured resistance [1]. The open circles in Fig. 4 indicate the crossing points where R_0 is the same for two different lengths at the same magnetic field. This crossing point moves towards $T = 0$ as the magnetic field is tuned to the critical point K_0^*. At the critical point, the $T \to 0$ resistance due to quantum fluctuations, is independent of the array length.

We have seen a magnetic-field-tuned transition from Josephson-like behavior to Coulomb

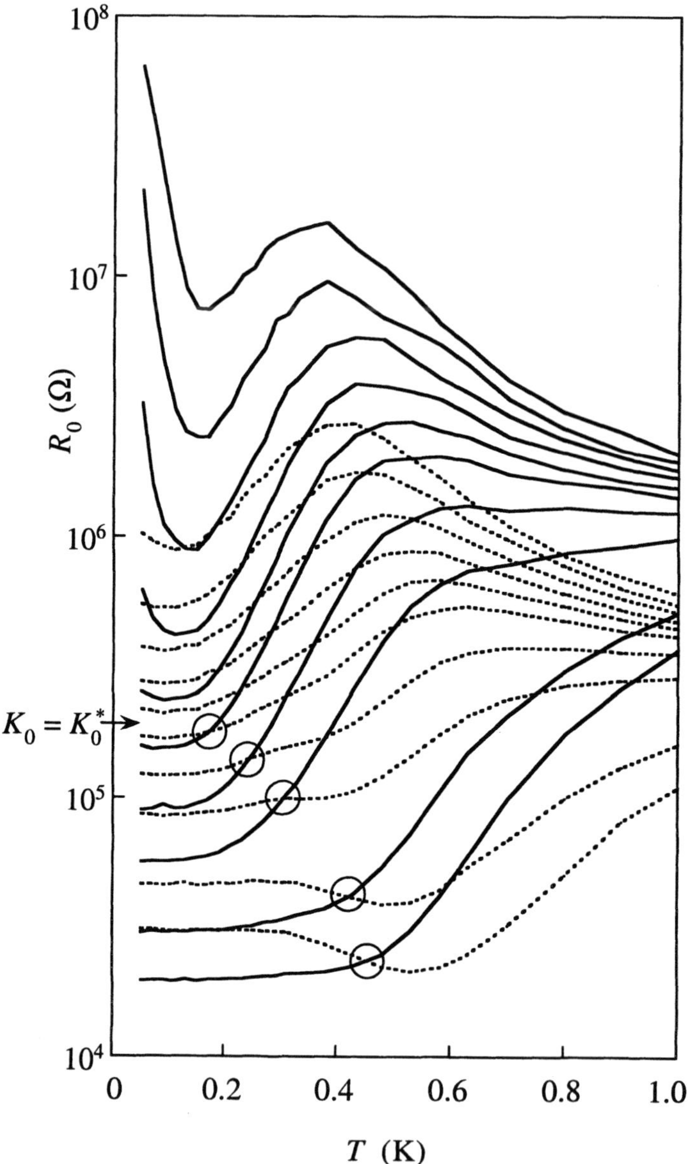

FIG. 4: Zero-bias resistance vs. temperature for two arrays having different number of junctions, N, but otherwise identical parameters. The set of solid curves are for $N = 255$ and the dashed curves are for $N = 63$, taken at the same magnetic fields between 0 and 7 mT. The open circles show where the measurements on the two arrays at the same magnetic field cross. At the magnetic field where $K_0 = K_0^*$, the two arrays have the same $T \to 0$ resistance, which is presumably independent of N.

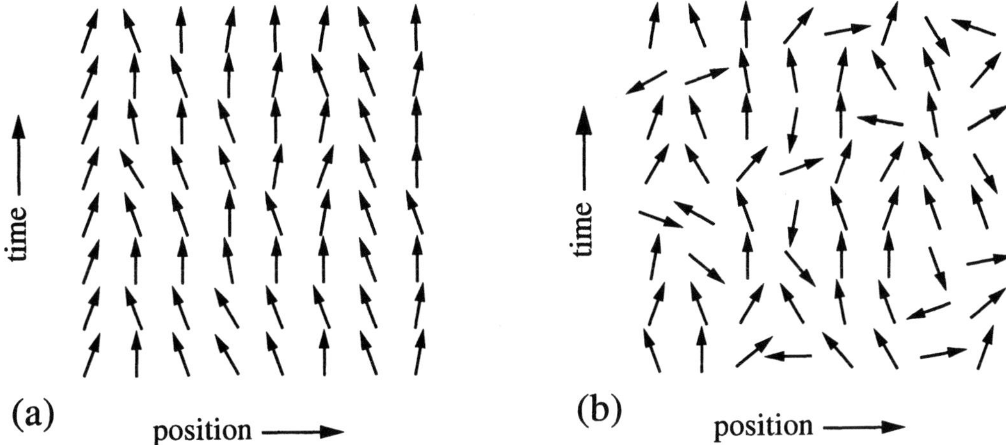

FIG. 5: Typical path or time history of a one-dimensional Josephson-junction array in (a) the superconducting phase and in (b) the insulating phase, respectively. The orientation of the arrows indicates the phase angles of the superconducting order parameter on the metallic elements connected by the Josephson junctions.

blockade, which may be called a superconductor-insulator transition, through the I-V curve (Fig. 3) and the temperature dependence of R_0 (Fig. 4). Moreover, the sharpness of the transition is strongly influenced by the length of the array. We can find qualitative explanation for this length dependence in a theoretical model of a quantum phase transition, which will be discussed in the following subsection.

B Mapping to the XY model

The SI transition can be described in an elegant theoretical framework as a quantum phase transition [1]. In these models, one can describe how a $T = 0$ property of a macroscopic quantum system with many degrees of freedom, will change as the complementary energies in the Hamiltonian of the system are adjusted. One can calculate the linear response, which in our case is the zero bias-resistance R_0, resulting from quantum fluctuations of the degrees of freedom. Within this framework, our 1D quantum system of Josephson junctions is mapped to the classical XY model of (1+1)D as sketched in Figs. 5a and 5b, the extra dimension being imaginary time, $i\hbar/k_BT$. Note that the role of temperature for the quantum system is to set the "size" of the system in the imaginary-time dimension. The (1+1)D classical XY model exhibits a Berzinski-Kosterlitz-Thouless phase transition [38, 39], from a disordered state (free vortices) to an ordered state (bound vortex pairs) as the strength of the dimensionless coupling constant, K_0 is increased. The quantum fluctuations of the phase of the superconducting wave function are thus described in

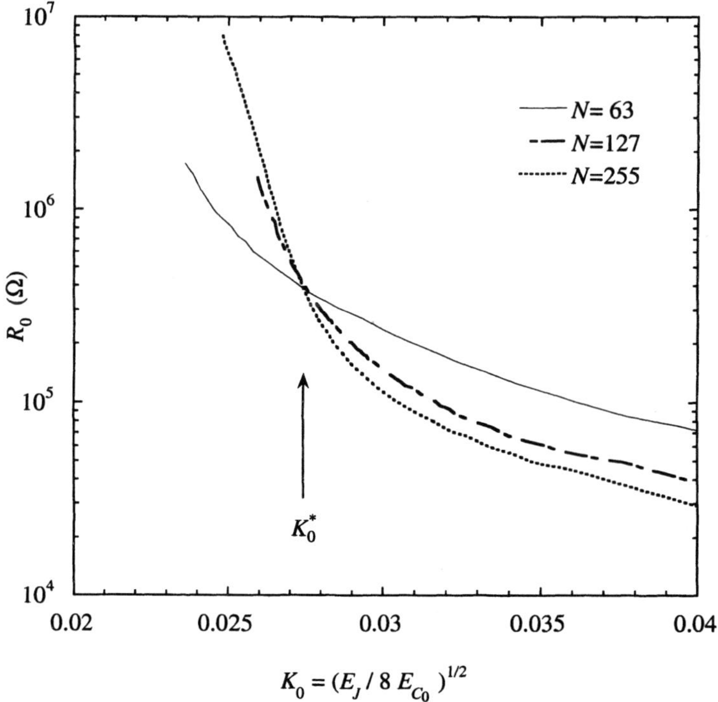

FIG. 6: Zero-bias resistance taken at $T = 0.05$ K as a function of the dimensionless coupling constant $K_0 = (E_J/8E_{C_0})^{1/2}$ for three arrays having different number of junctions.

terms of vortices, and in the insulating state (large quantum fluctuations) corresponds to the free-vortex state of the XY model [40]. The mapping to the isotropic XY model [15] can be done in the limit $C_0 \gg C$ which results in $K_0 = (E_J/8E_{C_0})^{1/2}$, where $E_{C_0} = e^2/2C_0$ is the charging energy associated with the stray capacitance of each electrode. Thus in this mapping, the junction capacitance, C is neglected.

Figure 6 shows a plot of R_0 measured at the lowest temperature, $T = 0.05$ K, vs. $K_0 = (E_J/8E_{C_0})^{1/2}$ for three arrays with $N = 255$, 127 and 63. We see in Fig. 6 that the three curves for different lengths cross at nearly the same point, $K_0^* = 0.027$. To the left of this crossing point we have the insulating state, where the resistance is larger for the longer arrays. To the right of this crossing point, we have the superconducting state, where the resistance is *smaller* for longer arrays. The arrays in the experiments have $C \gg C_0$, and thus we can not directly apply the theory. In our earlier work [23] we postulated that the effect of $C \gg C_0$ could be accounted for by "course graining" to the scale $\Lambda = (C/C_0)^{1/2}$, which would result in a coupling constant $J = (E_J/\Lambda E_C)^{1/2}$. Figure 4 of Ref. [23] shows that for this choice of the dimensionless coupling constant, the

R_0 vs. J curves do not cross at the same point, but in the region $J \in \{0.49, 0.55\}$. Choi *et al.* [20] have made a theoretical analysis of the role of a finite junction capacitance C by treating Λ as a small parameter. They found that the transition point should approach the limiting value $K_0^c = 2/\pi = 0.64$ when extrapolated to large Λ. The fact that the curves cross at one point in Fig. 6 would suggest that $K_0 = (E_J/8E_{C_0})^{1/2}$ is indeed the correct parameter for the transition. However, the experiment does not support the conclusions of Choi *et al.* in that the experimental critical point $K_0^* \approx 0.03 \ll K_0^c = 2/\pi$.

As we have seen earlier in Fig. 4, the resistance at $T > 0.6$ K is almost proportional to the array length, and in this sense, the arrays behave like classical 1D resistors. At lower temperatures, however, large deviations from this classical behavior occur. We can qualitatively understand these observations in the context of the (1 + 1)D XY model by considering the finite-size effect, which plays an important role in a real experiment. The zero-bias resistance of the array is determined by quantum fluctuations of the superconducting phase, which are described by the vortices in the (1+1)D XY model. As the temperature is lowered, the system becomes larger in the imaginary-time dimension, and at low enough temperatures, the system size in the real-space dimension, or the array length, determines the energy for free-vortex formation. The energy increases with increasing system size, and thus the probability of free-vortex formation is reduced in a longer array. This means that in a longer array, the superconducting state is favored. If we assume that the zero-bias resistance of the array is proportional to the probability of free-vortex formation in the isotropic XY model with the area of N^2 (i.e., N units in real space and N units in imaginary time), we obtain

$$R_0 \sim N^{2-\pi K_0} \tag{4}$$

by neglecting any renormalization effects [41]. For $K_0 > K_0^c = 2/\pi$, R_0 increases for decreasing N, as observed in Fig. 4 on the superconducting side of the transition.

IV COULOMB BLOCKADE IN SINGLE JUNCTIONS

A Theory for current-biased single Josephson junctions

The Hamiltonian of a single Josephson junction in an environment with sufficiently high impedance is written as

$$H = \frac{Q^2}{2C} - E_J \cos\phi, \tag{5}$$

where Q is the charge on the junction electrode, C is the capacitance of the junction, E_J is the Josephson energy, and ϕ is the Josephson-phase difference across the junction. The charge Q and $\hbar\phi/2e$ are quantum mechanically conjugate valuables, and a set of the eigenfunctions are Bloch waves of the form

$$\psi(\phi) = u(\phi)\exp(i\phi q/2e), \tag{6}$$

where q is called quasicharge and $u(\phi)$ is a periodic function,

$$u(\phi + 2\pi) = u(\phi). \tag{7}$$

The energy eigenvalue E plotted as a function of q has a band structure, and in all the allowed bands, it is $2e$ periodic. An example of the energy diagram for $E_J/E_C = 0.2$, where $E_C \equiv e^2/2C$ is the charging energy, is shown in Fig. 7a. Under constant current bias I_x, in the absence of quasiparticle or Cooper-pair tunneling, q increases uniformly in time according to

$$\frac{dq}{dt} = I_\mathrm{x} \tag{8}$$

so that the state of the system advances toward higher q within a given band as time goes on. The average voltage is given by

$$\langle V \rangle = \sum_{i_b, q} P(i_b, q) \frac{dE(i_b, q)}{dq}, \tag{9}$$

where i_b is the band index and $P(i_b, q)$ is the probability that the system is in the state (i_b, q). The probability $P(i_b, q)$ can be calculated by solving a set of coupled differential equations of the form

$$\frac{dP(i_b, q)}{dt} = \sum_{i'_b, q'} A(i_b, q, i'_b, q')\, P(i'_b, q') = 0\,, \tag{10}$$

where the matrix element $A(i_b, q, i'_b, q')$ describes the rate of transition between the states (i_b, q) and (i'_b, q'). The dominant process for $A(i_b, q, i'_b, q')$ depends on the magnitude of I_x.

An example of the theoretical I-V curve is shown in Fig. 7b. For sufficiently small I_x (region CB), the dominant process is stochastic quasiparticle tunneling, where q changes by e. This tunneling always occurs along the energy parabola $q^2/2C$, such that i_b changes by 0 or +1 if the initial state is in the lowest band ($i_b = 1$) and by ± 1 for all the other initial states ($i_b \geq 2$) [33]. The rate for the quasiparticle tunneling is given by

$$\Gamma(\Delta E) = \frac{\Delta E / e^2 R_\mathrm{qp}}{\exp(\Delta E / k_B T) - 1}, \tag{11}$$

where R_qp is the quasiparticle resistance and ΔE is the difference in energy between the initial (i_b, q) and final (i'_b, q') states,

$$\Delta E \equiv E(i'_b, q') - E(i_b, q). \tag{12}$$

At sufficiently low temperatures, the tunneling with $\Delta E > 0$ is extremely unfavorable, and the I-V curve is highly resistive (Coulomb blockade). For larger I_x (region BO in Fig. 7b),

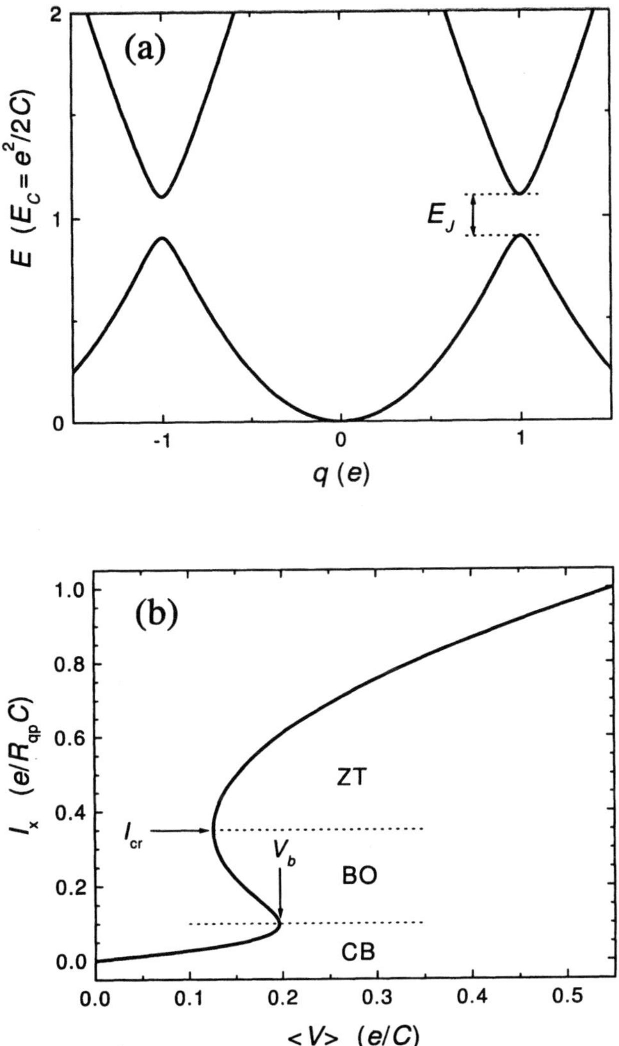

FIG. 7: (a) Energy diagram and (b) theoretical current-voltage characteristics for a single Josephson junction with $E_J/E_C = 0.2$ and $R_{\rm qp} = 200\,(h/\pi^2 e^2)$ at $k_BT/E_C = 0.2$, where E_J is the Josephson energy, $E_C \equiv e^2/2C$ is the charging energy, $R_{\rm qp}$ is the quasiparticle resistance, and k_BT is the thermal energy. (The energy diagram depends only on E_J/E_C.)

the quasicharge is frequently driven to the boundary of the Brillouin zone, $q = e$, then taken to $-e$ as a Cooper pair tunnels (Bloch oscillation). This process decreases $\langle V \rangle$, and as a result, the I-V curve has a region of negative differential resistance, or "back bending" in the low-current part. For still larger I_x (region ZT in Fig. 7b), Zener tunneling becomes important, and $\langle V \rangle$ increases again. In Zener tunneling, no quasicharge is transferred but the state of the system jumps from one band to another as it passes by the narrow gap between the bands. The probability of Zener tunneling from band i_b to $i_b + 1$ or vice versa is given by

$$P_Z = \exp\left[-\frac{\pi}{8}\frac{(\Delta E)^2}{i_b E_C}\frac{e}{\hbar I_x}\right]. \tag{13}$$

Following Ref. [42] which takes into account the above tunneling processes (quasiparticle, Cooper-pair, and Zener), we have calculate the I-V curve numerically. The parameters for the calculation are E_J/E_C, k_BT/E_C, and $\alpha \equiv h/\pi^2 e^2 R_{\rm qp}$. The current and the voltage are in units of $e/R_{\rm qp}C$ and e/C, respectively.

As we have seen in Fig. 7b, a typical I-V curve consists of three regions, so that it is characterized by the local voltage maximum, or blockade voltage V_b, and the local current minimum, or crossover current $I_{\rm cr}$. (Here, we have to mention that the back-bending feature is smeared out if k_BT/E_C or α is increased considerably.) Analytic expression of V_b and $I_{\rm cr}$ has been obtained theoretically for limiting cases [33]. The value of V_b is a function of E_J/E_C, and given by

$$V_b \approx \begin{cases} 0.25\, e/C & \text{for } E_J/E_C \ll 1, \\ \delta_0/e & \text{for } E_J/E_C \gg 1, \end{cases} \tag{14}$$

as $T \to 0$, where

$$\delta_0 = \frac{e^2}{C} 8 \left(\frac{1}{2\pi^2}\right)^{1/4} \left(\frac{E_J}{E_C}\right)^{3/4} \exp\left[-\left(8\frac{E_J}{E_C}\right)^{1/2}\right] \tag{15}$$

is the half width of the lowest energy band. As for $I_{\rm cr}$,

$$I_{\rm cr} \sim \left(I_Z \frac{e}{R_{\rm qp}C}\right)^{1/2} \tag{16}$$

is expected for $\alpha \ll (E_J/E_C)^2 \ll 1$ and $T \to 0$, where

$$I_Z \equiv \frac{\pi}{8}\frac{eE_J^2}{\hbar E_C} \tag{17}$$

is the Zener breakdown current. Note that $I_{\rm cr}$ is much smaller than I_Z. When we compare our experimental results with the theory, we need theoretical prediction for finite k_BT/E_C, and arbitrary E_J/E_C and α. For this reason we have done the numerical calculation. The measured V_b and $I_{\rm cr}$ will be compared with the calculation in Sec. IV.D.

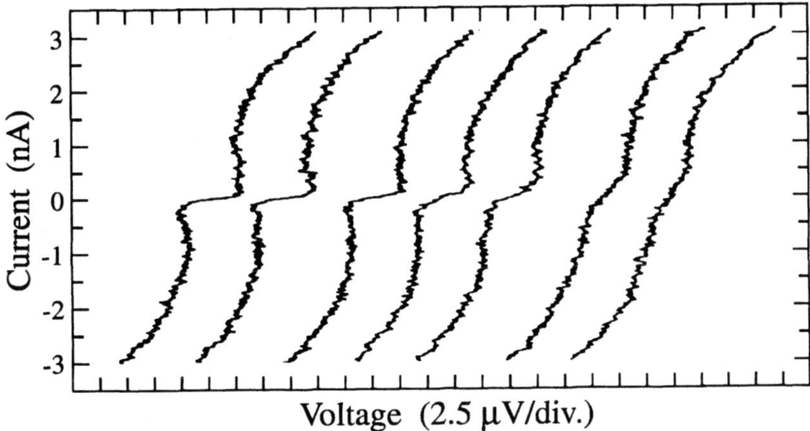

FIG. 8: Current-voltage characteristics of a single Josephson junction biased with thin-film resistors at several temperatures. From left to right, $T = 0.06$, 0.08, 0.11, 0.17, 0.23, 0.28, and 0.34 K, respectively. The origin of the voltage axis is displaced for each curve for clarity.

B Earier experiments on single junctions

In order to observe the Coulomb blockade in a single junction experimentally, the electromagnetic environment for the junction, or the measurement leads connected to the junction, should have a high impedance [25]. For this reason, thin-film resistors (NiCr alloy, AuPd alloy, Cr, and layered Ge/Pd) [30] and tunnel-junction ($Al/Al_2O_3/Al$) arrays [31, 32] were employed for the leads to bias a single junction ($Al/Al_2O_3/Al$).

Figure 8 shows the I-V curves of a single Josephson junction biased with thin-film resistors at several temperatures. A clear Coulomb blockade is seen at $T \leq 0.11$ K, and a "back bending" is also visible. In similar samples, high-frequency ($f = 0.4 - 10$ GHz) irradiation induced steps in the dc I-V curve at $I = \pm 2ef$ (not $\pm ef$), which can be explained as a phase locking of the externally applied signal to the Bloch oscillations [30].

Geerligs used 2D tunnel-junction arrays for the leads [31]. He claimed that in the normal state ($T > T_c$), a Coulomb blockade was visible in the single-junction I-V curve. In the superconducting state ($T < T_c$), however, neither the arrays nor the single junction develop clear charging effects, and the single junction showed a classic hysteretic I-V curve with an ordinary supercurrent. We believe that his arrays were in the superconducting side of the SI transition, and did not have high enough impedance below T_c.

Shimazu *et al.* biased a single junctions with 1D tunnel-junction arrays, and reported an increase of differential resistance around $V = 0$ in the normal state [32]. In the superconducting state, they measured the zero-bias resistance rather than the I-V curve. The zero-bias resistance in the superconducting state was higher than the normal-state resistance for some single junctions, which suggests the existence of a Coulomb blockade even in the superconducting state. However, this increase of the zero-bias resistance could also be due to simple quasiparticle tunneling, which is independent of any Coulomb blockade effects for Cooper pairs.

C Single Josephson junctions biased with SQUID arrays

We have employed 1D arrays of dc SQUIDs for the leads to bias a single Josephson junction [29]. The advantage of this SQUID configuration is that in contrast to the earlier experiments in Sec. IV.B, the impedance can be varied *in situ* by applying an external magnetic field at low temperatures. (See Fig. 4.) Thus, we can tune the electromagnetic environment for the single junction over a wide range. In Figs. 9 and 10, we show some results on a sample with $r_n = 17$ kΩ, $R_n = 1.4$ kΩ, and $N = 65$. The I-V curves for the single junction at several normalized magnetic fields, $\varphi \equiv BA_{\text{loop}}/\Phi_0$, are shown in Fig. 9. As φ is varied, the I-V curve develops a Coulomb blockade. We emphasize that the Josephson energy of the single junction is independent of φ, because it does not have a SQUID configuration and the field $\varphi\Phi_0/A_{\text{loop}} < 7$ mT applied here is much smaller than the critical field for Al films (≈ 0.1 T). The electromagnetic environment for the single junction (the SQUID array), however, is strongly varied with φ. The behavior of the single junction demonstrated in Fig. 9 does not result from the magnetic-field influencing the single-junction I-V curve, but rather from an environmental effect on the single junction. This experiment demonstrates in a direct way that the single-junction I-V curve is indeed sensitive to the electromagnetic environment.

The I-V curves of the two SQUID-array leads connected in series at $\varphi = 0.43$ ($R_0 =$ 0.61 MΩ), 0.46 ($R_0 = 3.2$ MΩ), and 0.49 ($R_0 = 43$ MΩ) are shown in Fig. 10. The I-V curves of the leads are nonlinear, and in general the SQUID array cannot be described by a liner impedance model [24]. However, we may characterize the environment by their R_0. Coulomb blockade is visible only when $R_0 \gg R_K$, which is consistent with the theoretical conditions for the clear observation of Coulomb blockade in single junctions [43]. For an arbitrary linear environment characterized by $Z_e(\omega)$, $\text{Re}[Z_e(\omega)] \gg R_K$ is required for the Coulomb blockade of quasiparticle tunneling and $\text{Re}[Z_e(\omega)] \gg R_K/4$ for that of Cooper-pair tunneling [43]. It is interesting to note that at $\varphi = 0.46$ (labeled "b"), the I-V curve of the leads is still "Josephson-like" (differential resistance is lower around $V = 0$), while that of the single junction is already "Coulomb-blockade-like". This feature becomes more distinct in samples with larger N [44].

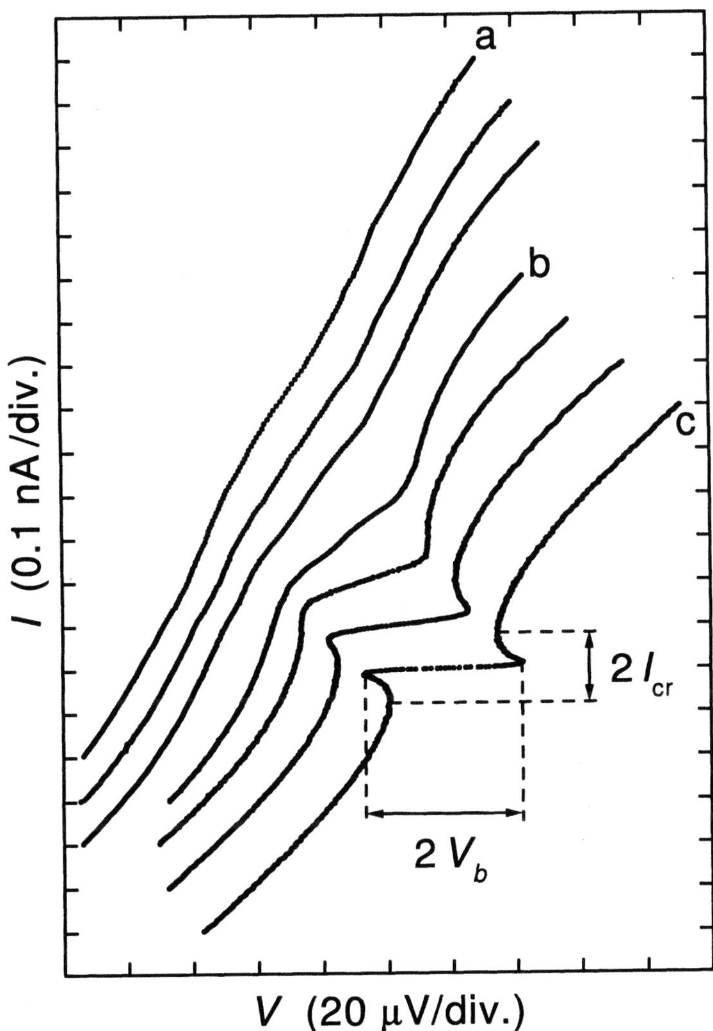

FIG. 9: Current-voltage (I-V) curves of the single junction at $T = 0.02$ K for a sample with $r_n = 17$ kΩ, $R_n = 1.4$ kΩ, and $N = 65$. From top left to bottom right, the normalized magnetic field $\varphi \equiv BA_{\rm loop}/\Phi_0$ is increased from 0.43 to 0.49 in steps of 0.01. The origin of each curve is offset for clarity. For the labeled curves, the I-V characteristics of the leads at the same φ are shown in Fig. 10

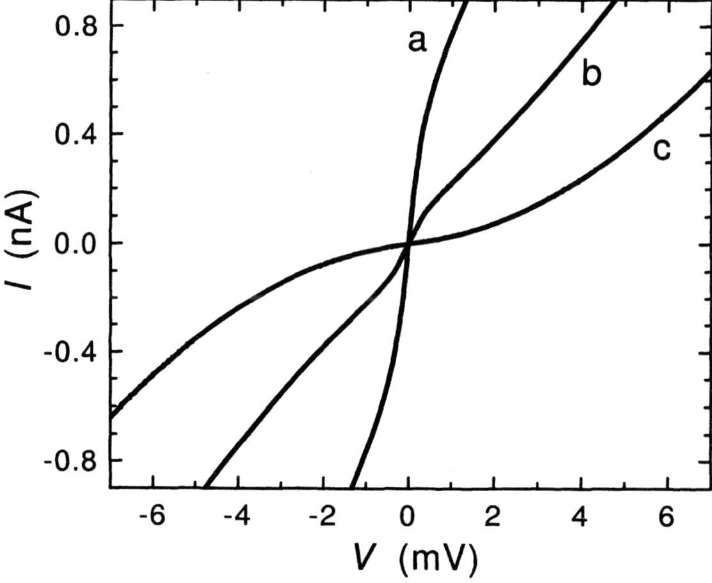

FIG. 10: Current-voltage curves of the two SQUID-array leads connected in series at $T = 0.02$ K for the same sample as in Fig. 9. From a to c, $\varphi \equiv BA_{\text{loop}}/\Phi_0$ is 0.43, 0.46, and 0.49, respectively.

D Comparison with the numerical calculation

The region of negative differential resistance seen in Fig. 9 when Coulomb blockade is well developed, is related to coherent tunneling of single Cooper pairs according to the theory [25,33] of a current-biased single Josephson junction in an environment with sufficiently high impedance. Following Ref. [42], we have calculated the blockade voltage V_b numerically as a function of E_J/E_C [45]. The measured V_b (See Fig. 9) for the samples having nominal junction area of 0.1×0.1 μm^2 is compared with the numerical calculation in Fig. 11. The boxes and circles represent the samples biased with thin-film resistors (Sec. IV.B) and with SQUID arrays (Sec. IV.C), respectively. For the sample shown in Fig. 9, we used the data at $\varphi = 0.49$ (curve c) in order to obtain V_b. At $\varphi = 0.49$ the voltage drop at the SQUID arrays is 10^2 times larger than that at the single junction, and the single junction is therefore considered to be current biased. Compare the voltage scale of Figs. 9 and 10. We calculated E_J from r_n, $E_J = h\Delta_0/8e^2r_n$. For E_C, we employed $c_s = 130$ fF/μm^2, and with this value the experimental data, especially those for the samples biased with SQUID arrays (the circles in Fig. 11), agree with the numerical calculation.

Actually, a smaller value, $c_s = 45\pm5$ fF/μm^2 [46], which was obtained for the junctions with 3×28 μm^2 and 7×54 μm^2, has been frequently employed [23,24,30,32,36]. Our

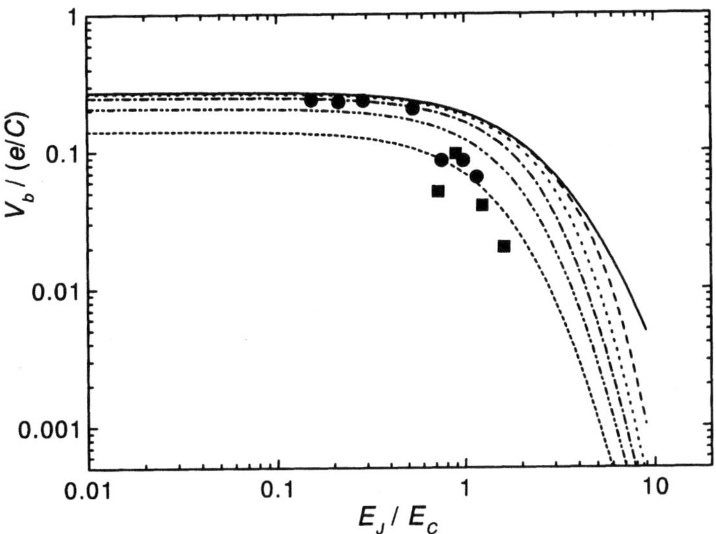

FIG. 11: Blockade voltage V_b divided by e/C as a function of E_J/E_C. From top to bottom, the curves represent the numerical calculations for normalized temperatures $k_BT/E_C = 0$, 0.02, 0.05, 0.1, 0.2, and 0.5, respectively.

apparently large c_s may be partly explained by distributed capacitance of the SQUID arrays or by residual environmental effects. We also note that the uncertainty in c_s seems to be large when the junction area is on the order of 0.01 μm^2 or smaller. For example, Fulton and Dolan measured samples with three junctions that share a common electrode, and obtained 0.20 − 0.23 fF for $(0.03 \pm 0.01\ \mu\text{m})^2 \times 3$ [47], i.e., $c_s = 42 - 192$ fF/μm^2. Geerligs *et al.* reported $c_s \approx 110$ fF/μm^2 for two-dimensional (190 × 60) junction arrays with the areas of 0.01 or 0.04 μm^2 [13]. More recently, Penttilä *et al.* studied resistively shunted single Josephson junctions with the area of 0.15 × 0.15 μm^2 [48]. The estimated C of their eight samples ranged between 0.8 and 6.6 fF, or $c_s = 36 - 293$ fF/μm^2.

When C is determined, it is possible to estimate R_{qp} of the samples from I_{cr}. We plot the measured I_{cr} for the single junctions biased with SQUID arrays (Sec. IV.C) as a function of E_J/E_C together with some theoretical curves based on our numerical calculation in Fig. 12. We obtain $R_{\text{qp}} = 10^0 - 10^1$ MΩ, or $R_{\text{qp}}/r_n = 10^2 - 10^3$.

V CONCLUSIONS

One-dimensional (1D) arrays of small-capacitance SQUIDs undergo a sharp transition, from Josephson-like behavior to the Coulomb blockade of Cooper-pair tunneling, as the effective Josephson coupling between nearest neighbors is tuned with an externally ap-

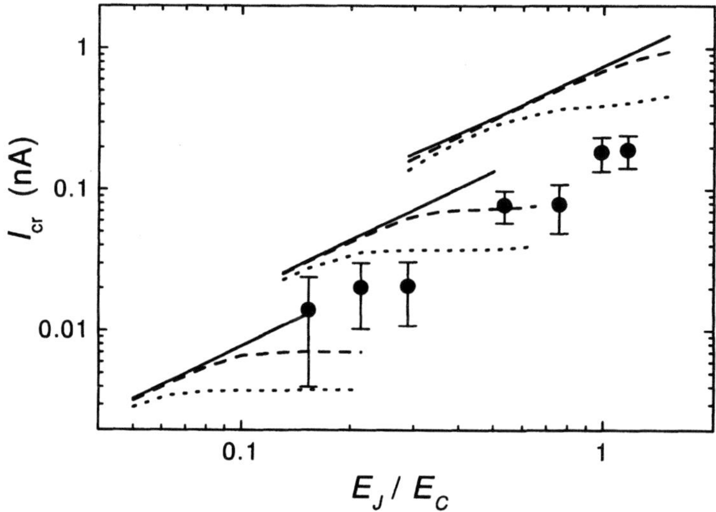

FIG. 12: Crossover current I_{cr} vs. E_J/E_C. The curves represent the numerical calculations for C = 1.3 fF, and from top to bottom $(\alpha \equiv h/\pi^2e^2R_{qp}, k_BT/E_C)$ = $(10^{-2}, 0)$, $(10^{-2}, 0.3)$, $(10^{-2}, 0.5)$, $(10^{-3}, 0)$, $(10^{-3}, 0.3)$, $(10^{-3}, 0.5)$, $(10^{-4}, 0)$, $(10^{-4}, 0.3)$, and $(10^{-4}, 0.5)$, respectively.

plied magnetic field. We have shown how length scaling of the zero-bias resistance of the array can be used to probe the superconductor-insulator quantum phase transition. The observed non-classical dependence of the zero-bias resistance on the length of the array, where the zero-bias resistance decreases with increasing length, can be supported qualitatively with a theoretical model which maps a 1D quantum system to the $(1+1)$D classical XY model.

We have also used the SQUID arrays as a tunable electromagnetic environment for a single small-capacitance Josephson junction, and demonstrated how the Coulomb blockade of Cooper-pair tunneling is induced in the single junction. When the Coulomb blockade is well developed, the measured current-voltage curve is consistent with the numerical calculation for a current-biased single Josephson junction.

ACKNOWLEDGMENTS

We are grateful to R. L. Kautz for great help in the numerical calculation, and to T. Kato and F. W. J. Hekking for fruitful discussions. This work was supported by Swedish NFR, and Special Postdoctoral Researchers Program and President's Special Research Grant of RIKEN. The samples for Sec. III were fabricated at the Swedish Nanometer Laboratory. M. W. would like to thank the Japan Society for the Promotion of Science (JSPS) and the Swedish Institute (SI) for financial support.

REFERENCES

[1] S. L. Sondhi, S. M. Girvin, J. P. Carini, and D. Shahar, Rev. Mod. Phys. **69**, 315 (1997).

[2] H. M. Jaeger, D. B. Haviland, B. G. Orr, and A. M. Goldman, Phys. Rev. B **40**, 182 (1989).

[3] A. F. Hebard and M. A. Paalanen, Phys. Rev. Lett. **65**, 927 (1990).

[4] R. P. Barber, Jr. and R. C. Dynes, Phys. Rev. B **48**, 10618 (1993).

[5] M. Watanabe, H. Shimada, S. Kobayashi, and Y. Ootuka, J. Phys. Soc. Jpn. **66**, 1419 (1997).

[6] Y. Liu, D. B. Haviland, B. Nease, and A. M. Goldman, Phys. Rev. B **47**, 5931 (1993).

[7] J. M. Valles, Jr., R. C. Dynes, and J. P. Garno, Phys. Rev. Lett. **69**, 3567 (1992).

[8] M.-C. Cha, M. P. A. Fisher, S. M. Girvin, M. Wallin, and A. P. Young, Phys. Rev. B **44**, 6883 (1991).

[9] A. P. Kampf and G. T. Zimanyi, Phys. Rev. B **47**, 279 (1993).

[10] E. Šimánek, Phys. Lett. A **177**, 367 (1993).

[11] A. van Otterlo, K.-H. Wagenblast, R. Fazio, and G. Schön, Phys. Rev. B **48**, 3316 (1993).

[12] M. Wallin, E. S. Sørensen, S. M. Girvin, and A. P. Young, Phys. Rev. B **49**, 12115 (1994).

[13] L. J. Geerligs, M. Peters, L. E. M. de Groot, A. Verbruggen, and J. E. Mooij, Phys. Rev. Lett. **63**, 326 (1989).

[14] C. D. Chen, P. Delsing, D. B. Haviland, Y. Harada, and T. Claeson, Phys. Rev. B **51**, 15645 (1995).

[15] R. M. Bradley and S. Doniach, Phys. Rev. B **30**, 1138 (1984).

[16] S. E. Korshunov, Euro. Phys. Lett. **9**, 107 (1989).

[17] P. A. Bobbert, R. Fazio, G. Schön, and A. D. Zaikin, Phys. Rev. B **45**, 2294 (1992).

[18] A. A. Odintsov, Phys. Rev. B **54**, 1228 (1996).

[19] L. I. Glazman and A. I. Larkin, Phys. Rev. Lett. **79**, 3736 (1997).

[20] M.-S. Choi, J. Yi, M. Y. Choi, J. Choi, and S.-I. Lee, Phys. Rev. B. **57**, R716 (1998).

[21] A. V. Herzog, P. Xiong, F. Sharifi, and R. C. Dynes, Phys. Rev. Lett. **76**, 668 (1996).

[22] F. Sharifi, A. V. Herzog, and R. C. Dynes, Phys. Rev. Lett. **71**, 428 (1993).

[23] E. Chow, P. Delsing, and D. B. Haviland, Phys. Rev. Lett **81**, 204 (1998).

[24] D. B. Haviland, K. Andersson, and P. Ågren, J. Low Temp. Phys. **118**, 733 (2000).

[25] D. V. Averin and K. K. Likharev, in *Mesoscopic Phenomena in Solids*, edited by B. L. Altshuler, P. A. Lee and R. A. Webb (Elsevier Science Publishers B. V., Amsterdam, 1991), chap. 6.

[26] E. Ben-Jacob, K. Mullen, and M. Amman, Phys. Lett. A **135**, 390 (1989).

[27] D. B. Haviland and P. Delsing, Phys. Rev. B **54**, R6857 (1996).

[28] K. K. Likharev and A. B. Zorin, J. Low Temp. Phys. **59**, 347 (1985).

[29] M. Watanabe and D. B. Haviland, Phys. Rev. Lett. **86**, 5120 (2001).

[30] D. B. Haviland, L. S. Kuzmin, P. Delsing, K. K. Likharev, and T. Claeson, Z. Phys. B **85**, 339 (1991).

[31] L. J. Geerligs, Ph.D thesis, Delft University of Technology (1990).

[32] Y. Shimazu, T. Yamagata, S. Ikehata, and S. Kobayashi, J. Phys. Soc. Jpn. **66**, 1409 (1997).

[33] G. Schön and A. D. Zaikin, Phys. Reports **198**, 237 (1990).

[34] D. B. Haviland, S. H. M. Persson, P. Delsing, and C. D. Chen, J. Vac. Sci. Technol. A **14**, 1839 (1996).

[35] V. Ambegaokar and A. Baratoff, Phys. Rev. Lett. **10**, 486 (1963); **11**, 104(E) (1963).

[36] D. B. Haviland, K. Andersson, P. Ågren, J. Johansson, V. Schöllmann, and M. Watanabe, Physica C **352**, 55 (2001).

[37] M. Watanabe, M. Morishita, and Y. Ootuka, Cryogenics **41**, 143 (2001).

[38] V. L. Berezinskii, Zh. Eksp. Teor. Fiz. **61**, 1144 (1971) [Sov. Phys. JETP **34**, 610 (1972)].

[39] J. M. Kosterlitz and D. J. Thouless, J. Phys. C **6**, 1181 (1973).

[40] G. Falci, R. Fazio, G. Schön, and A. Tagliacozzo, in *Macroscopic Quantum Phenomena and Coherence in Superconducting Networks*, edited by C. Giovanella and M. Tinkham (World Scientific, Singapore, 1996), p. 59.

[41] S. M. Girvin (private communication).

[42] U. Geigenmüller and G. Schön, Physica B **152**, 186 (1988).

[43] G.-L. Ingold and Y. V. Nazarov, in *Single Charge Tunneling*, edited by H. Grabert and M. H. Devoret (Plenum Press, New York, 1992), chap. 2.

[44] M. Watanabe and D. B. Haviland, J. Phys. Chem. Solids **63**, in press.

[45] M. Watanabe, D. B. Haviland, and R. L. Kautz, Supercond. Sci. Technol. **14**, 870 (2001).

[46] A. W. Lichtenberger, C. P. McClay, R. J. Mattauch, M. J. Feldman, S.-K. Pan, and A. R. Kerr, IEEE Trans. Magn. **25**, 1247 (1989).

[47] T. A. Fulton and G. J. Dolan, Phys. Rev. Lett. **59**, 109 (1987).

[48] J. S. Penttilä, Ü. Parts, P. J. Hakonen, M. A. Paalanen, and E. B. Sonin, Phys. Rev. Lett. **82**, 1004 (1999).

Barrier structure and electrical properties of interface-modified Josephson junctions

Y. Ishimaru , Y. Wu , M. Horibe , H. Wakana , S. Adachi , Y. Tarutani, and K. Tanabe

Superconductivity Research Laboratory, International Superconductivity Technology Center 1-10-13 Shinonome , Koto-ku, Tokyo, 135-0062 Japan

1. INTRODUCTION

To construct single flux quantum (SFQ) devices that employ high-Tc superconductors, ramp-edge type Josephson junctions have been most extensively studied. Recently, ramp-edge Josephson junctions using $YBa_2Cu_3O_{7-\delta}$(YBCO) with a barrier layer called an interface-engineered, interface-modified or surface-modified barrier have attracted considerable attention, because the junctions exhibit a high I_cR_n product and a smaller Ic spread than those for the junctions with a deposited artificial barrier layer. These junctions with different names are fabricated in almost the same way where the barriers are formed from an amorphous layer made by an ion bombardment process. There are some early studies of the junctions fabricated by using an ion bombardment process without artificial barrier layer deposition[1, 2]. Moeckly *et al.* first introduced the name of interface-engineered junction and reported remarkable results. Their junctions exhibited I_cR_n products at 20K of 0.5 ~ 3mV and the 1σ spread in Ic for 10 junctions lower than 8% [3, 4]. After this report, many researchers have studied this type of junctions[5~11]. At this stage, the outstanding performance of these junctions became obvious. Satoh *et al.* reported that the I_cR_n product at 4.2K was 1.5 ~ 2.5mV and the 1σ spread in Ic for 100 junctions was as low as 8% [5]. Soutome *et al.* reported the I_cR_n product at 4.2K of 2.2mV and the 1σ spread in Ic for 200 junctions of 10% [9]. These values, especially the 1σ spread in Ic of junction arrays, looked promising to many researchers studying oxide high-Tc Josephson junctions, because many researchers had been trying hard to decrease the 1σ spread value. Recently, frequent attempts to improve the 1σ spread in Ic have been undertaken. Soutome *et al.* have reported that the 1σ spread in Ic for 100 junctions can be reduced to 6.6% [12], and Wakana *et al.* also achieved the 1σ spread in Ic for 100 junctions as low as 6.5% [13]. Layered-type interface-engineered junctions have also been studied[14~17]

In this chapter, we call this type of junctions and barrier as interface-modified junctions and interface-modified barrier. As already mentioned, the interface-modified barrier is fabricated from the amorphous layer that is made by an ion bombardment process[18~19]. RF plasma discharge, Ar ion milling or electron cyclotron resonance (ECR) ion etching are usually used

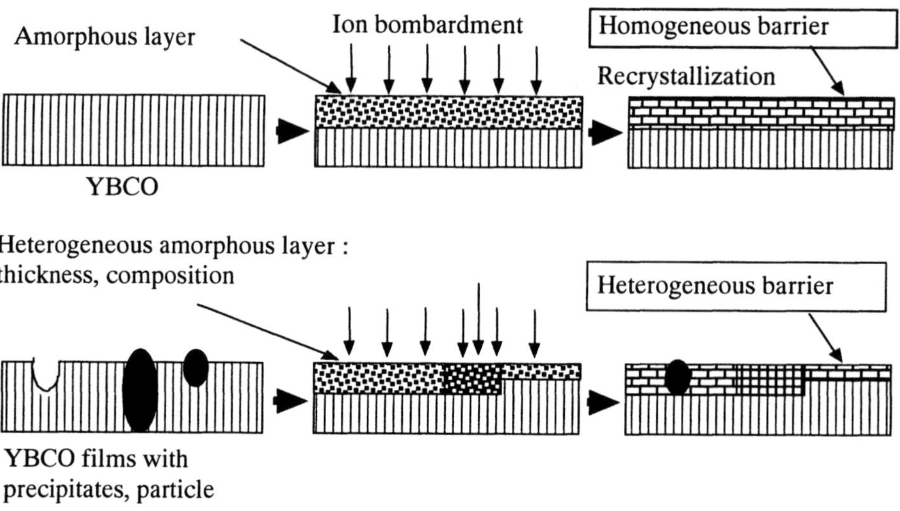

Fig. 1. Several courses of forming ununiform barrier structure.

for making the amorphous layer. This amorphous layer changes to a barrier through appropriate annealing and upper YBCO deposition processes. If the ion bombardment and recrystallization processes are done uniformly, formation of homogeneous barrier which can be expected as shown in Fig.1. However, actually, there are defects such as precipitates and particles in films, and there is spatial ununiformity in ion bombardment and recrystallization processes, that means ununiformity in ion energy and density distribution or spatial distribution of heating temperature. Thus, the actual interface-modified barrier becomes heterogeneous and is also shown in Fig.1. This heterogeneous barrier causes deterioration of 1σ spread in I_c of junction arrays. Although the electrical properties of this type of Josephson junctions have a strong dependence on the barrier structure or barrier recrystallization process, there has been controversy about the barrier structure. Wen *et al.* reported that the barrier on a real ramp edge surface is Ba-based cubic perovskite with an average lattice constant of 4.1Å, and the difference in the I-V characteristics between RSJ-type and flux-flow-type is caused by the different ratio of the cubic barrier region and the region showing a direct CuO_2 contact [19]. Moeckly *et al.* reported another type of face-centered cubic (fcc) barrier with a lattice constant of 5.1Å[4, 20]. From reflection high energy electron diffraction (RHEED) observation of the interface-modified barrier, existence of fcc-cubic Y_2O_3 with a lattice constant of 10.5Å was observed[10]. On the other hand, absence of a cubic layer at the interface region in real ramp edge junctions has been reported[21, 22]. Therefore, it is considered that there are several-types of interface-modified barriers and the barrier structures have a strong dependence on the recrystallization process which is influenced by , for example, the ion-bombardment conditions (ion acceleration voltage , ion species) and annealing conditions (temperature, time).

In this chapter, we report the results on observation of RHEED patterns in the recrystallization process of the amorphous layers made by ion bombardment. The variation of the RHEED patterns with the ion-bombardment and annealing conditions is investigated. Several types of barrier structures are observed by using cross-sectional transmission electron microscopy (TEM). The relation between the observed RHEED patterns and the barrier structures revealed by TEM is discussed. Moreover, we fabricated the interface-modified ramp-edge Josephson junctions by pulsed laser deposition (PLD). There are several important factors which deteriorate the 1σ spread values of junction arrays. We will report several factors including the uniformity of ion beam for making an amorphous layer, and the base-electrode film quality such as the density of outgrowth or particles and the roughness of the ramp surface. We will also show that the recrystallization of an interface-modified barrier from an amorphous layer strongly depends on the kinetic energy from laser plume plasma or deposited particles as well as the thermal energy from substrate heating. Decreasing the shadowing effect by laser plume can make junction arrays with a low 1σ spread. Finally, we will report TEM observation results for real ramp-edge junctions and the relation between barrier structure and I-V characteristics of the junctions.

2. RHEED and TEM observation

2.1 Sample preparation

For the purpose of RHEED observation of the interface-modified barrier recrystallization process, we used a-axis oriented YBCO films. Fig.2 shows the spatial relation between the

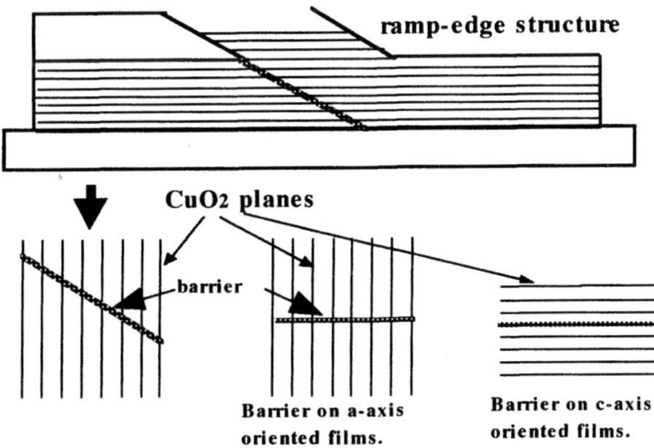

Fig. 2. Spatial relations between the barrier and CuO_2 planes of two-sided YBCO in ramp-edge junctions, and in junctions fabricated using a-axis and c-axis oriented YBCO films.

barrier and CuO_2 planes of YBCO on both sides in ramp-edge junctions. The barrier crosses to the CuO_2 planes with a typical angle of 20~30° . However, since RHEED observation of real ramp-edge surface is impossible, we have to choose a-axis or c-axis oriented YBCO films. We chose a-axis films, because the growth orientation of the barrier in ramp-edge junctions is similar to that on a-axis films rather than c-axis films.

a-Axis oriented YBCO films were deposited on $SrTiO_3$(=STO)(100) substrate by PLD. A KrF excimer laser (λ=248nm) with a power density of about 2J/cm^2 on a stoichiometric high-density target and a repetition rate of 5Hz was used. About 200nm thick films were grown in an oxygen pressure of 200mTorr at a substrate temperature of 580℃. The deposition, recrystallization process and all the RHEED observations were done in a multi-chamber vacuum system without exposuring the samples to air as shown in Fig. 3.

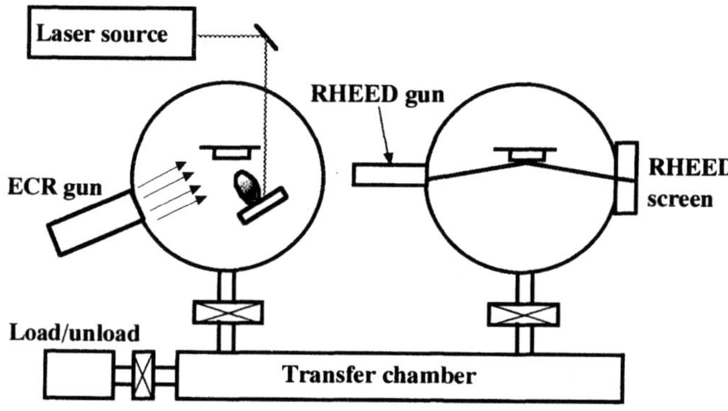

Fig. 3. Schematic illustration of experimental equipments used for barrier recrystallization process observation.

Fig. 4 illustrates the flow chart of the RHEED pattern observation process. All of the RHEED patterns in this paper were observed toward the direction where the electron beam was parallel to the [100] of STO substrates. First, the RHEED patterns of a-axis oriented YBCO films were observed at room temperature. After this observation, ion bombardment was applied to the YBCO film surfaces at room temperature using an ECR type ion source with substrate rotation and an incident angle of 30° from the direction parallel to the film surface. The ion acceleration voltages ranging from 200 to 900V were examined and pure Ar or Ar and O_2 mixed gas was used as a process gas. Then, the second observation of RHEED patterns was carried out. The samples were annealed to make a interface-modified barrier from the layer damaged by the ECR ion bombardment. The substrate temperature was linearly increased from room temperature to 660℃ or 700℃ in 30 minutes in 200mTorr oxygen, maintained at these temperature for 30 minutes in the same atmosphere, then , decreased to room temperature. The third RHEED observation was carried out after the annealing process was finished. In addition, to observe the initial

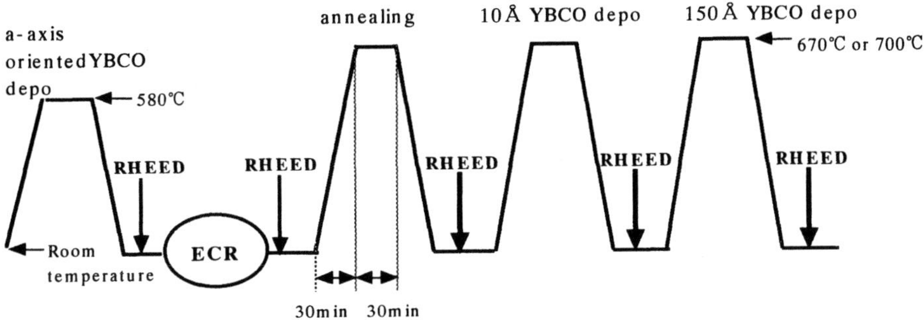

Fig. 4. Flow chart of RHEED pattern observation process.

process of YBCO film growth on the recrystallized surface, the fourth and fifth observations were done at room temperature after depositing 10Å and 150Å thick YBCO. The substrate temperature and the rise time were the same as in the annealing process. After RHEED observation , barrier structures of some typical samples were observed by cross-sectional TEM. Fig.5 shows the X-ray diffraction patterns of the used films showing only (h00) peaks of YBCO which confirm that they are purely a-axis oriented films.

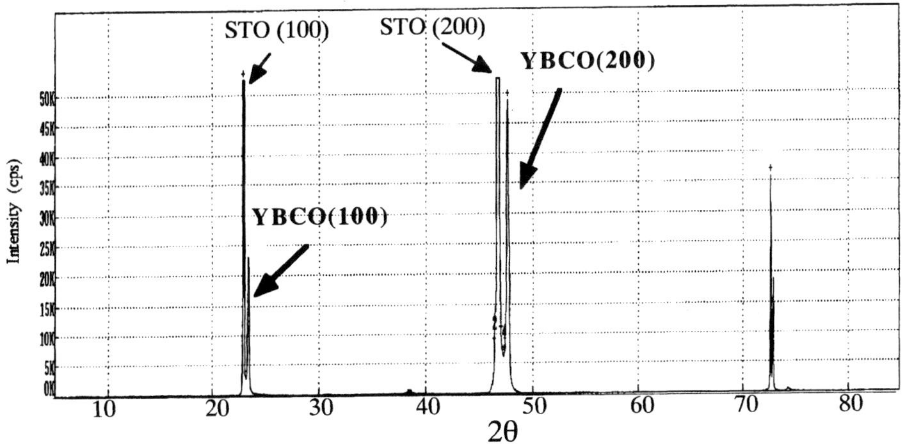

Fig. 5. XRD pattern of an as-deposited a-axis oriented YBCO film on STO substrate.

2.2 RHEED and TEM observation results

(First, we show the experimental data for the ion acceleration voltage varying from 350 to 700V.) Fig. 6 shows the RHEED patterns of the as-deposited YBCO film surface and the surface after ECR ion bombardment. A typical pattern of a tri-layered perovskite structure which confirms the existence of a-axis oriented YBCO can be clearly observed. After ion bombardment, this tri-layered perovskite structure disappears and a halo pattern showing existence of an amorphous layer appears. Similar halo patterns were always observed under almost all experimental conditions, i.e., for use of pure Ar or mixed gas of Ar and oxygen, at the ion acceleration voltage (V_{acc}) varying from 350 to 700V. However, for the case of V_{acc} below 230V, a weak trace of the perovskite pattern was sometimes observed in addition to the halo pattern. This halo pattern was changed to the typical three types of patterns after annealing, depending on the ECR and annealing conditions.

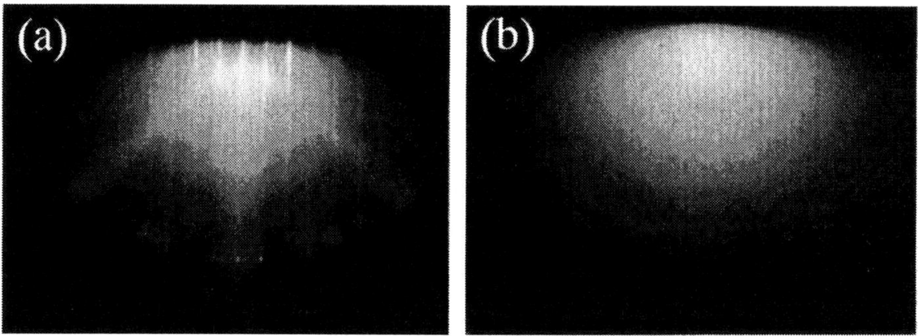

Fig. 6. RHEED patterns from an as-deposited YBCO film surface (a) and a surface after ECR ion bombardment (b).

Figs. 7~9 show the illustrations of these three types of RHEED patterns[23]. All of the RHEED patterns are shown not by photographs but by illustrations, because it is easier to understand the features and variations of the patterns. The RHEED pattern illustrations in this report were drawn by picking the qualitative features from RHEED photographs. For the case of ECR ion acceleration voltage of 700V, the amorphous pattern changes to a tri-layered perovskite structure pattern independently of the kind of process gas after annealing at 660℃ and 700℃ (Fig.7). The tri-layered perovskite structure pattern grows stronger and stronger with increasing the YBCO deposition thickness. In this case, a high substrate temperature results in only more rapid recrystallization to a-axis oriented YBCO. The important result shown in Fig.7(d) is that the tri-layered perovskite structure pattern can reappear only by the annealing without YBCO deposition.

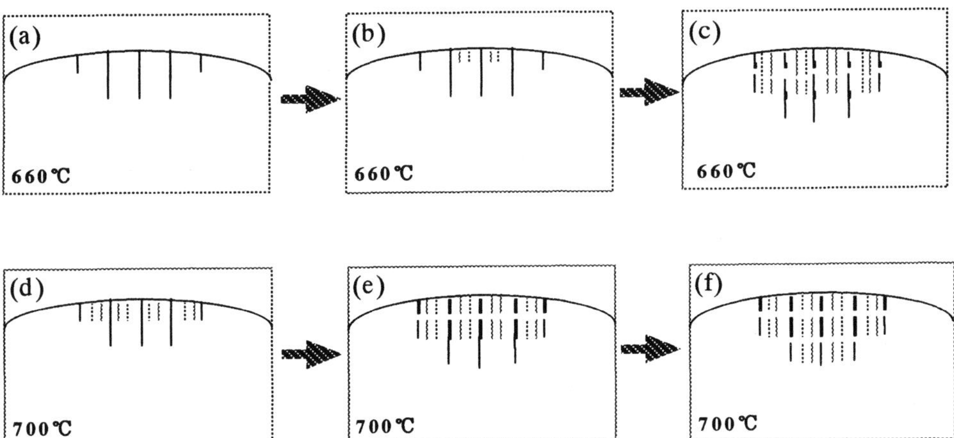

Fig. 7. RHEED patterns in recrystallization process for the ECR ion acceleration voltage of 700V. The variation of RHEED pattern is independent of the kind of process gas. The process temperatures for (a), (b), (c) and (d), (e), (f) are 660 and 700°C, respectively. (a), (d) are the patterns after annealing. (b), (e) are the patterns after 10Å thick YBCO deposition, and (c), (f) are the patterns after 150Å thick YBCO deposition.

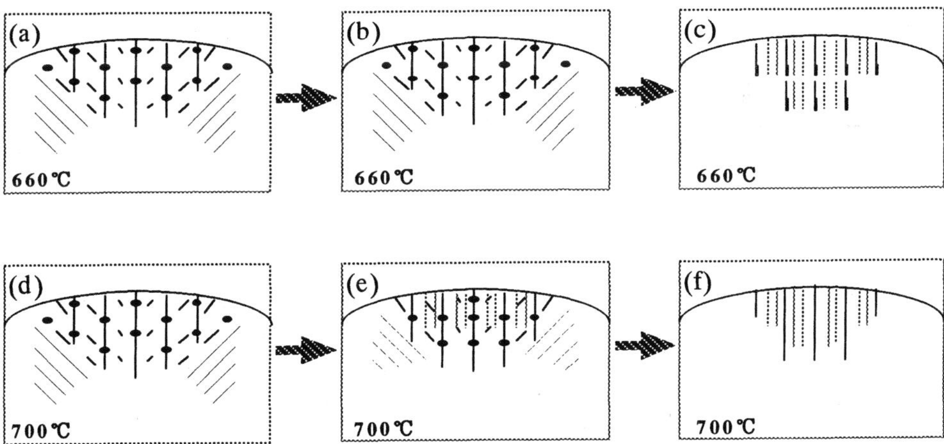

Fig. 8. RHEED patterns in recrystallization process for the ECR treatment in Ar with the acceleration voltage of 500 or 350V. The process temperatures for (a), (b), (c) and (d), (e), (f) are 660 and 700°C, respectively. (a), (d) are the patterns after annealing. (b), (e) are the patterns after 10Å thick YBCO deposition, and (c), (f) are the patterns after 150Å thick YBCO deposition.

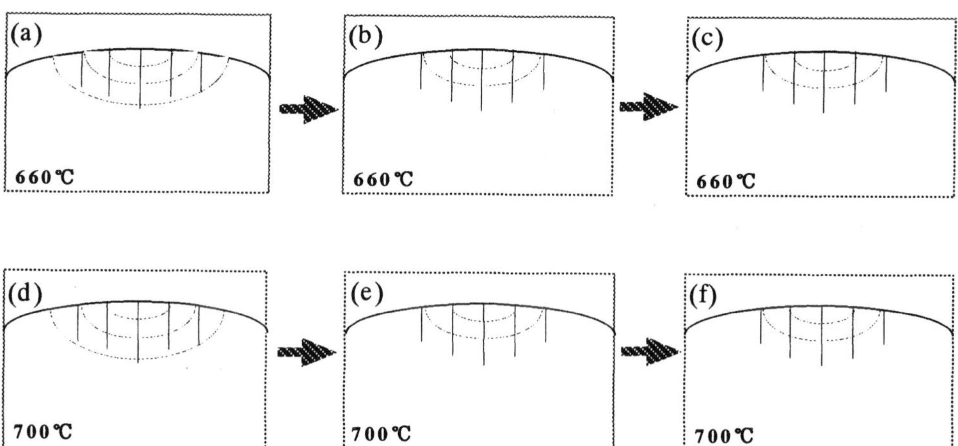

Fig. 9. RHEED patterns in recrystallization process for the ECR treatment in a mixture of Ar and oxygen with the acceleration voltage of 500 or 350V. The process temperatures for (a), (b), (c) and (d), (e), (f) are 660 and 700°C, respectively. (a), (d) are the patterns after annealing. (b), (e) are the patterns after 10Å thick YBCO deposition , and (c), (f) are the patterns after 150 Å thick YBCO deposition.

For the cases of lower acceleration voltages of 350V and 500V, the amorphous pattern changes to different patterns strongly depending on the kind of process gas. When pure Ar gas was used, complex RHEED patterns shown in Fig. 8 appeared after annealing and after deposition of a 10Å thick YBCO film. However, the tri-layered perovskite structure pattern reappeared after depositing 150Å thick YBCO. This means that a-axis oriented YBCO films can grow on the layer showing this complex RHEED pattern. On the other hand, when Ar and O_2 mixed gas was used, weak ring patterns without a tri-layered perovskite structure pattern as shown in Fig.9 was observed after annealing, and deposition of 10Å or 150Å thick YBCO.

Fig. 10 shows the relation between RHEED patterns and cross-sectional TEM images of the samples[23~25]. Fig. 10(d) shows a cross-sectional TEM image of the sample which was subjected to ECR treatment in Ar and oxygen mixed gas with the acceleration voltage of 700V and the annealing temperature of 700℃. There is no cubic or any other distinct barrier structure at the interface between the upper and lower YBCO. This TEM image is consistent with the variation of the RHEED pattern which shows reappearance of a tri-layered perovskite structure pattern after annealing.

Fig. 10(e) shows a cross-sectional TEM image of the sample which was subjected to ECR treatment in Ar with the acceleration voltage of 500V and the annealing temperature of 660 ℃. About 3nm thick second phase or barrier layer can be clearly observed at the interface. The detailed structure of the barrier is shown and analyzed in Fig. 11(a). The regions containing a fcc pseudo-cubic structure can be clearly observed and the average lattice constant of this pseudo-cubic was found to be about 5.3Å which is similar to the value

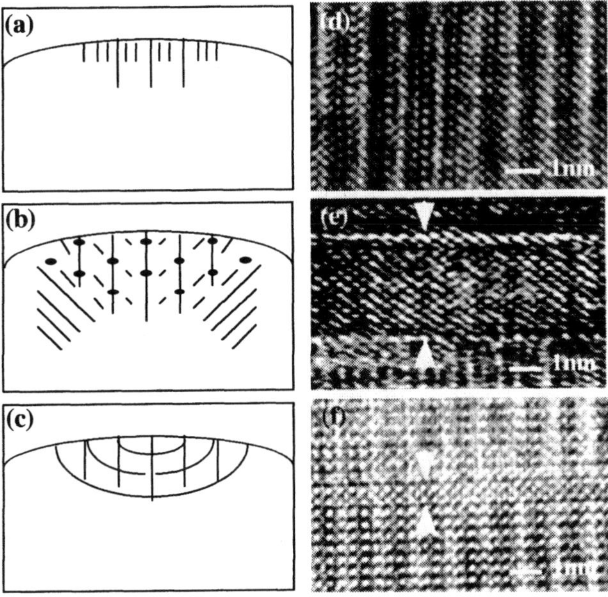

Fig. 10. Relation between RHEED patterns and cross-sectional TEM images of the sample. (a),(d) were subjected to ECR treatment in Ar and oxygen mixed gas with the acceleration voltage of 700V and the annealing temperature of 700℃. (b),(e) were subjected to ECR treatment in Ar with the acceleration voltage of 500V and the annealing temperature of 660℃. (c),(f) were subjected to ECR treatment in a mixture of Ar and oxygen at the acceleration voltage of 350V and the annealing temperature of 700℃.

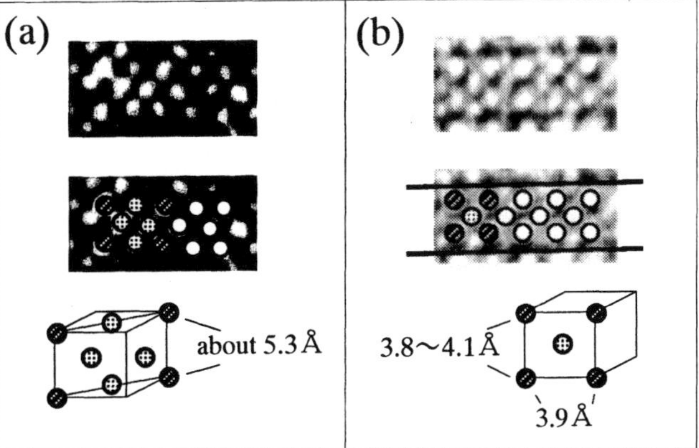

Fig. 11. Crystal structures of cubic or pseudo-cubic layer observed in Figs.10(e) and 10(g). (a) is a fcc cubic structure viewed along [110] direction shown in Fig.10(e). (b) is a fcc pseudo-cubic structure viewed along [100] direction shown in Fig.10(g).

reported by Moeckly *et al.* [4, 20]. The <110> axis direction of this pseudo-cubic layer is parallel to <100> of the lower YBCO layer.

Fig. 10(g) shows a cross-sectional TEM image of the sample which was subjected to ECR treatment in a mixture of Ar and oxygen at the acceleration voltage of 350V and annealing temperature of 700℃. A thin uniform barrier layer about 0.8nm can be clearly observed at the interface. The detailed structure of the barrier is shown and analyzed in Fig. 11(b). The regions containing a fcc pseudo-cubic structure can be clearly observed and the lattice constant of this pseudo-cubic phase was found to be about 3.8Å~4.1Å. The average lattice constant value of about 3.95Å is not similar to the value of Ba-based perovskite reported by Wen *et al.* , but similar to the value of Cu-based perovskite [18]. The <100> of this pseudo-cubic layer is the parallel to <100> of the lower YBCO layer. However , there still remains inconsistency between this TEM image and the RHEED patterns with weak rings shown in Fig. 9 , because the ring pattern means existence of a polycrystal-like surface. One of the explanations may be different sizes of the observed region between TEM and RHEED. The spot size of the focused electron beam for RHEED (20~50μm) is much larger than that for TEM observation. It was also found from low magnification TEM image that the interface is not fully but partially covered by a pseudo-cubic layer. Thus, there are possibly many small pseudo-cubic grains with various crystal orientations at the interface.

Fig. 12 shows the RHEED patterns after annealing at 660 or 700℃ of the samples subjected to Ar ion bombardment with various ion acceleration voltages[25]. The patterns for the samples annealed at 660 or 700℃ are qualitatively almost the same, though there is a slight difference. The RHEED patterns for the acceleration voltage from 350 to 700V are the same that are shown before. Ring-shape RHEED patterns were observed at a higher

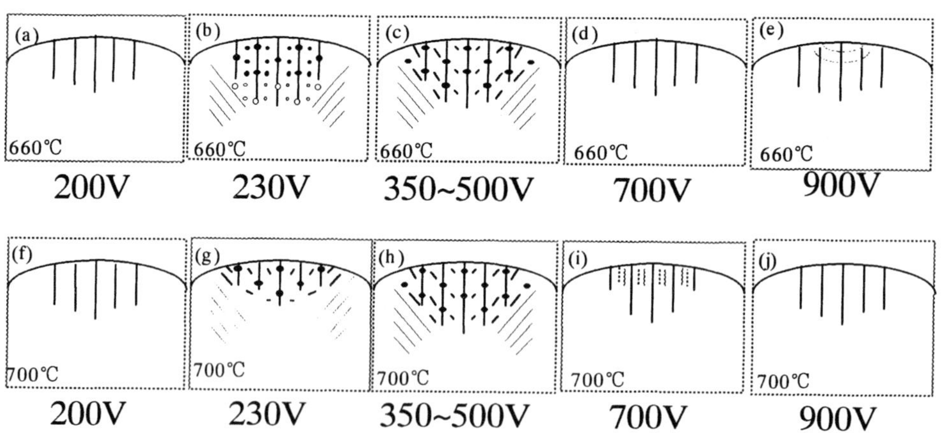

Fig. 12. RHEED patterns after annealing without YBCO deposition by using Ar ion bombardment with various ion acceleration voltages. (a)~(e) are patterns from the samples annealed at a temperature of 660℃, and (f)~(j) are patterns from the samples annealed at a temperature of 700℃.

voltage of 900V. Complex RHEED patterns similar to those for the cases of 350~500V are also observed at 230V after 660℃ annealing. These patterns suggest a smaller fluctuation of the cubic structure, because the patterns are closer to those from the <110>-axis oriented fcc cubic surface. It is considered that these several types of RHEED patterns reflect different barrier microstructures which are possibly caused by substantial deviation of the atomic composition in the amorphous layer from that of as-deposited YBCO[10, 19, 21]. The sputtering yields for the constituent elements by ion bombardment are expected to vary depending on ion acceleration voltage. The ion acceleration voltage dependence of the atomic composition in the amorphous layer was actually reported[26].

To check the thermal stability of the barrier, RHEED patterns after the first 660℃ 30 minutes annealing are compared with those after additional second 660℃ 180 minutes annealing. The Ar ion acceleration voltages of 200, 230, 500V were employed. Fig. 13 shows the RHEED patterns after the first and additional second annealing for the samples subjected to ion bombardment with various ion acceleration voltages. There is no change in the pattern after the second annealing for each ion acceleration voltage. Thus, the several-type barrier structures exhibiting the RHEED patterns as in Fig.13 have good thermal stability, if the process temperature is not higher than the annealing temperature for the barrier recrystallization.

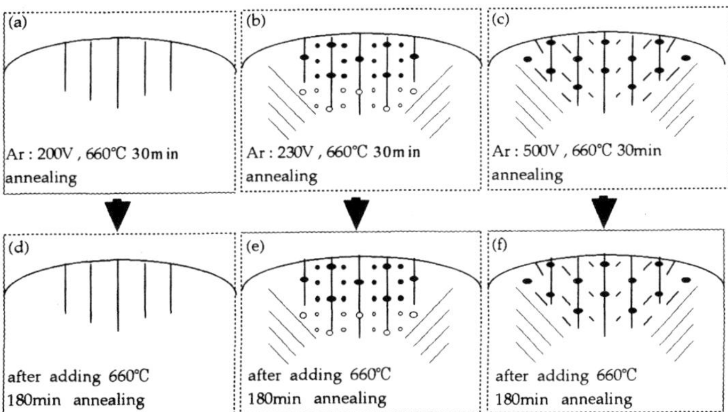

Fig. 13. Thermal stability of cubic layers examined by RHEED patterns. (a)(b)(c) are the RHEED patterns after 30min annealing at 660℃ for the samples bombarded with the Ar ion acceleration voltage of 200,230,500V respectively. (d)(e)(f) are the RHEED patterns after further 180min annealing at 660℃ for the samples in (a)(b)(c), respectively.

3. Interface-modified ramp-edge junction properties

3.1 Junction fabrication process

Fig.14 shows the schematic illustrations of the junction fabrication process[27~28]. We firstly deposited 200nm thick YBCO films and 500nm thick LaSrAlTaOx (LSAT) films on LSAT(100) substrates at a substrate temperature of 780℃ by PLD. YBCO films were deposited in 400mTorr O_2 at a KrF excimer laser power density of about 1.5J/cm^2 and the substrate-target distance of 44mm. LSAT films were deposited in 100mTorr O_2 at a laser power density of about 1.4J/cm^2 and the substrate-target distance of 39mm. Next, about 430nm thick LSAT layer was etched by using photolithography and Ar ion milling, leaving a thin LSAT layer on the YBCO base electrode[5]. Then, a ramp-edge structure was fabricated by using ECR ion bombardment with an incident angle of 30 degree and substrate rotation. We employed two steps for this etching process. First, a base electrode of YBCO was patterned by using the LSAT layer as an etching mask at an acceleration voltage of

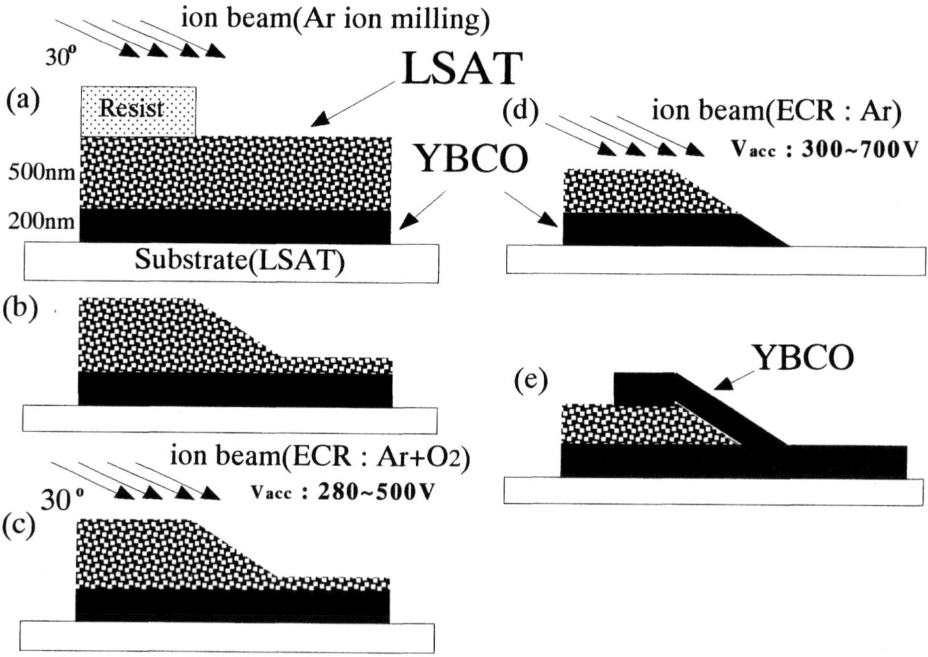

Fig. 14 . Schematic illustrations of junction fabrication process.

280~500V in pure-Ar or Ar and oxygen mixed gas[27]. Then, Ar ion bombardment at an appropriate acceleration voltage for 1 minute was added to make an important amorphous layer that is expected to form an interface-modified barrier after annealing and YBCO deposition processes. A counter YBCO layer was deposited at a substrate temperature of 670~720℃ in 200~400mTorr O_2 at a laser power density of about 1.5J/cm^2. The substrate-target distance was changed from 39 to 61mm. The deposition rate of YBCO was about 40 to 80nm/min. A 450nm thick Au film was sputtered on the counter YBCO layer and patterned. Finally, the counter YBCO layer was patterned to a width of 5μm by photolithography and Ar ion milling. The typical Tc of the base and counter YBCO layers in ramp-edge junction structures were about 87K and 83K, respectively. The temperature and magnetic field dependence of I-V characteristics were measured by a standard four-probe method in a magnetic shield.

In the junction fabrication process, we used two types of ion sources, one is an ECR ion source with a small ion beam diameter, and the other is a KAUFMAN-type ion source with a large ion beam diameter. Three configurations shown in Fig. 15 were examined and 1σ spread of Ic values were compared. We also used two kinds of YBCO films for base electrode, one was PLD films and the other was the films deposited by off-axis magnetron sputtering. In some cases we used a polishing technique for making a smooth base electrode surface. We used these films illustrated in Fig. 16 to compare 1σ spread of junction Ic values[29~30].

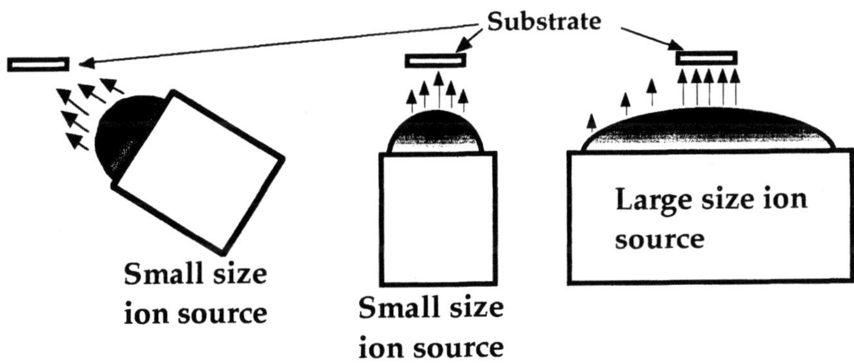

Fig. 15 . Schematic illustrations of ion bombardment configurations for making an important amorphous layer.

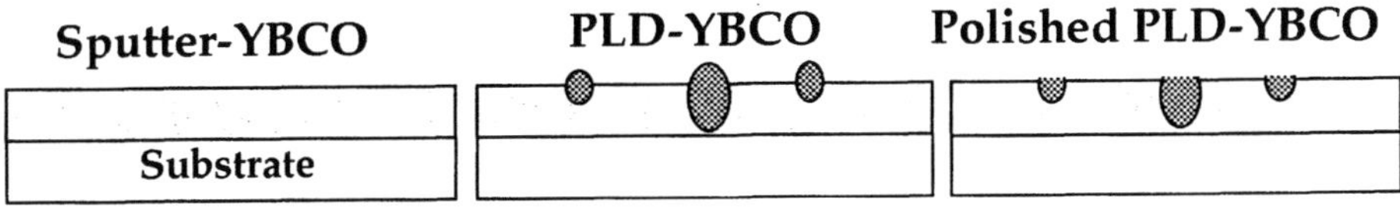

Fig. 16 . Schematic illustrations of three types of base YBCO films.

3.2 Junction properties

Fig. 17 shows typical I-V characteristics of the junctions. The junctions fabricated under appropriate conditions showed RSJ or RCSJ like I-V curves with typical IcRn products of about 0.5 to 3mV at 4.2K. Fig. 18 shows the Jc dependence of IcRn product of our junctions. This relation is similar to the correlation between Jc and IcRn previously reported by several researchers [31, 33~35]. The maximum IcRn values reported for interface-modified ramp-edge junctions are almost the same about 3~4mV at 4.2K[31, 33~35]. Recently, the highest IcRn value of 5.16mV at 4.0K was reported for the junctions with substantially higher Jc values [36~37].

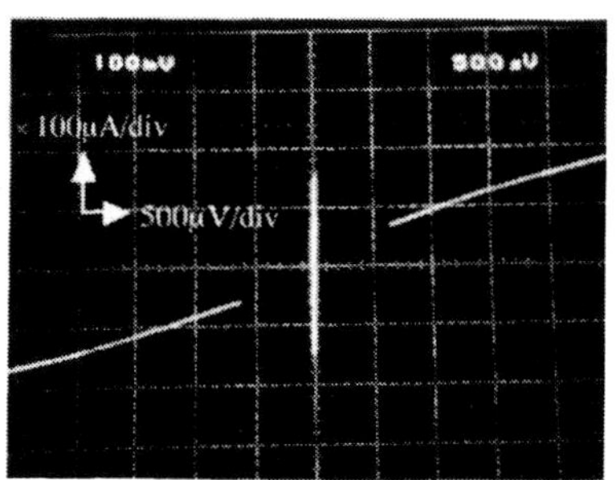

Fig. 17. I-V characteristics of the junction at 4.2K.

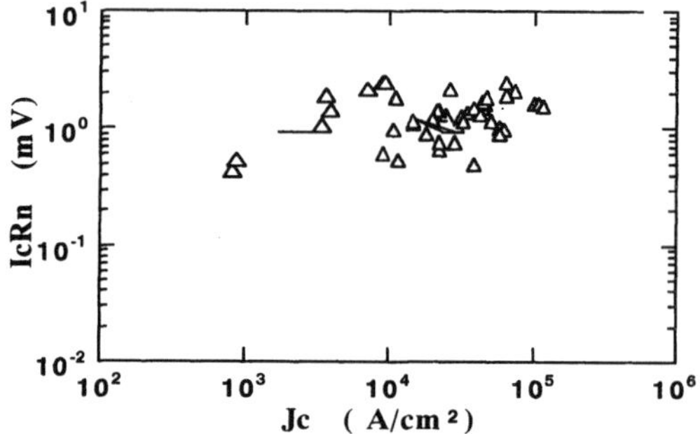

Fig. 18. Jc dependence of IcRn of the junction at 4.2K.

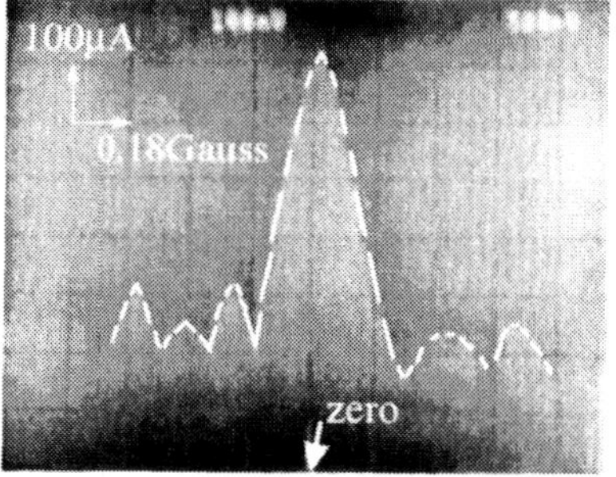

Fig. 19. Magnetic field dependence of Ic of the junction at 4.2K.

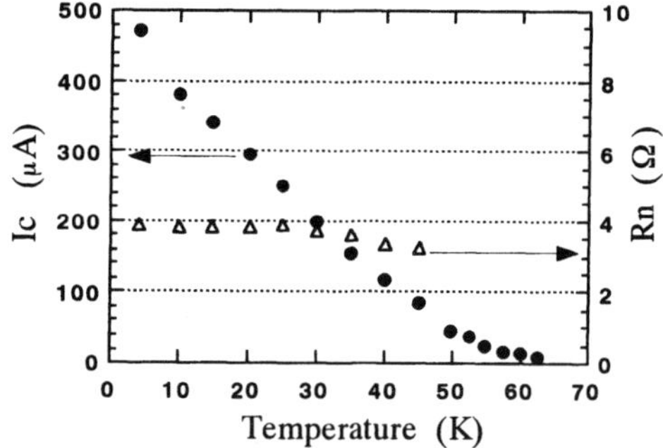

Fig. 20. Temperature dependence of Ic and Rn for the junction.

Fig. 19 shows the magnetic field dependence of Ic at 4.2K of the typical junction. A Fraunhofer-like pattern is observed. The modulation depth of Ic is about 80%, implying the existence of only small excess current. All the samples with RSJ or RCSJ-type I-V characteristics which we examined (about 40 samples) exhibit Fraunhofer-like magnetic field dependence of Ic patterns with the Ic modulation depth of about 70 to 100%.

Fig. 20 shows the temperature dependence of Ic and Rn values for the typical junction. Ic quasi-linearly decreases with increasing temperature, while Rn is almost constant below 45K. This tendency is similar to those previously reported for interface-modified junctions[3, 5, 7, 28, 32, 38, 39].

It is well known that the I-V characteristics of the interface-modified junctions are significantly changed by varying the annealing temperature and upper-YBCO deposition temperature. The Jc of the junctions increases with increasing the annealing or deposition temperature[31,35,40], and higher-Jc samples usually show flux-flow type I-V characteristics. These facts mean that a higher recrystallization energy in the barrier fabrication process leads to a stronger superconducting coupling between both sides of the barrier. Thus, if there are other factors contributing to a higher recrystallization energy in the barrier fabrication process, I-V characteristics of the junctions may also be significantly influenced.

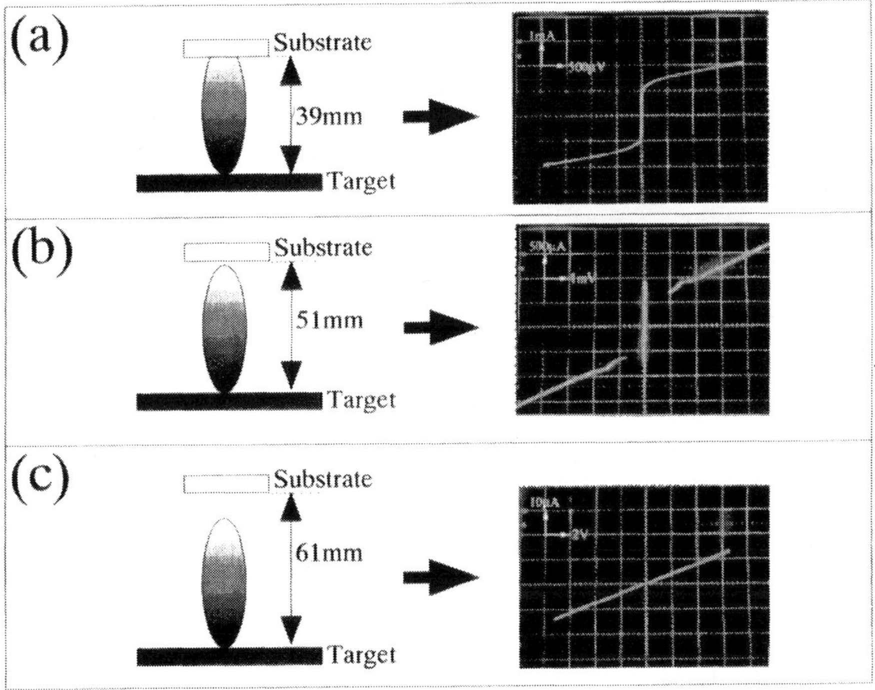

Fig. 21. I-V characteristics at 4.2K for the junctions fabricated at the substrate-target distance of 39mm (a), 51mm (b), and 61mm (c). The relative position of the substrate to the laser plume is also illustrated.

We examined the relation between the substrate-target distance for the deposition of counter-YBCO and the I-V characteristics of the junctions. Figs. 21(a), (b), and (c) show the I-V characteristics for the junctions fabricated at the substrate-target distance of 39, 51, and 61mm, respectively, and also schematically illustrate the relative position of the substrate to the laser plume. The conditions except the substrate-target distance, for example, the conditions of ECR ion bombardment and the deposition temperature for the counter YBCO layer were not changed. Tc of counter YBCO films deposited with different substrate-target distance were almost the same, about 83K. Surprisingly, a large difference is found in the

I-V characteristics. The junction fabricated with the shortest distance exhibits flux-flow-type I-V characteristics as in Fig. 21(a). On the other hand, the junctions fabricated with a much longer distance of 61mm exhibits a high resistance (500kΩ) linear I-V curve without a superconducting current as in Fig. 21 (c). The junctions fabricated with the intermediate distance only shows resistively shunted junction (RSJ) or resistively and capacitively shunted junction (RCSJ) like I-V curves with typical I_cR_n products of 1.0~3.2mV at 4.2K. These results mean that the recrystallization of the interface-modified barrier from an amorphous layer is strongly influenced by not only the energy from substrate heating, but also the kinetic energy from laser plume plasma or deposited particles[28, 41~42].

Now we discuss the relation between the I-V characteristics of the junctions and the barrier microstructures revealed by RHEED and TEM observations. It was found that there are several different microstructures of barriers which are reproducibly fabricated depending on the ion acceleration voltage and ion species in the ion bombardment process. However, if we assume the configuration of ion bombardment as illustrated in Fig. 22(a), one side of the ramp-edge surface is completely different from the opposite side, i.e., one is bombarded by higher ion energy or momentum and the other is bombarded by lower ion energy or momentum. Actually, this configuration occurs for interface-modified junctions fabricated through ion bombardment process with substrate rotation, in particular, when the ion etching rate is relatively large. Fig. 22(c) shows the I-V characteristics of a 25JJ series-array at 4.2K. This sample was fabricated through the ion bombardment process with substrate rotation. A series of 12 voltage jumps with lower I_c values are clearly separated from the other 13 voltage jumps with higher I_c values. The numbers of JJs in the series with lower and higher I_c just correspond with the numbers of the ramp edge surfaces bombarded by ions with a higher and lower energy, respectively. This experimental result strongly supports the above idea that the barrier properties have a strong dependence on the actual bombardment energy of ions incident on the ramp surface[25, 43~44]. This result also seems consistent with the results of RHEED and TEM observations that the barrier microstructures have a strong dependence on the ion acceleration voltage. A similar experimental fact that the I_c values of the junctions depend on the direction of the ramp-edge surface in a chip was reported, and this was explained in the same manner as mentioned above.[31~32] Further to confirm the hypothesis, we fabricated the samples employing the ion bombardment configuration without substrate rotation as shown in Fig. 22(b). In this case, the right and left ramp edge surfaces are bombarded with the same fixed incident angle, leading to formation of the same barrier structure on both sides. Fig. 22(d) shows the I-V characteristics at 4.2K of a 25JJ series-array fabricated employing this configuration with other junction fabrication parameters unchanged. The I-V characteristics of the array is drastically changed, i.e., the I_c separation into two groups as shown in Fig. 22(c) disappears.

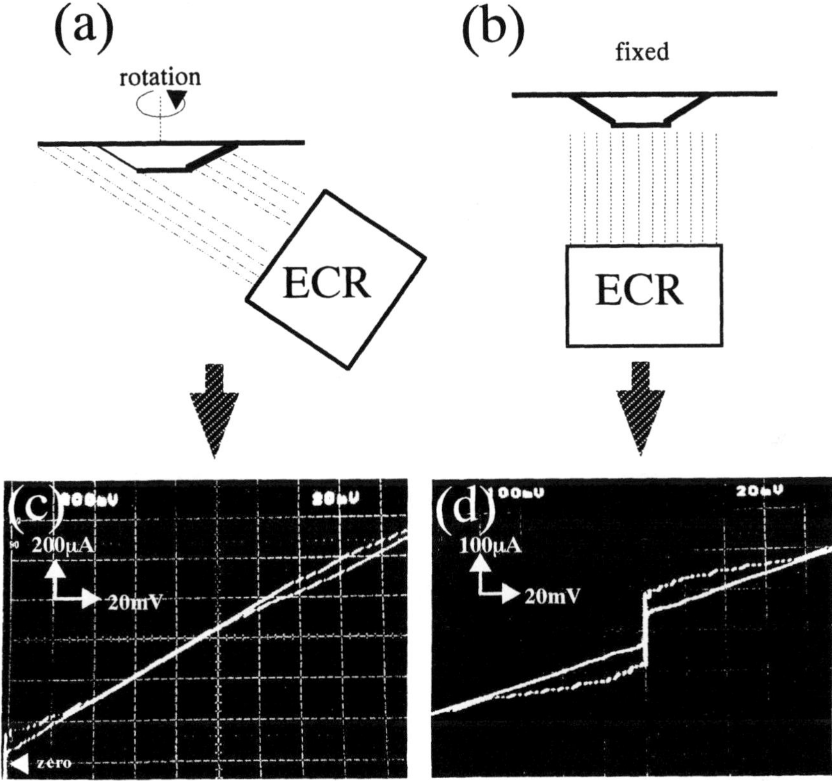

Fig. 22. Schematic illustrations of the relation between ramp-edge surfaces and the ECR gun for ion bombardment. The substrate is rotated in the case of (a), and the substrate is fixed without rotation in (b). I-V characteristics at 4.2K of a 25JJ series-arrays fabricated with substrate rotation (c), and without substrate rotation (d).

From the previous experimental results, it is speculated that if the used ion beam is not uniform spatially, the local barrier structure at different positions in a junction array may be slightly different, as a result, the Ic values of the junction array may have a larger spread. To confirm this hypothesis, we fabricated two series of samples employing a smaller size ion source with lower ion beam uniformity and a larger size ion source with higher ion beam uniformity. Moreover, two types of base electrode YBCO films were used for fabricating junction arrays. Fig. 23 shows the experimental results. In all the cases, the 1σ spread of the junction arrays increases with decreasing the average Ic value. The important fact is that the 1σ spreads of the samples fabricated by using the larger ion source are smaller than the values for the samples fabricated by using the smaller ion source. Another important point is that the junction arrays using sputtered YBCO films for base electrode have smaller 1σ spread values than the junctions fabricated using PLD YBCO films. This means that the base electrode film quality such as the density of precipitates and particles is also important for making junction arrays with smaller 1σ spreads.

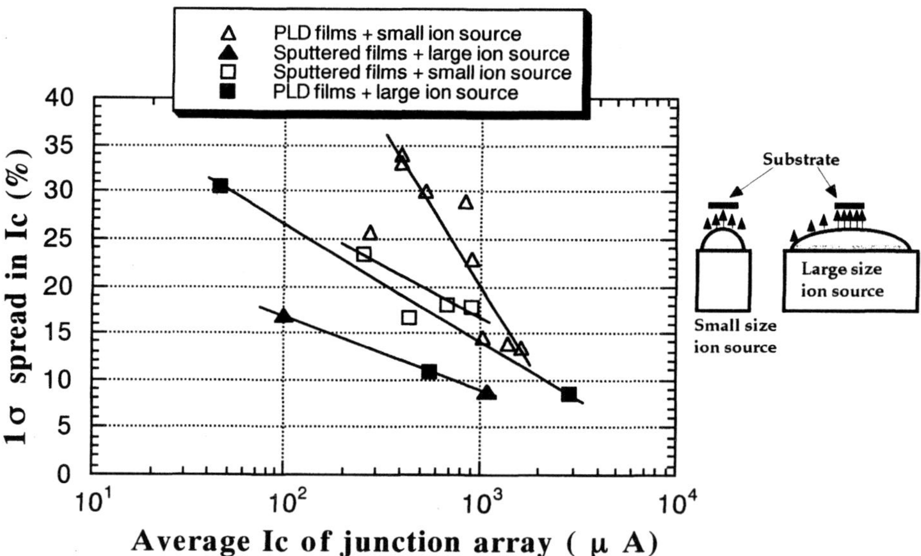

Fig. 23. Relation between average Ic and 1σ spread value of junction arrays fabricated by using two types of ion sources and two types of base YBCO films. Four solid lines are guides for the eye. The 1σ spreads of the samples fabricated by using the larger ion source are smaller than the values of the samples fabricated by using the smaller ion source. The junction arrays using sputtered YBCO films for base electrode have smaller 1σ spread values than the junctions fabricated using PLD YBCO films.

Fig. 24 shows the dependence of the Ic spread for the junction arrays fabricated by using as-grown PLD-YBCO , polished PLD-YBCO and as-grown sputtered-YBCO films for base electrodes. In this experiment, the ECR ion source was used. The 1σ spread values for polished PLD-YBCO base electrode are lower than those for as-grown PLD-YBCO. And, the 1σ spread values for polished PLD-YBCO are almost the same as those for as-grown sputtered-YBCO films. These results indicate that the roughness of the base-electrode films is also important to decrease the Ic spread of the junction arrays.

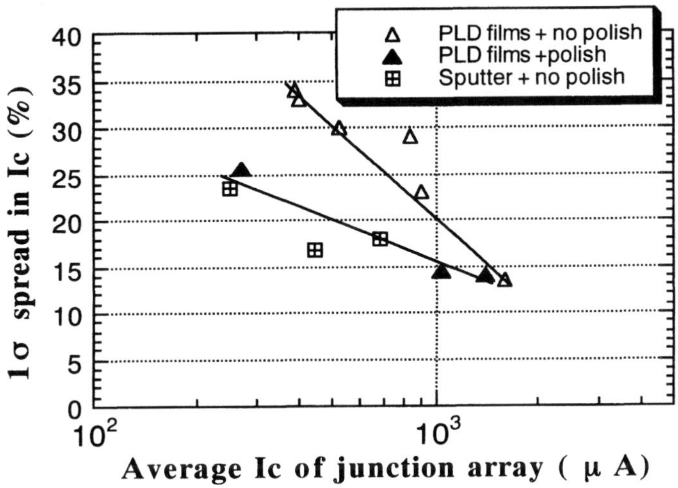

Fig. 24. Relation between average Ic and 1σ spread value of junction arrays fabricated by using three types of base YBCO films. Two lines are guides for the eye.

Fig. 25 show the optical microscope images of base-electrode YBCO film surfaces without polishing and after polishing, and AFM images of ramp-edge surfaces fabricated by using these films as base-electrodes[29~30]. There are many particles on the film surface without polishing. In this case, these particles make shadows when the ramp-edge is fabricated by ion beam etching, so the ramp-edge surface have larger roughness than the one fabricated by using the polished base electrode. This ramp-edge surface roughness is important to decrease the 1σ spread of junction arrays[45].

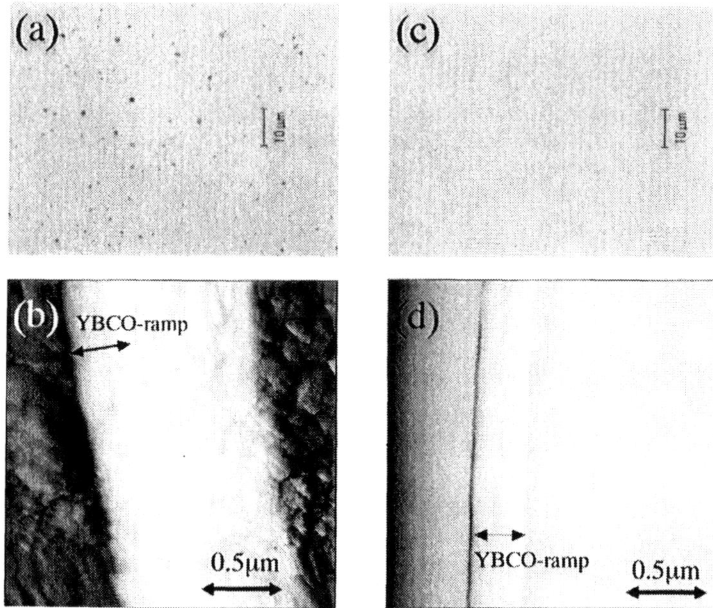

Fig. 25. Optical scope images of base-electrode YBCO film surfaces without polishing(a) and with polishing(c). AFM images of ramp-edge surfaces fabricated by using the as-grown YBCO film(b) and the polished YBCO film(d) as base electrodes.

We fabricated a 100 JJ series-array employing the combination of the large-size ion source and polished PLD-YBCO base electrode. The 1σ Ic spread of 8.6% shown in Fig. 26(a) was obtained for this series-array. We fabricated a 1000 JJ series-array employing the combination of the large size ion source and sputtered base electrode. We also developed an automatic evaluating system for Ic spread values of the junction arrays[46]. The junction array was fabricated in a rectangular region of 0.5 × 1.7 mm^2, and the 1σ Ic spread of 17% in Fig. 26(b) was obtained for this series-array with a relatively small average Ic of 100μA.

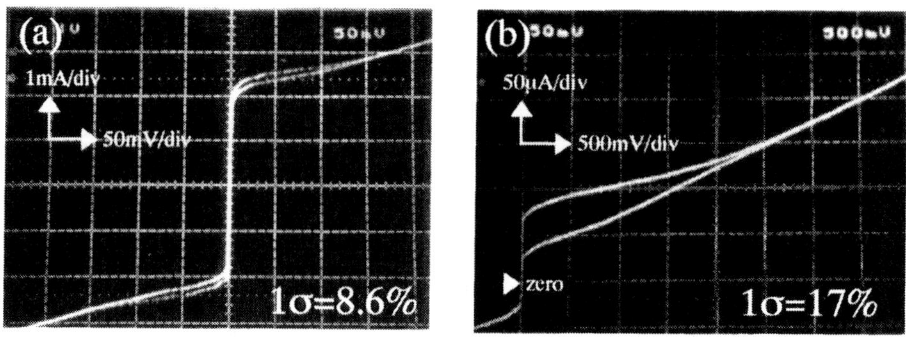

Fig. 26 I-V curves and 1σ spread values at 4.2K for 100JJs array (a) and 1000JJs arrays (b).

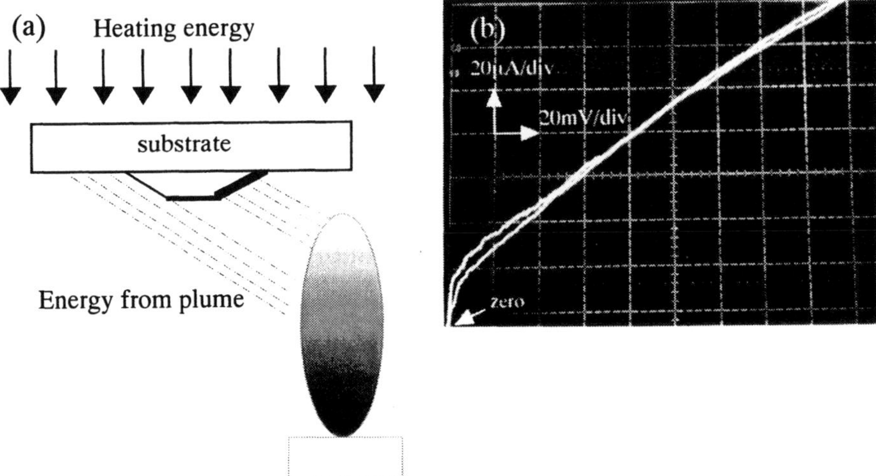

Fig. 27 Schematic illustrations of the relation between ramp-edge surfaces and laser plume(a). I-V characteristics at 4.2K of a 100JJ series-array fabricated by using this configuration for counter-electrode deposition(b).

We reported and discussed one of the important factors for 1σ spread of JJ arrays before. We mention again that for the configuration of ion bombardment as illustrated in fig. 22(a), one side of the ramp-edge surface is completely different from the opposite side , i.e., one is bombarded by higher ion energy or momentum and the other is bombarded by lower ion energy or momentum. As a result, Ic separation into two groups shown in Fig. 22(b) appears in the I-V characteristics of series-arrays. This is due to the shadowing effect of ion bombardment. The similar effect is observed in another point of the junction fabrication process, that occurred in the counter-electrode deposition process. If PLD method is used to deposit counter-electrode films, the same configuration as illustrated in fig. 27(a) is generated, where one side of the ramp-edge surface is completely different from the opposite side , i.e., one receives higher energy particles and the other receives lower energy particles[41]. We mentioned that the total energy from heating and deposited particles influences the junction properties, as revealed in fig. 21[28, 42]. Thus, if the heating energy is supplied uniformly but the energy from plume-plasma or deposited particles is not equal under this configuration, the junction properties may be different between both sides of the ramp. Fig. 27(b) is I-V characteristics of the sample which was fabricated by using this configuration in the counter-electrode deposition process. An Ic separation into two groups clearly appears in the I-V characteristics of the series-array. This shadowing effect usually occurs more or less in the counter-electrode deposition process. If this shadowing effect could be weaken, the 1σ spread of JJ arrays would be decreased. Wakana *et al.* have recently successfully decreased this shadowing effect and obtained excellent 1σ spread values for 100JJs and 1000JJs of 6.5% and 8.5 %, respectively [13, 41~42].

3.3 Barrier structure of ramp-edge junctions

In this section, we show the TEM observation results for real interface-modified ramp-edge junctions. Fig. 28 shows a cross-sectional TEM image of the junction which exhibits RSJ-type I-V characteristics. The junction was fabricated by using the Ar-ion acceleration voltage of 700V to make an amorphous layer and the annealing and upper-YBCO deposition temperature of 670℃. There is no cubic or pseudo-cubic structures at the interface. Any other structures from second phases are not observed[21, 22, 35, 44], though the contrast at the interface shown by the arrows indicates existence of strain. This barrier structure without cubic or pseudo-cubic phase is consistent with the results of RHEED and TEM observation for a-axis YBCO films fabricated with almost the same ion bombardment conditions.

Figs. 29~31 show cross-sectional TEM images of the junction arrays which exhibit RSJ-type I-V characteristics shown in Figs. 32~34, respectively. These junctions were fabricated by using the Ar-ion acceleration voltage of 300V to make an amorphous layer and almost same the annealing and upper-YBCO deposition temperature of about 700℃. However, these junction arrays have different average I_c values by about two orders of magnitude(1000μA~10μA). In these TEM images, clear cubic or pseudo-cubic structures at the interface regions can be observed[18~19, 47~48]. This barrier structure with cubic or pseudo-cubic phase is consistent with the results of RHEED and TEM observation for a-axis YBCO films fabricated with almost the same ion bombardment conditions.

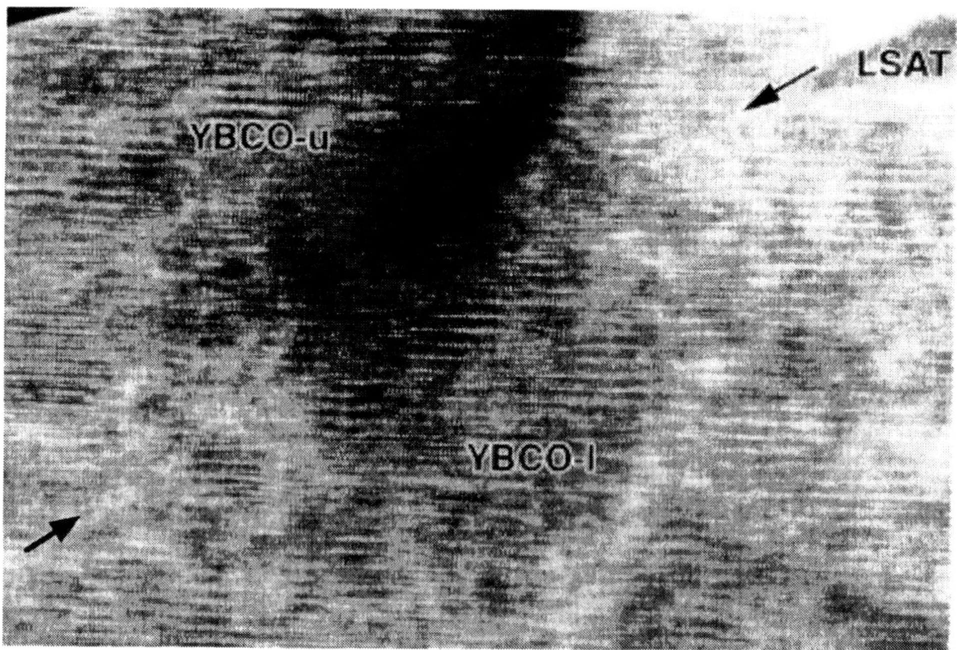

Fig. 28 Cross-sectional TEM image of the ramp-edge junction fabricated by using Ar-ion acceleration voltage of 700V and the annealing and upper-YBCO deposition temperature of 670℃.

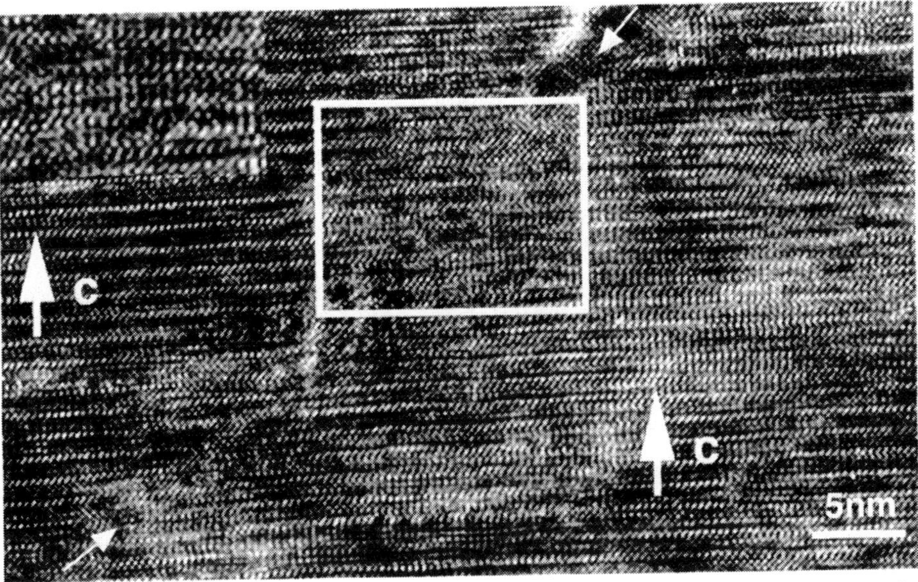

Fig. 29. Cross-sectional TEM image of the ramp-edge junction fabricated by using Ar-ion acceleration voltage of 300V and etching time of 4min 30sec. The annealing and upper-YBCO deposition temperature is 700℃.

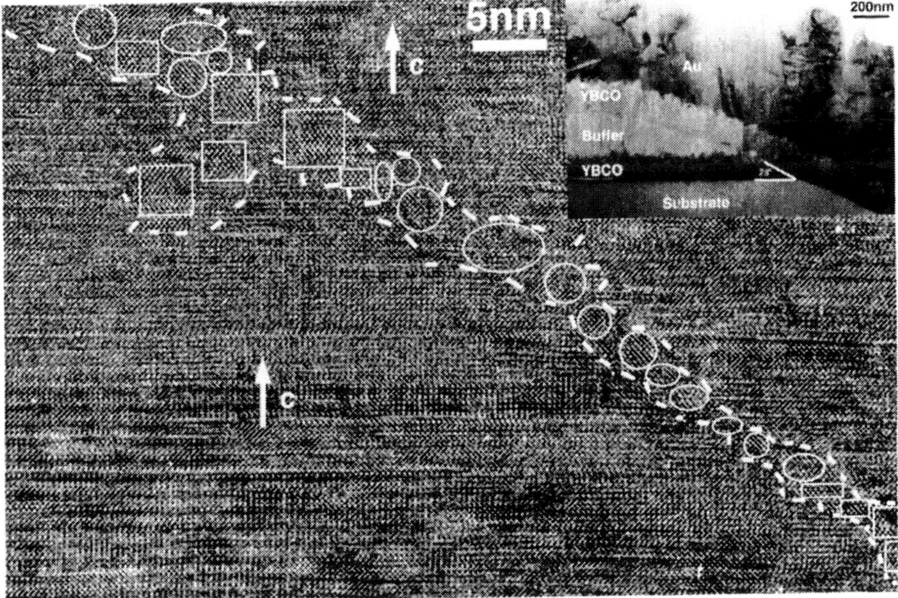

Fig. 30. Cross-sectional TEM image of the ramp-edge junction fabricated by using Ar-ion acceleration voltage of 300V and etching time of 15min. The annealing and upper-YBCO deposition temperature is 697℃.

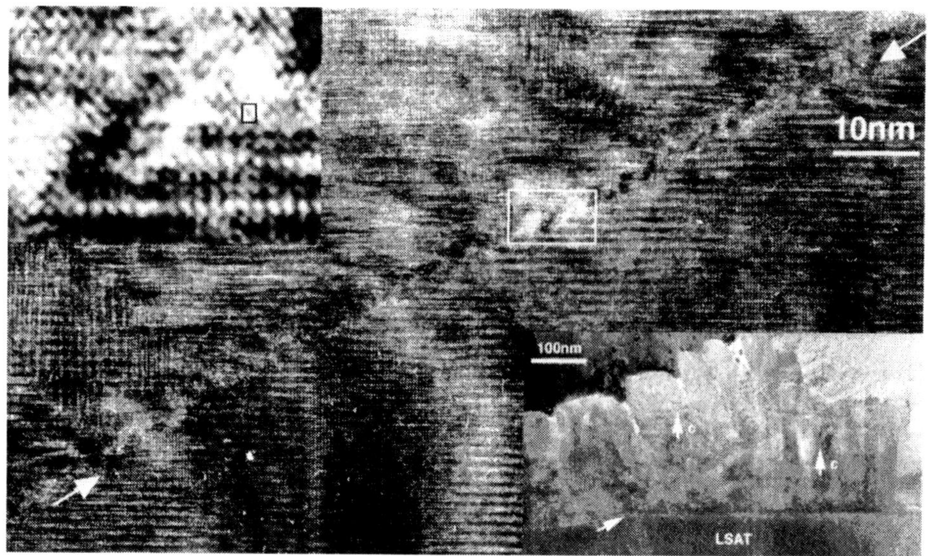

Fig. 31. Cross-sectional TEM image of the ramp-edge junction fabricated by using Ar-ion acceleration voltage of 300V and etching time of 15min. The annealing and upper-YBCO deposition temperature is 700°C.

Fig. 29 shows the ramp-edge TEM image of the sample with the average It value of about 900μA for the junctions array. There is a little cubic or pseudo-cubic region at the interface between the lower and upper YBCO films, and large regions without a cubic phase are observed at the interface in the low magnification TEM image. I-V characteristics of this sample are shown in Fig. 32(a) and a schematic illustration of the barrier is exhibited in Fig. 32(b). Fig. 30 shows the ramp-edge TEM image of the sample with the average Ic value of about 200μA for the junctions array. Clear cubic or pseudo-cubic regions shown by circles or rectangulars can be observed at the interface. However, these cubic or pseudo-cubic regions do not continuously cover the interface, and the lower magnification image shows that large regions without cubic phases still remain at the interface. The image shows that a barrier, with three types of crystal structures different from YBCO, disturbs the 3-times-unit-cell ordered (along c-axis) perovskite pseudo-cubic structure of YBCO. The thickness of the barrier is 1 to 3 nm. Due to the slight lattice mismatch between pseudo-cubic phases and YBCO, there are some misfit dislocations at the interface. Even in the thin barrier region, there is lattice parameter variation. The cubic structure observed by Huang *et al.* [20], a phase similar to the Ba-based perovskite-like structure observed by Wen *et al.*[18~19] and another phase, coexist in the barrier. I-V characteristics of this sample are shown in Fig. 33(a) and a schematic illustration of the barrier is exhibited in Fig. 33(b). Fig. 31 shows the ramp-edge TEM image of the sample with a small average Ic value (~10μA) for the junctions array. Cubic or pseudo-cubic regions continuously cover the interface, and the thickness of this barrier is distributed from about 2 to 5 nm. I-V characteristics of this sample are shown in Fig. 34(a) and a schematic illustration of the barrier is exhibited in Fig. 34(b). These TEM

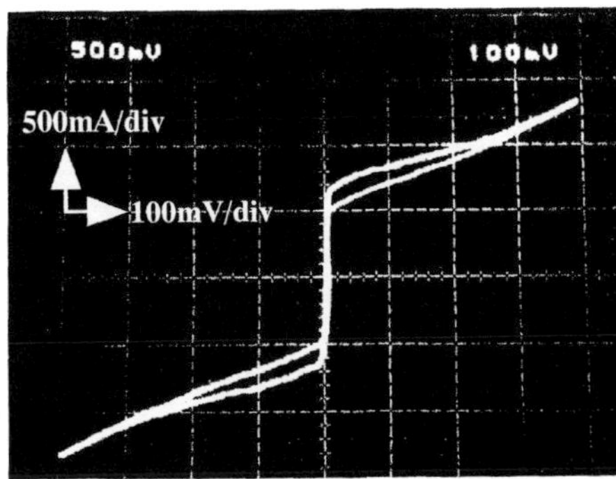

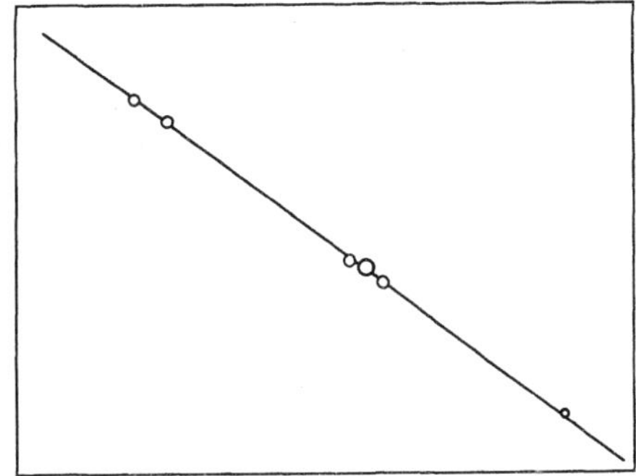

Fig. 32 I-V characteristics of a 120JJ array with the average Ic value of 895μA and the 1σ spread value of 14.0% at 4.2K(a). (b) is the schematic illustration of the barrier structure at the interface observed in the TEM image of Fig. 29.

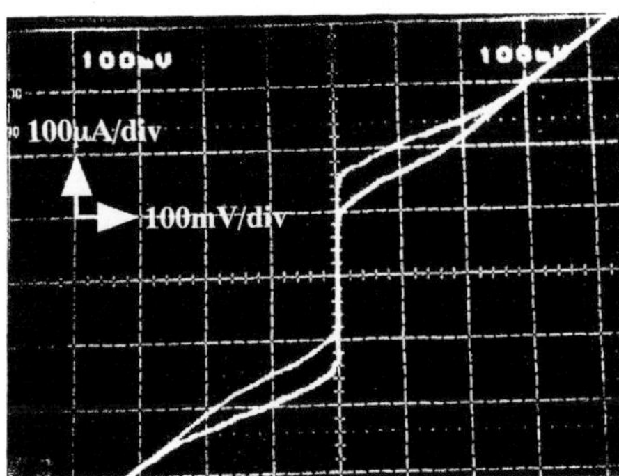

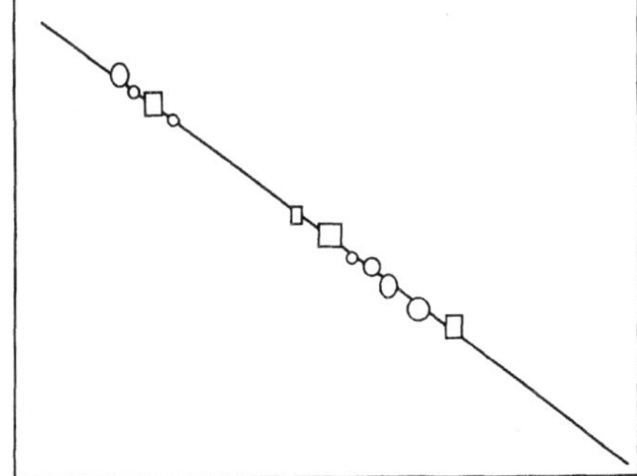

Fig. 33 I-V characteristics of a 120JJ array with the average Ic value of 206μA and the 1σ spread value of 16.3% at 4.2K(a). (b) is the schematic illustration of the barrier structure at the interface observed in the TEM image of Fig. 30.

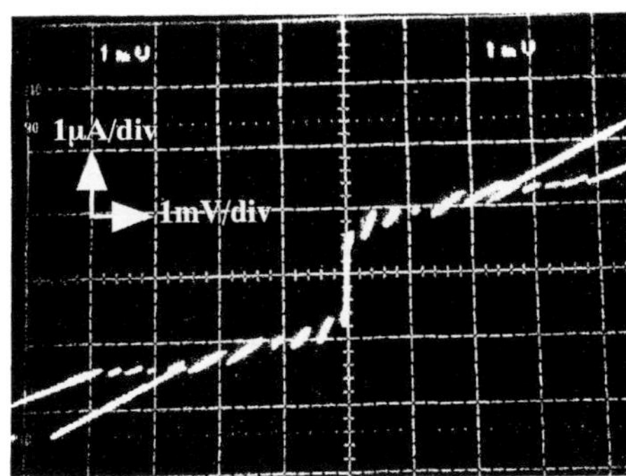

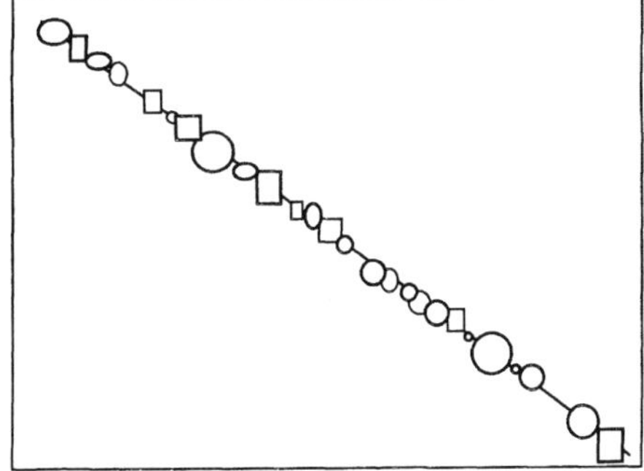

Fig. 34 A part of I-V characteristics of a 100JJ array. The average Ic and the 1σ spread values can not be calculated. (b) is the schematic illustration of the barrier structure at the interface observed in the TEM image of Fig. 31.

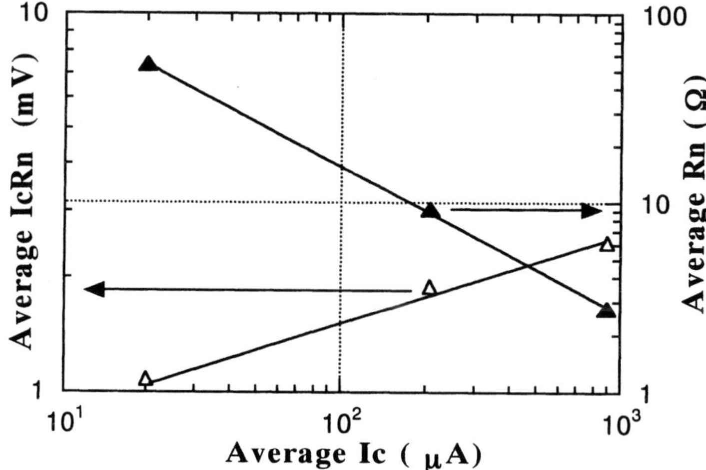

Fig. 35 Average Ic dependence of average Rn and average IcRn values . This plot was used by the junction data of figs. 32,33,34.

images show a clear relation between the average Ic value and the ratio of cubic or pseudo-cubic region to the region without these phases. For the sample with a large average Ic, no cubic region is dominant. The average Ic decreases with increasing the cubic or pseudo-cubic region at the interface. We calculated the average Rn values of these junction arrays by using the I-V curves in the high bias region. Fig. 35 shows the average Ic dependence of average Rn and average IcRn values which are plotted using the I-V data shown in Figs. 32-34. From this plot, the relation of $IcRn \propto Jc^{0.2}$ is clearly obtained. Similar relations with the N values in $IcRn \propto Jc^{N}$ ranging from about 0.2 to 0.3 have been reported by several researchers[7, 31, 33 , 38]. This relation gives an important suggestion. If the cubic or pseudo-cubic regions are insulating and do not allow the supercurrent flow, the junction Ic values have a strong dependence on the thickness of other barrier region which means the strained or disordered region at the interface. If this thickness is constant and only the ratio of the cubic or pseudo-cubic region changes, the IcRn product shuold be constant. Though the Ic linealy decreases and the Rn linealy increases with increasing the ratio of the cubic area, the IcRn product is not constant. Thus, higher Ic junctions should have a thinner strain or disordered region at the interface, and lower Ic junctions have a thicker one. The thickness of the strained or disordered region seems essential for the interface-modified barrier[47].

Fig. 36 shows a cross-sectional TEM image of the junction which exhibits RSJ-type I-V characteristics with a large Ic(>2mA). The junction was fabricated by using a La-dope YBCO base-electrode film and the Ar-ion acceleration voltage of 500V to make an amorphous layer . The annealing and upper-YBCO deposition temperature of 700°C was used. Only this sample was used the combination between La-dope YBCO base-electrode and YBCO counter-electrode. All of the other samples and dates in this chapter were made and obtained from the combination between YBCO base-electrode and YBCO counter-electrode. There is no cubic or pseudo-cubic structures at the interface shown in Fig. 36.

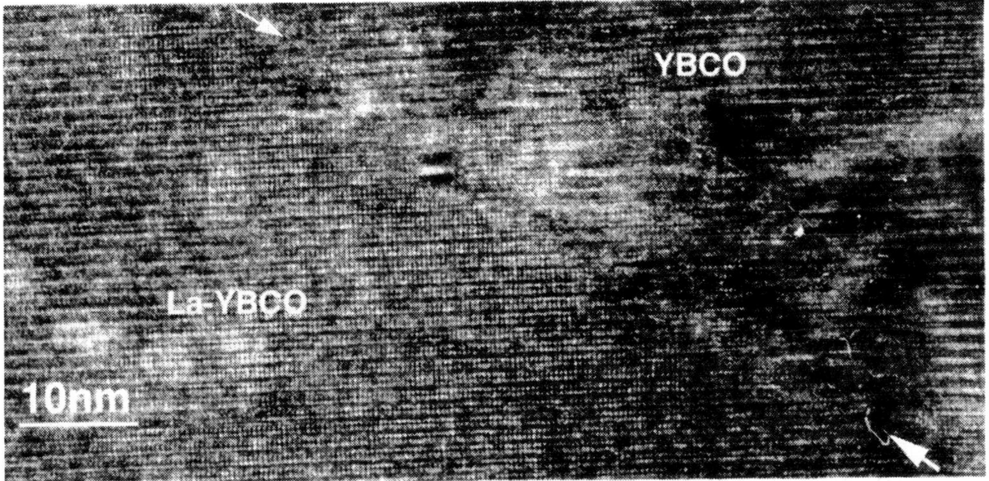

Fig. 36. Cross-sectional TEM image of the ramp-edge junction fabricated by using the Ar-ion acceleration voltage of 500V and etching time of 2min 30sec. The annealing and upper-YBCO deposition temperature is 700℃.

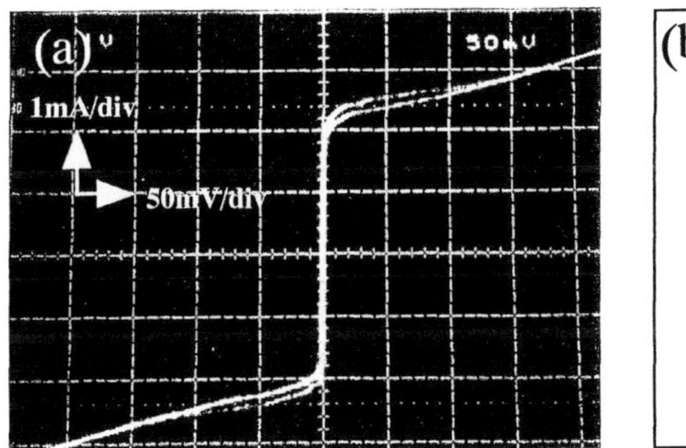

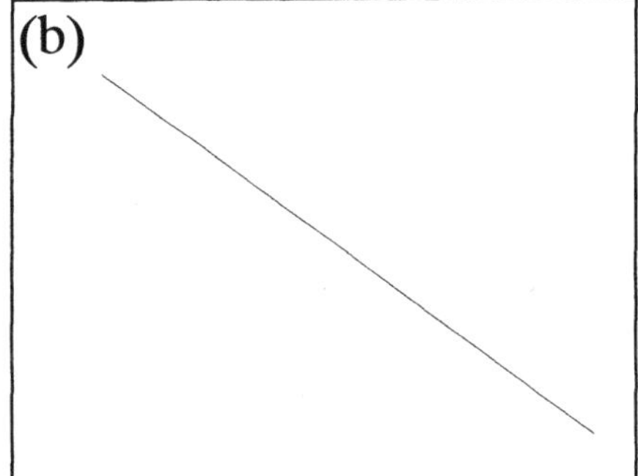

Fig. 37 I-V characteristics of a 100JJ array with the average Ic value of 2.85mA and1σ spread value of 8.6% at 4.2K(a). (b) is the schematic illustration of the barrier structure at the interface observed in the TEM image of Fig. 36.

The contrast region at the interface shown by the arrows indicates existence of strain which is almost the same as in the TEM image of Fig. 28. This barrier structure without cubic or pseudo-cubic phases is not consistent with the results of RHEED and TEM observation for a-axis YBCO films fabricated with almost the same ion acceleration voltage. However, the barrier structures for the cases of higher and lower acceleration voltage cases are consistent with those for the a-axis YBCO films. Thus, the ion acceleration voltage of 500V may be near the critical point at which makes the cubic structures appear or not. Soutome *et al.* also reported the same structure by using ion acceleration voltage of 500V[10].

4 Summary

We have demonstrated that RHEED can be well applied to observing the recrystallization process of the interface-modified barrier. A halo pattern showing existence of an amorphous layer is changed to three types of RHEED patterns after annealing by changing the ion acceleration and annealing conditions. A high ion acceleration voltage leads to recrystallization to YBCO. For lower voltages, complex or ring-shape RHEED patterns different from that of YBCO are observed. In these cases, cubic or pseudo-cubic structure regions exist at the interface, as reveled by TEM. RHEED patterns showing existence of cubic or pseudo-cubic structure are stable and do not change by further annealing at the same or lower temperature than that for the first annealing temperature to make a barrier from the amorphous layer.

YBCO interface-modified junctions have been fabricated. The junctions fabricated under optimum conditions exhibited RCSJ-like I-V curves with a typical I_cR_n product at 4.2K of 1.0 ~ 3.2mV. The experimental results which we reported here suggest that different electrical properties (such as I_c values) of the junctions are caused by different barrier microstructures, which largely depend on the fabrication and recrystallization conditions of the amorphous layer. One of the important factors is the actual bombardment energy of ions incident on the ramp surface, thus using uniform ion beam with symmetrical configuration is important to decrease the 1σ spread value. Moreover, the base-electrode quality which means the density of particles and the surface roughness are also important to decrease the 1σ spread value. We also show that the recrystallization of the interface-modified barrier from an amorphous layer strongly depends on the kinetic energy from laser plume plasma or deposited particles, as well as the thermal energy from substrate heating. Decreasing the shadowing effect enables fabrication of junction arrays with a low 1σ spread.

Finally, we report the interface-modified barrier structure in the real ramp-edge junctions, and discuss the relation between the barrier structure and I-V characteristics. Two types of barrier structures are clearly observed. One contains cubic or pseudo-cubic phases at the interface, and the other contains no cubic phases. Clear RSJ-type junction properties can be obtained for both types of interface-modified barriers. These barrier structures are almost consistent with the experimental results of RHEED and TEM observation obtained by using a-axis oriented films. For the cases of the samples with cubic phases, the average I_c of junction array decreases with increasing the ratio of cubic or pseudo-cubic regions at the interface. However, not the cubic region, but the thickness of the strained or disordered region may be the essential as a junction barrier. It is considered that higher-I_c junctions have a thinner strained or disordered region and lower-I_c junctions have a thicker one. This strained or disordered region possibly works as an actual junction barrier. In Fig. 38, we summarize the relation between the variation of interface-modified barrier and the tendency of junction properties.

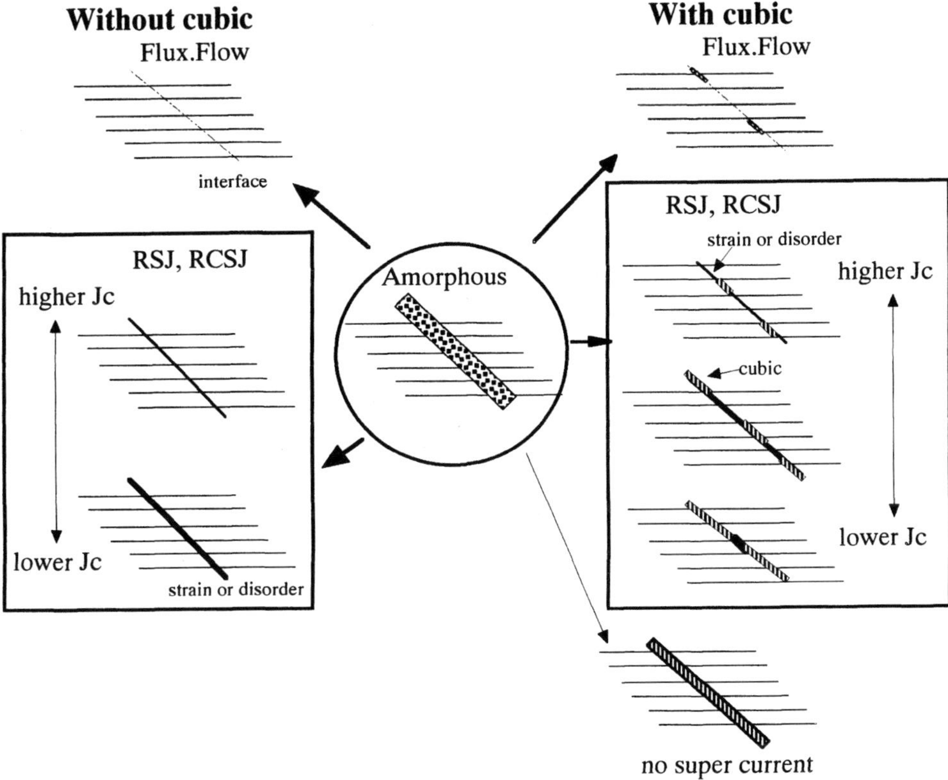

Fig. 38 Relation between the variation of interface-modified barrier and tendency of junction properties. There are two types of barrier structures, one contains cubic or pseudo-cubic phases at the interface, the other contains no cubic phases. The thickness of strained or disordered region may be essential for the interface-modified barrier .

Acknowledgements

The authors would like to thank O. Horibe for preparation of thin films and H. Tano, T. Suzuki, Y. Yoshida, Y. Oshikubo and U. Kawabe for their helpful support. They also would like to thank H. Sugiyama and M. Iiyama for helpful discussion and S. Tanaka for his encouragement. This work was supported by the New Energy and Industrial Technology Development Organization (NEDO) as Collaborative Research and Development of Fundamental Technologies for Superconductivity Applications.

References

[1] R.B. Laibowits, R.H. Koch, A. Gupta, G. Koren, W.J. Gallagher, V. Foglietti, B. Oh and J.M. Viggiano , Appl. Phys. Lett 56 (1990) 686.
[2] K. Harada, H. Myoren and Y. Osaka , Jpn.J.Appl.Phys 30(1991)L1387.
[3] B.H. Moeckly and K.Char , Appl. Phys. Lett 71 (1997) 2526.
[4] B.H. Moeckly , K. Char , Y.Huang and K.L.Merkle , IEEE Trans. Appl. Supercond 9 (1999) 3358.
[5] T. Satoh , M. Hidaka and S. Tahara , IEEE Trans. Appl. Supercond 9 (1999) 3141.
[6] B.D. Hunt, M.D. Forrester, J. Talvacchio and R.M. Young , IEEE Trans. Appl. Supercond 9 (1999) 3362.
[7] A. Fujimaki , K. Kawai , N. Hayashi , M. Horibe , M. Maruyama and H. Hayakawa , IEEE Trans. Appl. Supercond 9 (1999) 3436.
[8] R.Dittmann , J.-K.Heinsohn , A.I.Braginski , C.L.Jia , IEEE Trans. Appl. Supercond 9 (1999) 3440.
[9] Y. Soutome , T. Fukazawa , A. Tsukamoto , Y. Tarutani , K. Takagi , 1999 Program and Extended Abstract Int. Workshop on Superconductivity (Kauai Island, Hawaii) (Tokyo : ISTEC) p 113.
[10] S. Inoue , T. Nagano , H. Sugiyama and J. Yoshida , 1999 Program and Extended Abstract Int. Workshop on Superconductivity (Kauai Island, Hawaii) (Tokyo : ISTEC) p 103.
[11] T. Makita , G. Alvarez , K. Toma and K. Tanabe , *Proc 12th International Symposium on Superconductivity (ISS 99)* (Eds : T.Yamashita , K.Tanabe), (Springer Verlag Tokyo, 2000), p 999.
[12] Y.Soutome , T.Fukazawa , A.Tsukamoto, A.Tukamoto and K.Takagi , to be published in IEICE TRANS. ELECTRONICS.
[13] H.Wakana : in preparation for publication.
[14] K.Yoshida, T.Furutani, M.Maruyama, Y.Inagaki, M.Horibe, A.Fujimaki and H.Hayakawa , *Proc 12th International Symposium on Superconductivity (ISS 99)* (Eds : T.Yamashita , K.Tanabe), (Springer Verlag Tokyo, 2000), p 1005.
[15] M.Maruyama, K.Yoshida, T.Furutani, Y.Inagaki, M.Horibe, M.Inoue, A.Fujimaki and H.Hayakawa , Jpn.J.Appl.Phys 39(2000)L205.
[16] B.H.Moeckly , Appl. Phys. Lett 78 (2001) 790.
[17] H.Sato, A.Kaneko, T.Kaneda, T.Yamada, H.Yamamoto, K.Hohkawa and H.Akoh , to be published in Physica C.
[18] J.G. Wen, T. Satoh, M. Hidaka, S. Tahara, N. Koshizuka and S. Tanaka , *Proc 12th International Symposium on Superconductivity (ISS 99)* (Eds : T.Yamashita , K.Tanabe), (Springer Verlag Tokyo, 2000), p 984.
[19] J.G. Wen, N. Koshizuka, S. Tanaka, T. Satoh, M. Hidaka and S. Tahara , Appl. Phys. Lett 75 (1999) 2470.
[20] Y. Huang, K.L. Merkle , B.H. Moeckly , K. Char , Physica C 314(1999)36.
[21] Y. Soutome , T. Fukazawa , A. Tsukamoto , Y. Tarutani and K. Takagi , *Proc 12th International Symposium on Superconductivity (ISS 99)* (Eds : T.Yamashita , K.Tanabe), (Springer Verlag Tokyo, 2000), p 990.

[22] J.-K.Heinsohn, R.Dittmann, J.R.Contreras, J.Scherbel, A.Klushin, M.Siegel, C.L.Jia , S.Golubov, M.Y.Kupryanov , J.Appl.Phys 89(2001)3852.
[23] Y. Ishimaru, Y.Wu, O.Horibe, Y.Tarutani and K. Tanabe , Physica C 357-360 (2001)1432.
[24] Y. Wu, Y. Ishimaru, K. Tanabe , Physica C 366(2001)51.
[25] Y. Ishimaru , Y. Wu , O. Horibe , H. Tano , T. Suzuki ,Y. Yoshida , M. Horibe , H. Wakana , S. Adachi , Y. Takahashi, Y. Oshikubo , H. Sugiyama , M. Iiyama , Y. Tarutani and K. Tanabe , to be published in Physica C.
[26] M. Horibe , Y. Inagaki , K. Yoshida , G. Matsuda , N. Hayashi , A. Fujimaki and H. Hayakawa , Jpn.J.Appl.Phys 39(2000)L284.
[27] T. Makita , K. Toma, K. Ishikawa, H. Zama, T. Utagawa, U. Kawabe and K. Tanabe , IEEE Trans. Appl. Supercond 11 , No1, (2001) 155.
[28] H. Tano, Y. Ishimaru, T. Suzuki, Y. Tarutani, U. Kawabe and K. Tanabe , Physica C 357-360 (2001)1428.
[29] Y.Ishimaru , S.Miura, F.Wang, N.Tanaka, A.Yoshida, T.Morisita and N.Yokoyama , *Proc 11th International Symposium on Superconductivity (ISS 98)* (Eds : N. Koshizuka , S. Tajima), (Springer Verlag Tokyo, 1999), p 1035.
[30] T.Hato, Y.Ishimaru, H.Aso, A.Yoshida and N.Yokoyama , *Proc 12th International Symposium on Superconductivity (ISS 99)* (Eds : T.Yamashita , K.Tanabe), (Springer Verlag Tokyo, 2000), p 1008.
[31] T. Satoh , J.G. Wen , M. Hidaka , S. Tahara , N. Koshizuka and S. Tanaka , Supercond.Sci.Technol 13(2000)88.
[32] J.K. Heinshon, R.H. Hadfield, R. Dittmann , Physica C 326-327 (1999)157.
[33] B.H.Moeckly , *Proc 11th International Symposium on Superconductivity (ISS 98)* (Eds : N. Koshizuka , S. Tajima), (Springer Verlag Tokyo, 1999), p 1141.
[34] J. Yoshida, S. Inoue, H. Sugiyama and T. Nagano , IEEE Trans. Appl. Supercond 11 , No1, (2001) 784.
[35] Y. Soutome , R.Hanson , T. Fukazawa , K.Saitoh , A. Tsukamoto , Y. Tarutani and K. Takagi , IEEE Trans. Appl. Supercond 11 , No1, (2001) 163.
[36] H. Shimakage, R.H.Ono, L.R.Vale and Z.Wang , IEEE Trans. Appl. Supercond 11 , No2, (2001) 4032.
[37] H. Shimakage, R.H.Ono, L.R.Vale, Y.Uzawa and Z.Wang , Physica C 357-360 (2001)1416.
[38] J. Yoshida, S. Inoue, H. Sugiyama and T. Nagano , Physica C 335 (2000)226.
[39] T. Makita , K.Toma , K.Ishikawa, H.Zama, T.Utagawa , U.Kawabe and K. Tanabe , IEEE Trans. Appl. Supercond 11 , No1, (2001) 155.
[40] M.Horibe , T.Ito , Y.Inagaki , G.Matsuda , A. Fujimaki and H. Hayakawa , IEEE Trans. Appl. Supercond 11 , No1, (2001) 159.
[41] H. Wakana, S. Adachi, M. Horibe, Y. Ishimaru, O. Horibe, Y. Tarutani and K. Tanabe, to be published in Jpn.J.Appl.Phys.
[42] H. Wakana, S. Adachi, M. Horibe, Y. Ishimaru, O. Horibe, Y. Tarutani and K. Tanabe, to be published in Physica C.
[43] Y. Ishimaru , Y.Wu , O.Horibe , H.Tano , T.Suzuki , Y.Tarutani , U.Kawabe and K.Tanabe , to be published in Physica C.

[44] Y. Ishimaru , Y.Wu , O.Horibe , H.Tano , T.Suzuki , Y.Tarutani , U.Kawabe and K.Tanabe , to be published in Jpn.J.Appl.Phys.

[45] M. Horibe, T. Suzuki, H. Tano, H. Wakana, Y. Ishimaru , S. Adachi, O. Horibe, Y. Tarutani and K. Tanabe , to be published in Physica C.

[46] Y. Tarutani, Y. Ishimaru, H. Wakana, M. Horibe, O. Horibe, H. Sugiyama, M. Iiyama , S. Adachi, Y. Oshikubo, H. Tano, T. Suzuki and K. Tanabe , to be published in Physica C.

[47] Y. Wu, Y. Ishimaru, H. Wakana, S. Adachi, Y. Tarutani and K. Tanabe : in preparetion for publication.

[48] Y. Wu, H. Wakana, S. Adachi and K. Tanabe: in preparetion for publication.

Paramagnetic Meissner Effect in Granular Superconductors Studied Through the Magnetic Properties of 2D-Josephson Junction Arrays

F. M. Araujo-Moreira*
Department of Physics
Multidisciplinary Center for Development of Ceramic Materials-MCDCM
Universidade Federal de São Carlos
Caixa Postal 676, São Carlos SP, 13565-905, Brazil

P. Barbara
Physics Department, Georgetown University, Box 571228
Washington DC, 0057-1228, USA

A. B. Cawthorne
Neocera, Inc.
10000 Virginia Manor Road, Suite 300
Beltsville, MD, 20705-4215, USA

C. J. Lobb
Center for Superconductivity Research
Department of Physics, University of Maryland at College Park
College Park MD, 20742-4111, USA

I. Introduction: JJA and Granularity

Granular superconductors can be considered as a collection of superconducting grains embedded in a weakly superconducting - or even normal - matrix. Granularity often occurs in high-temperature superconductors (HTS) manifested by a two-part transition. The first one represents the intragranular contribution, associated to the grains exhibiting ordinary superconducting properties. The other component originates from intergranular material, and is associated to the weak-link structure, thus, to the Josephson junctions network (Fig. 1). From this picture, intragranular

* Corresponding author; e-mail address: faraujo@df.ufscar.br

properties would be intrinsic, while intergranular, on the contrary, would be extrinsic, generating processing dependent effects. This particular feature started to be used more than thirty years ago in low temperature superconductors (LTS) to create disordered networks of defects in a controlled way [1-3].

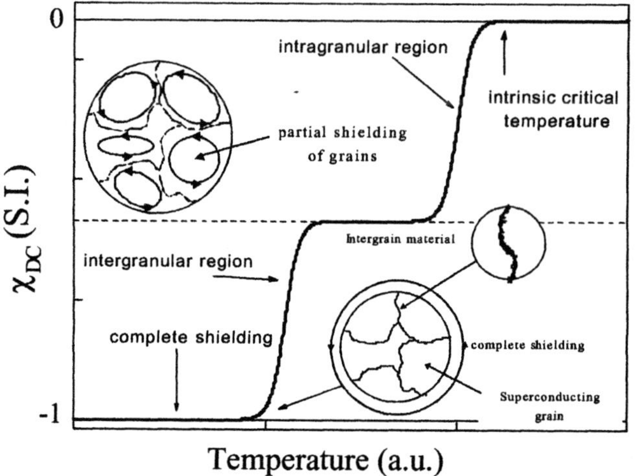

Figure 1 - Sketch of both inter and intragranular regions in a GSS; obtained from a DC magnetization experiment as a function of temperature with an external magnetic field, H. As usual, we have defined the magnetic susceptibility as $\chi_{DC} = \partial M/\partial H$.

Also, granular LTS systems have been employed since 1980 to fabricate Josephson networks [4], in an attempt to obtain experimental data to be compared with simulation results for two-dimensional (2D-JJA). Today, processing dependent effects allow the study of the enhancement of the critical current as a consequence of additional pinning centers in HTS and LTS.

For granular samples in the macroscopic scale, the fraction f_g, is a measure of the normalized superconducting volume. For single-crystals and other nearly-perfect structures, granularity is a more subtle feature that can be envisaged as the result of symmetry breaking. Thus, one might have granularity in the nanometer scale, generated by localized defects like impurities, oxygen deficiency, vacancies, atomic substitutions and the genuinely intrinsic granularity associated with the layered structure of perovskites. On the micrometer scale, granularity results from the existence of extended defects, such as grain and twin boundaries. From this picture, granularity could have many contributions, each one with a different volume fraction. In transport and magnetic experiments, this multigranular feature, is recognized through the presence of different plateaus[5-8], as shown in Fig. 2.

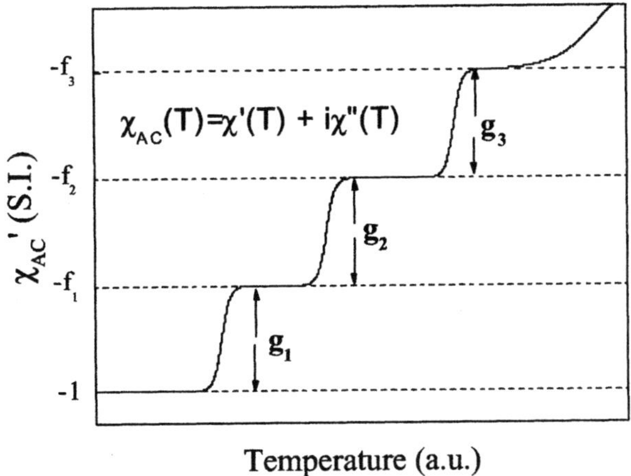

Figure 2 - Multilevel granular structure observed from a magnetic susceptibility experiment as a function of temperature; each family of defects, g, has a typical volume fraction, f_i.

The small coherence length of HTS implies that any imperfection may contribute to both the weak-link properties and the flux pinning. This leads to many interesting peculiarities and anomalies, many of which have been tentatively explained over the years in terms of the granular character of HTS materials.

By contrast, artificial Josephson junction arrays consist of superconducting islands arranged on a symmetrical lattice (Fig. 3), coupled by Josephson junctions (Fig. 4), where is possible to introduce a controlled degree of disorder. In this case, a JJA with disorder can be considered as the limiting case of an extreme non-homogeneous type-II superconductor, allowing its study in samples where the disorder is nearly exactly known.

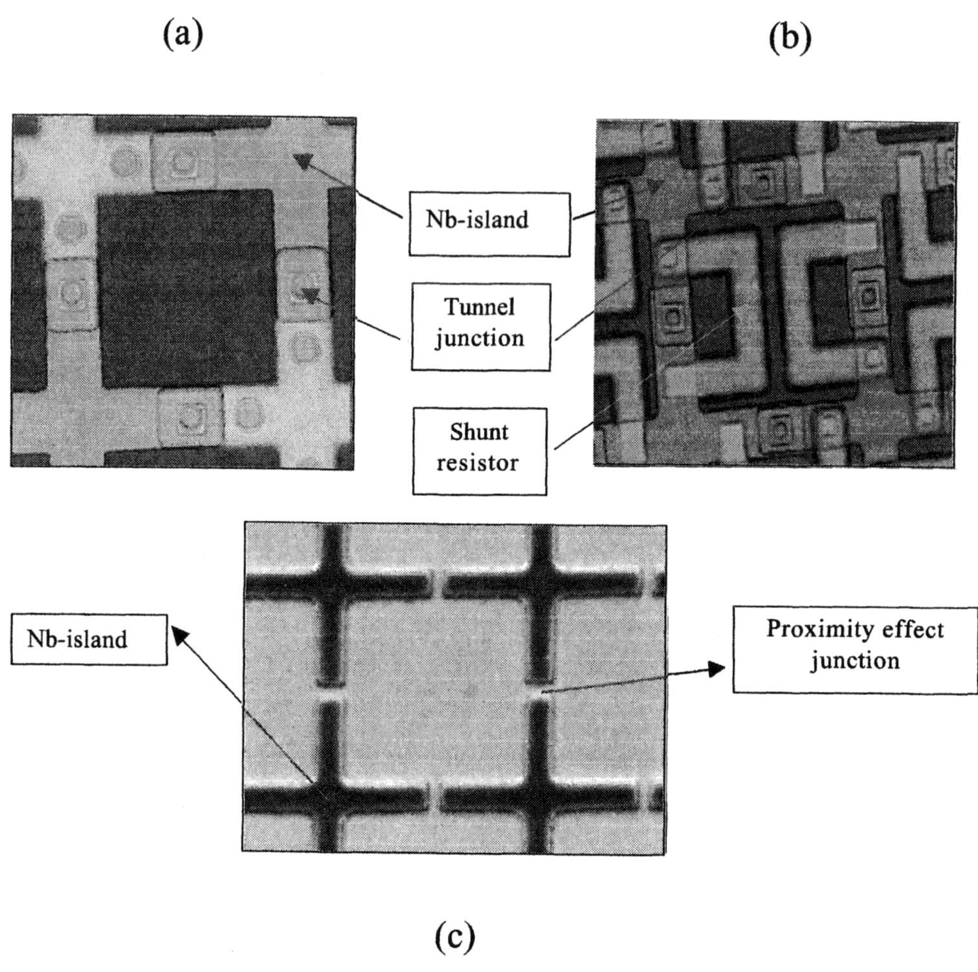

Figure 3 - For completeness, we show here the photograph of different types of Josephson junction arrays: (a) SIS/unshunted, (b) SIS/shunted, and (c) SNS.

Since JJA are artificial structures, they can be very well characterized. Their discrete nature, together with the very well-know physics of the Josephson junction, allows the numerical simulation of their behavior. The details of the physical properties of two dimensional Josephson junction arrays have been recently extensively discussed by Newrock *et al.*[9], and Martinoli *et al.*[10]

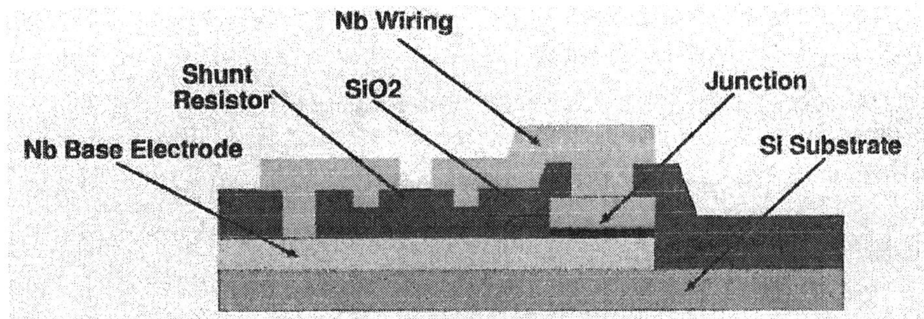

Figure 4 - Cross section of a shunted JJA showing superconducting regions coupled by the Josephson junction.

II. Magnetic Response of Granular Superconductors and JJA

We consider first the magnetic response of a granular superconductor in either an AC or DC field of small magnitude. This field should be weak enough to guarantee that the critical current of the intergranular material is not exceeded at low temperatures. After a zero-field cooling process (ZFC), which consists in cooling the sample from above its critical temperature (T_C) with no applied magnetic field, the magnetic response to the application of a magnetic field is that of a perfect diamagnet. In this case, the intragranular screening currents prevent the magnetic field from entering the grains, whereas intergranular currents flow across the sample to ensure a null magnetic flux throughout the whole specimen. This temperature dependence of the magnetic response gives rise to the well-known double-plateau behavior for the DC-susceptibility, χ_{DC}, and the corresponding double-drop/double-peak for the complex AC magnetic susceptibility, χ_{AC} [5-8,11].

Cooling the sample in the presence of a magnetic field, by following a field-cooling process (FC), the screening currents are, at temperatures immediately below T_c, restricted to the intragranular contribution, a situation that remains until the temperature reaches T_{inter}, below which $J_{c,intra}(T)$ is no longer zero. Intergranular currents, which develop below T_{inter}, might contribute a signal that can be either paramagnetic or diamagnetic. Figure 5-a depicts a circumstance where the intergranular contribution χ_{inter} is positive, although smaller than $|\chi_{intra}|$. In this situation, the resulting susceptibility χ_{sample} reenters between T_{inter} and T_C, yet remaining negative for all temperatures. An extreme situation is shown in Figure 5-b, for which $\chi_{inter} > |\chi_{intra}|$. In this latter condition, χ_{sample} also reenters and changes sign close to T_{inter}. As we will see later, this competition between diamagnetic and paramagnetic contributions is of fundamental importance to the correct interpretation of the magnetic response of JJA.

All these possibilities about the signal and magnitude of χ_{sample} have been extensively reported in the literature, involving both LTS and HTS materials [12-15]. The reentrant behavior mentioned before is also known in the literature as the Paramagnetic Meissner effect, PME. As we will show later, we have reported its occurrence as a reentrant behavior in χ_{AC} (T) measurements of two-dimensional JJA[16,17], which are, as we showed before, particularly ordered granular systems.

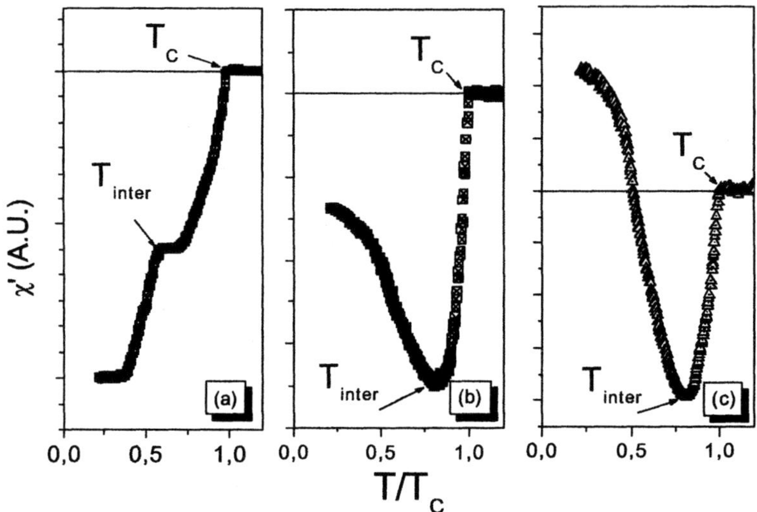

Figure 5 - Situation where (a) both components, χ_{inter} and χ_{intra}, are negative; (b) χ_{inter} is paramagnetic, although smaller than $| \chi_{intra} |$; (c) extreme situation where $\chi_{inter} > |\chi_{intra}|$. In this case χ_{sample} reenters and changes sign close to T_{inter}.

Complex AC magnetic susceptibility is a powerful low-field technique to determine the magnetic response of many systems, like granular superconductors and Josephson junction arrays. It that has been successfully used to measure several parameters such as critical temperature, critical current density and penetration depth in superconductors. To measure samples in the shape of thin films, the so-called screening method has been developed. It involves the use of primary and secondary coils, with diameters smaller than the dimension of the sample. When these coils are located near the surface of the film, the response, i.e., the complex output voltage V, does not depend on the radius of the film or its properties near the edges. In the reflection technique [18], an excitation coil (primary) coaxially surrounds a pair of counter-wound pick up coils (secondaries). When there is no sample in the system, the net output from these secondary coils is close to zero, since the pick up coils are close to identical in shape, but are wound in opposite directions. The sample is positioned as close as possible to the set of coils, to maximize the induced signal on the pick up coils (Fig. 6).

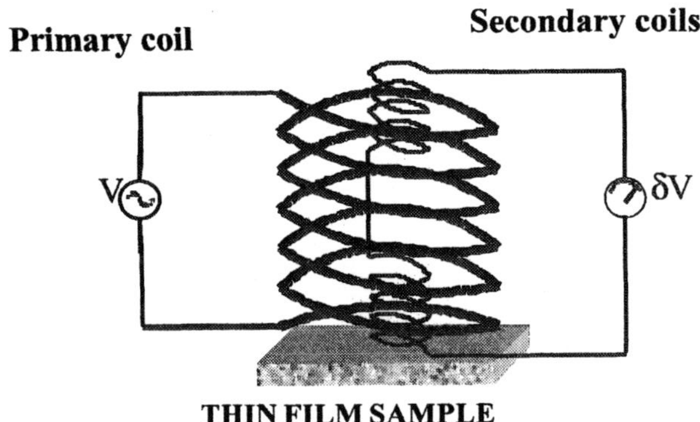

Figure 6 - Screening method in the reflection technique, where an excitation coil (primary) coaxially surrounds a pair of counter-wound pick up coils (secondaries).

An alternating current sufficient to create a magnetic field of amplitude h_{AC} and frequency f is applied to the primary coil. The output voltage of the secondary coils, V, is a function of the complex susceptibility, $\chi_{AC} = \chi' + i\ \chi''$, and is measured through the usual lock-in technique. If we take the current on the primary as a reference, V can be expressed by two orthogonal components. The first one is the inductive component, V_L (in phase with the time-derivative of the reference current) and the second one the quadrature resistive component, V_R (in phase with the reference current). This means that V_L and V_R are correlated with the average magnetic moment and the energy losses of the sample, respectively.

We used the screening method in the reflection configuration to measure $\chi_{AC}(T)$ of Josephson junction arrays [16,17]. Measurements were performed as a function of the temperature T (1.5K < T < 15K), the amplitude of the excitation field h_{AC} (1 mOe < h_{AC} < 10 Oe), and the external magnetic field H_{DC} (0 < H_{DC} < 100 Oe) parallel with the plane of the sample (Fig. 7). The frequency in the experiments reported here was fixed at f = 1.0 kHz. The susceptometer was positioned inside a double wall μ-metal shield, screening the sample region from Earth's magnetic field.

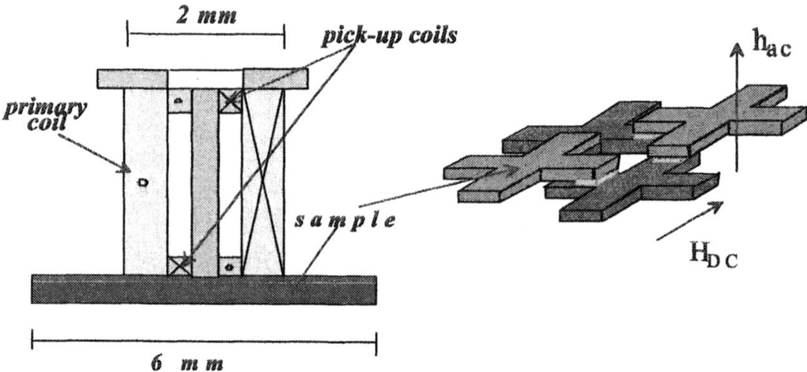

Figure 7 - Sketch of the experimental setup, where the excitation field h_{AC} and the external magnetic field H_{DC} are respectively perpendicular and parallel to the plane of the sample.

Let us now study the relation between the measured complex voltage, $V = V_L + iV_R$, and the components of the AC magnetic susceptibility, χ' and χ". We assume that the current in the drive coil (primary) is given by $I_D e^{i\omega t}$, which creates at the sample an average magnetic field $H_D e^{i\omega t}$. Looking at a section of the sample as a simple loop, we model its response as an impedance Z_s in series with a geometrical inductance, L_g. The impedance depends on the material parameters as well as the size of the loop. For a normal metal sample, $Z_S = 2\pi\rho(rt)\Delta r$, with ρ the resistivity of the material, r the radius of the loop, t the thickness of the sample, and Δr the width of the loop. We can obtain analogous equations for the specific case of a superconducting material. The equation relating the drive field to the current response I_S of the loop is given by:

$$-\frac{\partial \Phi_{ext}}{\partial t} = -i\omega\mu_0 H_D e^{i\omega t} A = I_S(Z_S + i\omega L_g) \qquad [1]$$

where A is the area of the loop. Taking $Z_S = X = iY$, Eq. [1] reduces to:

$$I_S = \frac{-iA\omega\mu_0 H_D e^{i\omega t}}{X + i(Y + \omega L_g)} \qquad [2]$$

The induced voltage in the pick-up coil is given by:

$$-M_{SP}(i\omega I_S) = V_P \qquad [3]$$

where M_{SP} is the mutual inductance between the sample and the pickup coil. Combining Eqs. [2] and [3] we obtain:

$$V_P = -\frac{\omega^2 A M_{SP} \mu_0 H_D e^{i\omega t}}{X + i(Y + \omega L_g)} \qquad [4]$$

To obtain the magnetic susceptibility we first find a relationship between the effective magnetization <M> of the loop and I_S. Since $B = \mu_0 (H + M)$ we may write:

$$\mu_0[<H> + <M>]A = <B> A = \Phi \qquad [5]$$

From this, we identify the magnetic flux due to the current in the sample as being proportional to the average magnetization:

$$\mu_0 <M> A = L_g I_S \qquad [6]$$

Combining Eqs. [2] and [6], gives:

$$-\frac{i\omega L_g H_D e^{i\omega t}}{X + i(Y + \omega L_g)} = <M> = (\chi' - i\chi'')H_D e^{i\omega t} \qquad [7]$$

Where we have neglected higher harmonics considering the response of the loop given by the average magnetization:

$$<M> = (\chi' - i\chi'')H_D e^{i\omega t} \qquad [8]$$

Since the pickup coil is counter wound, it only responds to dM/dt, so that:

$$V_P \propto -\frac{\partial M}{\partial t} \propto (-\omega\chi'' - i\omega\chi')H_D e^{i\omega t} \qquad [9]$$

From Eqs. [2], [6], [7] and [8], we obtain:

$$\frac{\mu_0 M_{SP} A\omega}{L_g}(-\chi'' - i\chi')H_D e^{i\omega t} = V_P = V_P' + iV_P'' \qquad [10]$$

which agrees with Eq.[9]. From Eq. [7] we can write:

$$\chi' = \frac{\omega L_g Y + \omega^2 L_g^2}{X^2 + (Y + \omega L_g)^2} \qquad [11\text{-a}]$$

$$\chi'' = \frac{\omega L_g X}{X^2 + (Y + \omega L_g)^2} \qquad [11\text{-b}]$$

To get the complete response of a real sample, these equations should be integrated over the whole specimen. For the special case of a superconducting loop far below T_C, where we can neglect the normal channel in a two-fluid model, the induced EMF in a magnetic field $H_D e^{i\omega t}$ is still given by $\varepsilon = -i\omega A\mu_0 H_D e^{i\omega t}$. The loop has now a kinetic inductance L_K as well as a geometrical inductance L_g so that the current is given by $-i\omega A\mu_0 H_D e^{i\omega t} = i\omega(L_K + L_g)I_S$, or $I_S = -(A\mu_0 H_D e^{i\omega t})/(L_K + L_g)$. Eq. [6] implies that the magnetization is:

$$< M > = -\frac{L_g}{L_K + L_g} H_D e^{i\omega t} \qquad [12]$$

or, alternatively, that:

$$\chi' = -\frac{L_g}{L_K + L_g} \qquad [13]$$
$$\chi'' = 0$$

which agrees with Eqs. [11] setting $X = 0$ and $Y = \omega L_K$.

This means, the components of the complex AC magnetic susceptibility can be associated with the measured voltage through:

$$\chi' \propto V_L \qquad [14\text{-a}]$$
$$\chi'' \propto V_R \qquad [14\text{-b}]$$
$$\chi'' \propto V_R$$

as we stated in the beginning.

III. NUMERICAL SIMULATIONS OF THE MAGNETIC BEHAVIOR OF JJA

We have found that all our experimental results obtained from the magnetic properties of JJA can be qualitatively explained by analyzing the dynamics of a single unit cell in the array [16, 17].

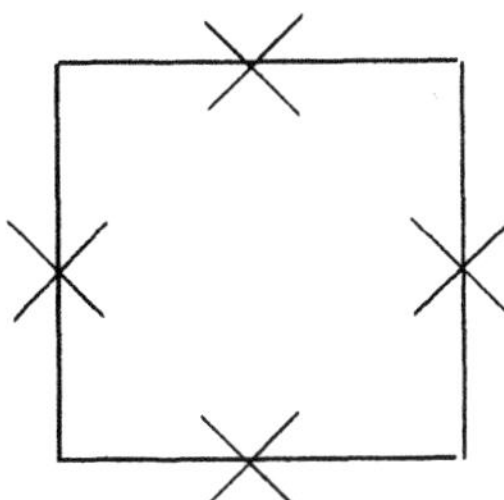

Figure 8 – Unit cell of the array, containing a loop with four identical junctions.

In our experiments, the unit cell is a loop containing four junctions (Fig. 8) and the measurements correspond to ZFC AC magnetic susceptibility. We model a single unit cell as having four identical junctions, each with capacitance C_J, quasi-particle resistance R_J and critical current I_C. We apply an external field of the form:

$$H_{ext} = h_{AC}\cos(\omega t) \qquad [14]$$

The total magnetic flux, Φ_{TOT}, threading the four-junction superconducting loop is given by:

$$\Phi_{TOT} = \Phi_{EXT} + LI \qquad [15]$$

where $\Phi_{EXT} = \mu_0 a^2 H_{EXT}$ with μ_0 being the vacuum permeability, I is the circulating current in the loop, L is the inductance of the loop and Φ_{EXT} is the flux related to the applied magnetic field. Therefore de current is given by:

$$I = I_C \sin\gamma_i + \frac{\Phi_0}{2\pi R_J}\frac{d\gamma_i}{dt} + \frac{C_J\Phi_0}{2\pi}\frac{d^2\gamma_i}{dt^2} \qquad [16]$$

Here, γ_i is the superconducting phase difference across the ith junction and I_C is the critical current of each junction. In the case of our model with four junctions, the fluxoid quantization condition, which relates each γ_i to the external flux, is:

$$\gamma_i = \frac{\pi}{2}n - \frac{\pi}{2}\frac{\Phi_{TOT}}{\Phi_0} \qquad [17]$$

where n is an integer and, by symmetry, we assume:

$$\gamma_1 = \gamma_2 = \gamma_3 = \gamma_4 = \gamma_i \qquad [18]$$

In the case of an oscillatory external magnetic field of the form of Eq. [14], the magnetization is given by:

$$M = \frac{LI}{\mu_0 a^2} \qquad [19]$$

It may be expanded as a Fourier series in the form:

$$M(t) = h_{AC}\sum_{n=0}^{\infty}[\chi_n^{'} \cos(n\omega t) + \chi_n^{''} \sin(n\omega t)] \qquad [20]$$

We calculated χ' and χ'' through this equation. Both Euler and fourth-order Runge-Kutta integration methods provided the same numerical results. In our model we do not include other effects (such as thermal activation) beyond the above equations. In this case, the temperature-dependent parameter is the critical current of the junctions, given to good approximation by [19]:

$$I_C(T) = I_C(0)\sqrt{1-\frac{T}{T_C}}\tanh\left[1.54\frac{T_C}{T}\sqrt{1-\frac{T}{T_C}}\right] \qquad [21]$$

We calculated χ_1 as a function of T. χ_1 depends on the parameter β_L, which is proportional to the number of flux quanta that can be screened by the maximum critical current in the junctions, and the parameter β_C, which is proportional to the capacitance of the junction:

$$\beta_L(T) = \frac{2\pi L I_C(T)}{\Phi_0} \qquad [22]$$

$$\beta_C(T) = \frac{2\pi I_C C_J R_J^2}{\Phi_0} \qquad [23]$$

By using the equations above, we can simulate the magnetic behavior of a particular JJA with specific parameters β_C, β_L, and $I_C(T)$. This last parameter gives the temperature dependency of the simulated properties.

IV. PME IN GRANULAR SUPERCONDUCTING SYSTEMS

The paramagnetic Meissner effect (PME) measured in high T_C granular superconductors has been attributed to the presence of π-junctions between the grains. Here we present measurements of complex AC magnetic susceptibility from two-dimensional arrays of conventional (non π) Nb-AlO_x-Nb Josephson junctions. We measured the susceptibility as a function of the temperature T, the AC amplitude of the excitation field, h_{AC}, and the external magnetic field, H_{DC}. The experiments show a strong paramagnetic contribution from any multi-junction loop, which manifests itself as a reentrant screening at low temperature, for values of h_{AC} higher than 50 mOe. The highly simplified model described before, based on a single loop containing four junctions, accounts for this paramagnetic contribution and the range of parameters in which it appears. This model offers an alternative explanation of PME which does not involve π-junctions.

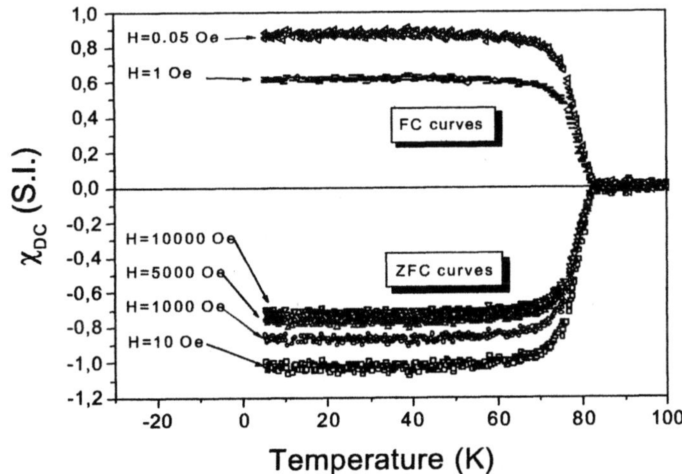

Figure 9 - Paramagnetic FC (field cooled) response obtained from Bi-based polycrystalline superconducting samples, similar to those obtained by Braunisch *et al*[20,21].

An experimental study of the paramagnetic response in Bi-based polycrystalline superconductors was first reported by Braunisch *et al.*[20,21]. They measured a paramagnetic DC susceptibility at values of temperature lower than the critical temperature T_C of their superconducting samples. We have obtained similar results in 2212-BSCCO samples (Fig. 9).

This paramagnetic response was in striking contrast to the usual diamagnetic Meissner effect, where the magnetic field is excluded from superconductors. The PME appeared systematically under specific experimental conditions and depended on sample preparation and morphology:

(1) The samples had to be cooled below T_C in the presence of small magnetic field, H < 1 Oe; by increasing the value of H in their field cooled (FC) experiments, they observed a crossover of the DC magnetic susceptibility to diamagnetic values.

(2) The PME was strongly dependent on the granular structure. Grinding the samples into small powder (a process that substantially weakens the contact between the grains) suppressed or even destroyed PME.

(3) Weak links with rather high critical currents were essential for the occurrence of PME: only melt-processed samples, more densely packed and with higher critical currents with respect to the sintered ones, showed PME.

Braunisch *et al.* also found that, after cooling the same samples following a zero external field procedure (ZFC), the measured susceptibility was diamagnetic. The authors attributed PME to the occurrence of spontaneous currents, flowing in direction opposite to ordinary Meissner screening currents. They proposed that anomalous Josephson junctions between the grains may be responsible for the existence of such currents. In these junctions (π-junctions) the Cooper pairs acquire a phase shift equal to π in the tunneling process and the Josephson current has direction opposite to conventional junctions. π-junctions may be the consequence of magnetic impurities in the junction[22,23], or non-s wave pairing symmetry[24].

Regardless of the origin of π-junctions, Dominguez *et al.*[25] have modeled a granular superconductor by considering a network where the nodes represent the grains and the links represent the coupling between the grains. This network was a mixture of normal junctions and π-junctions. In this case, the low-temperature and low field ZFC susceptibility was of the order of -1 (SI), while the FC susceptibility was paramagnetic for some values of magnetic field, reproducing qualitatively the experimental data obtained from Bi-based superconductors.

Other models based on networks of conventional junctions could explain the PME experimental results. Auletta *et al.*[26,27] found that numerical simulations of a two-dimensional array of conventional Josephson junctions, made of concentric multi-junction loops, lead to positive FC magnetic susceptibility, qualitatively similar to the experimental PME.

V. Experimental and Numerical Results

Our experiments on the AC magnetic susceptibility of two-dimensional Josephson junction arrays, in ZFC experiments, are in agreement with this last picture, i.e. they show that networks of conventional Josephson junctions can give a paramagnetic contribution to the measured susceptibility. We use the simple multi-junction loop model showed in before to explain how, in spite of the very different experimental conditions, our experiment can provide an alternative explanation for PME.

Our samples consist of 100×150 unshunted tunnel junctions, similar to those shown in Fig. 3a. The unit cell had square geometry with lattice spacing $a = 46\ \mu m$ and a junction area of $5 \times 5\ \mu m^2$ (Fig. 10). From these dimensions, we estimated that the inductance of each loop was about 64 pH. The critical current density for the junctions forming the arrays was about 600 A/cm^2 at 4.2K, giving $I_C = 150\ \mu A$ for each junction.

We have performed four different types of experiments:

(1) $\chi_{AC}(T)$, for different fixed values of h_{AC} and $H_{DC} = 0$;

(2) $\chi_{AC}(T)$ for a fixed value of h_{AC}, and different values of H_{DC};

(3) $\chi_{AC}(h_{AC})$ for fixed values of the temperature, and $H_{DC} = 0$;

(4) $I_C(H_{DC})$ for a fixed temperature, from I x V characteristics.

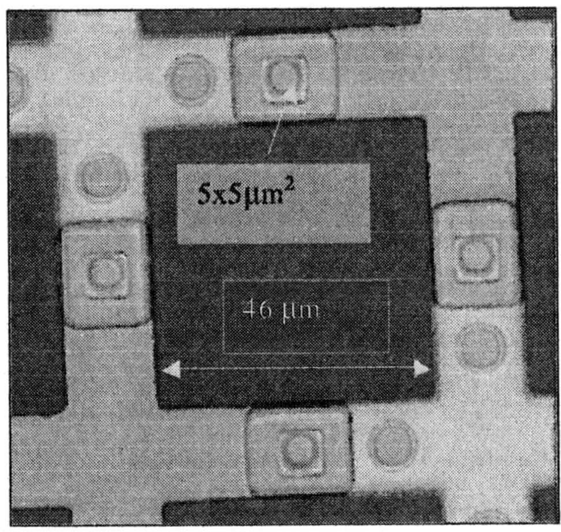

Figure 10 - The unit cell of the JJA with square geometry with lattice spacing $a = 46\ \mu m$ and a junction area of $5 \times 5\ \mu m^2$.

Figure 11 shows the results for $\chi_{AC}(T)$, obtained from ZFC experiments, for h_{AC}= 10, 80, 96, 144 and 320 mOe, and with H_{DC}= 0. For h_{AC} smaller than about 50 mOe, the behavior of both components of $\chi(T)$ is quite similar to typical superconducting samples[11]. The real component, $\chi'(T)$, which is a measure of the screening current, becomes more negative at lower temperatures, indicating stronger superconductivity through the Meissner effect. The imaginary component $\chi''(T)$ has a peak, indicating a maximum in the losses, around the critical temperature, T_C. Notice that $\chi' \approx -0.7$ (SI) for h_{AC} = 10 mOe, at low temperature. The sample can only partially screen the external magnetic field. As we can see in Figure 8, the screening of the array is weaker compared to screening in a thick (500 nm) niobium film, taken as a reference. We therefore define the array to be in a Meissner-like state for h_{AC}< 50 mOe. This partial screening will be qualitatively explained in the following section, through the single-loop picture.

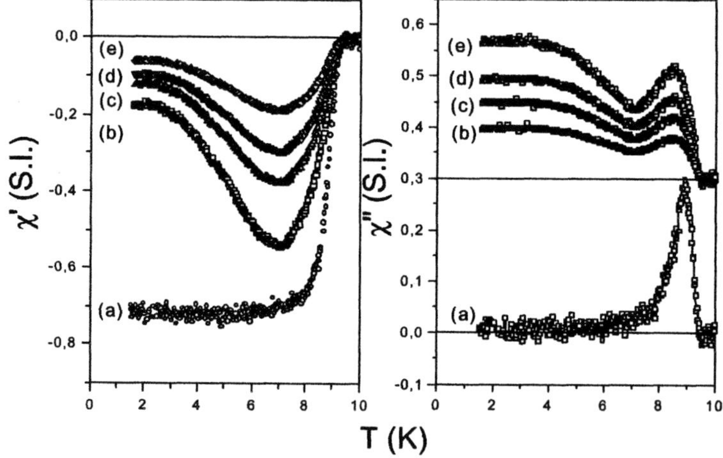

Figure 11 - Curves of $\chi_{AC}(T)$ for h_{AC}=(a)10 mOe, (b) 80 mOe, (c) 96 mOe, (d) 144 mOe and (e) 320 mOe with no applied external magnetic field.

Outside the Meissner-like regime, for values of h_{AC} > 50 mOe, $\chi'(T)$ is reentrant. It first increases in modulus as the temperature is lowered from the critical temperature T_C, then decreases at a lower temperature. The minimum in $\chi'(T)$ appears at T ≈ 7.0 K. For all the temperatures, at a fixed value of T, the modulus of $\chi'(T)$ decreases by increasing h_{AC}. The out-of-phase component, $\chi''(T)$, is correlated with the reentrance observed in $\chi'(T)$, showing increasing losses as the screening decreases, indicating an apparent weakening of the order parameter at low temperatures.

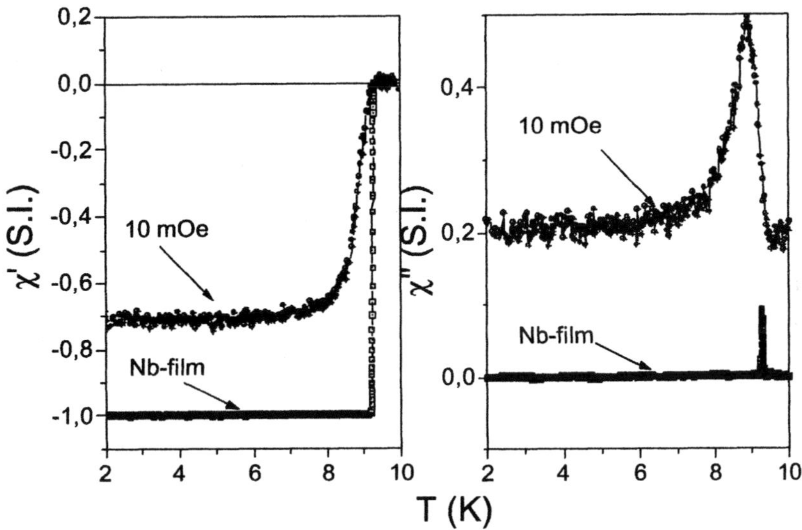

Figure 12 - The JJA can screen the external magnetic field only partially, in a weaker way than that of a thick (500 nm) niobium film, taken as a reference for the total Meissner state ($\chi'=-1$). In this figure we compare the Nb response with that from the JJA for h_{AC}= 10 mOe. The imaginary component, χ'', has been shifted of 0.2 S.I. units for clarity.

To experimentally investigate the origin of the reentrance, we have measured $\chi_{AC}(T)$ at a fixed value of the amplitude of the excitation field, h_{AC}= 96 mOe, for different values of H_{DC}, as shown in Figure 13. As already mentioned in Section III, the external magnetic field H_{DC} is parallel to the plane of the array (Fig. 7). For our sample geometry, this parallel field suppresses the critical current I_C of each junction, while inducing a negligible flux into the "holes" of the array (we estimate an alignment to the parallel direction within 0.1°. The measurements show that the position of the reentrance is being tuned by H_{DC}. We also observe that the value of temperature T_m at which of $\chi'(T)$ has a minimum shifts towards lower temperatures as we raise H_{DC}, down to $T_m \approx 6$ K at H_{DC} = 33 Oe. Further increase of H_{DC} shifts T_m back to higher temperature.

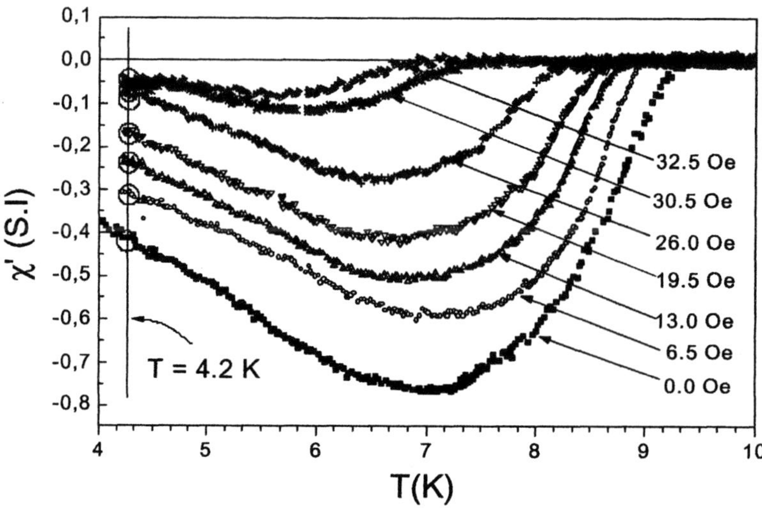

Figure 13 - Curves for $\chi_{AC}(T)$ for h_{AC}= 96 mOe, for H_{DC}=0, 6.5, 13, 19.5, 26, 30.5, and 32.5 Oe. The circles indicate the values of χ' at T=4.2 K to be used in Fig. 10.

This non-monotonic behavior is similar to the dependence of the Josephson junction critical current on a magnetic field applied in the plane of the junction[28] (Fraunhofer pattern). We measured $I_C(H_{DC})$ from transport current-voltage characteristics, at different values of H_{DC} and at T = 4.2K (see Fig. 14). We find that $\chi'(T=4.2K)$, obtained from the isotherm T = 4.2K (see Fig. 13), shows the same Fraunhofer-like dependence on H_{DC} as the critical current I_C of the junctions forming the array. This gives further proof that only the junction critical current is varied in this experiment. This also indicates that the screening currents at low temperature (i.e., the reentrant region) are proportional to the critical currents of the junctions. Furthermore, this shows an alternative way to obtain $I_C(H_{DC})$ in big arrays.

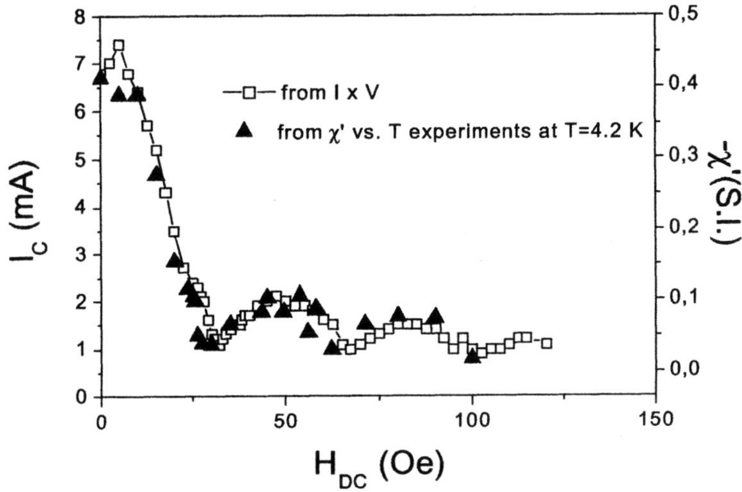

Figure 14 - Critical current I_C(open squares) and χ' (triangles) as a function of the external magnetic field, H_{DC}, for T = 4.2K.

We have simulated $\chi_1'(T)$ and $\chi_1''(T)$ for different values of h_{AC} by following the equations developed in a preceding section. The obtained results are shown in Fig. 15. For values of h_{AC} smaller than 47 mOe (corresponding to $\mu_0 a^2 h_{AC} \approx 5\Phi_0$), $\chi_1'(T)$ decreases with decreasing temperature, and $\chi_1''(T)$ is close to zero. By increasing h_{AC} above 47 mOe, reentrance at low temperature clearly appears and the screening becomes weaker in all the temperature range. This is consistent with the experiment, where the magnitude of $\chi'(T)$ decreases with increasing h_{AC}. Note that at these high values of h_{AC} the simulated $\chi_1''(T)$ increases significantly, i.e. the simulation reproduces the dramatic increase of losses at low temperature that is found in the experiment (see Figs. 11 and 12).

In the simulated $\chi_1'(T)$ the reentrance appears as a paramagnetic region below 5.0 K. Another paramagnetic region is present above 7.5 K. These regions correspond to the measured decrease of screening below and above the minimum value of $\chi_1'(T)$ in Figure 11. The simulated $\chi_1'(T)$ is either paramagnetic or diamagnetic, depending on the temperature range. This surprising result can be understood by calculating the curves Φ_{TOT} vs. Φ_{EXT} at different temperatures. As an example, in Figure 16 we plot these curves for the same parameters used for the Figure 15-c, so that:

$$\Phi_{EXT} = \mu_0 h_{AC} a^2 \cos\omega t = 7\Phi_0 \cos\omega t$$

At low values of temperature, these curves are very hysteretic, showing multiple branches (see Fig. 16-c). The hysteresis decreases with increasing the temperature (and eventually disappears at T ≈ 8.5 K).

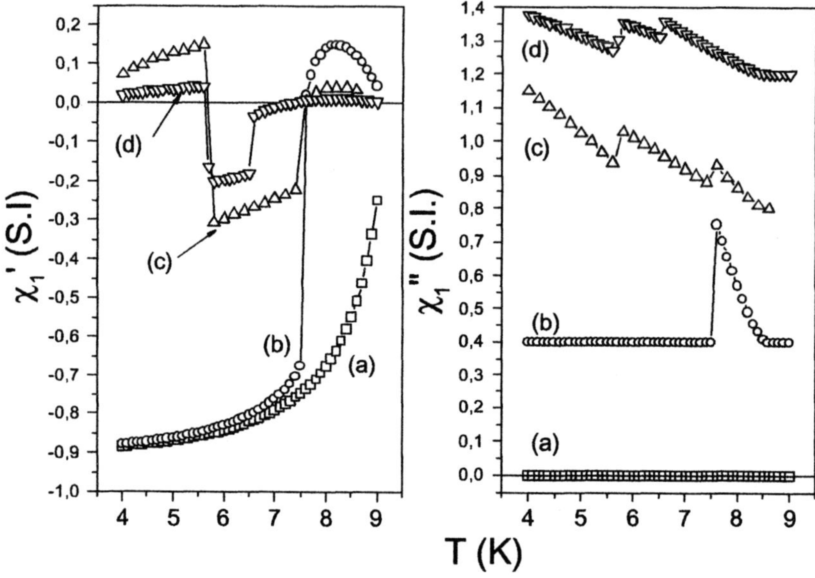

Figure 15 - Curves for the simulated $\chi_1'(T)$ and $\chi_1''(T)$ for (a) h_{AC}= 5 mOe, (b) 29 mOe, (c) 69 mOe, (d) 118 mOe. In all cases $\beta_L(T=4.2K)=30$ and $\beta_C(T=4.2K)=60$. Curves for $\chi_1''(T)$ have been vertically shifted by 0.4 (SI) for clarity.

The important aspect of these curves is that they contain both paramagnetic and diamagnetic states. In Figure 16, the line $\Phi_{TOT}=\Phi_{EXT}$ marks the boundary between diamagnetic (DIA) states and paramagnetic (PARA) states. For clarity, we shaded the diamagnetic areas in the graph Φ_{TOT} vs. Φ_{EXT}, while clear areas correspond to paramagnetic states $\Phi_{TOT}>\Phi_{EXT}$.

At a fixed value of temperature and h_{AC}, the value of $\chi_1(T)$ is a time average of all the magnetic states that the system transverses during one cycle of H_{EXT}. In other words, $\chi_1(T)$ is either diamagnetic or paramagnetic, depending on the shape of the part of the hysteresis curve that is spanned during one cycle of H_{EXT}.

The shape of the curve Φ_{TOT} vs. Φ_{EXT}, changes with temperature. For example, Figure 15-a, at T = 7.6 K, has three stable branches at positive values of Φ_{EXT}. Note that one branch is completely paramagnetic. This causes the average response to be paramagnetic, making $\chi'_1(T)$ positive. This scenario corresponds to the positive values of $\chi'_1(T)$ at T > 7.5 K, shown in Figure 15-c.

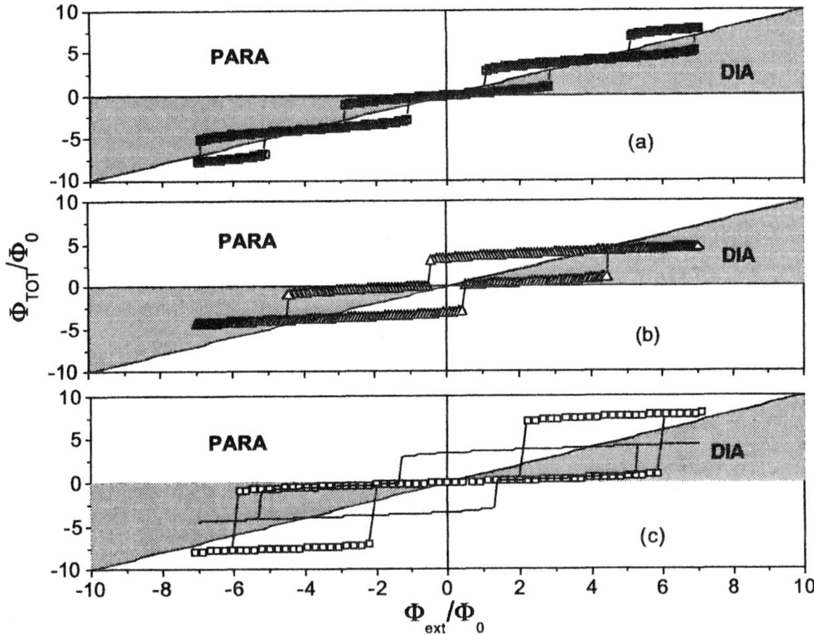

Figure 16 - Simulated curves of Φ_{TOT} vs. Φ_{EXT} for (a) T = 7.6 K, (b) 6.0 K, (c) 4.0 K; for all curves $\beta_L(T=4.2K)=30$; for curve (c) we show curves for $\beta_C(T=4.2K)=60$ (open squares) and $\beta_C(T=4.2K)=2$ (continuous solid line).

By lowering the temperature, the length of all the branches increases until, at about T = 7.5K, the second branch extends in the diamagnetic region up to the highest values of Φ_{EXT} ($7\Phi_0$) and the third branch becomes inaccessible. In fact, for T < 7.5 K, the curve Φ_{TOT} vs. Φ_{EXT}, consists of two branches, as shown in Figure 16-b. In this region, the average is diamagnetic, i.e. $\chi'_1(T)$ is negative. (Note that the branch crossing $\Phi_{TOT} = 0$

is all diamagnetic, while the second branch is diamagnetic at high values of Φ_{EXT} and paramagnetic at low values of Φ_{EXT}: the average turns out to be diamagnetic). This scenario corresponds to the negative values of $\chi'_1(T)$ at 5.0 <T< 7.5 K, shown in Figure 16-c.

By further lowering the temperature, below T ≈ 5.0 K, the first branch extends to higher values of Φ_{EXT}, where the third branch becomes stable and the second unstable (at $\Phi_{TOT}/\Phi_0 = 6$ there is a crossover of stability between the second and the third branch). Therefore, the system switches directly from the first branch to the third, which is fully paramagnetic (see Fig. 16). In this case, the average response is paramagnetic and $\chi'_1(T)$ is positive. This scenario corresponds to the positive values of $\chi'_1(T)$ at T < 5.0 K, shown in Fig. 15-c). Our simulations show that different values of β_C affect the switching point (i.e. the length) of the branches in the low temperature range. For example, we show that the curve at T = 4.2 K for $\beta_C = 2$ (solid line in Fig. 16-c) is qualitatively very similar to the curve at T = 6.0 K (Fig. 16-b). This happens because of the early switch from the first to higher branches occurring at lower values of Φ_{EXT}, i.e. at values of Φ_{EXT} where the second branch is still stable. For small values of β_C, our simulations show no reentrance at low temperature. We have also calculated the susceptibility spectra, $\chi'_1(h_{AC})$ at T = 4.2 K (Fig. 17). The jump in $\chi'_1(T)$ corresponds to magnetic flux entering the loop at the switch from the first branch to higher branches. The calculated value of external field (47 mOe) corresponding to this jump is very close to the measured value (50 mOe). As we will show coming works, this is strongly related to vortex avalanche and a critical state phenomena in JJA.

In the simulated $\chi_{AC}(T)$ the minimum value of $\chi'_1(T)$ is about -0.9 (see Fig. 16), meaning that the sample can only provide partial screening. This is due to the fact that the diamagnetic branch crossing the value $\Phi_{EXT} = 0$ (see Fig. 16) has a non-zero slope, i.e. some flux penetrates the sample even in the Meissner-like regime. What distinguishes the Meissner-like regime from the reentrant regime is the fact that the Meissner-like regime is reversible, while the reentrant regime is not. In fact, the reentrant regime involves switching to higher branches, which introduces pinning and hysteresis, in other words, irreversibility.

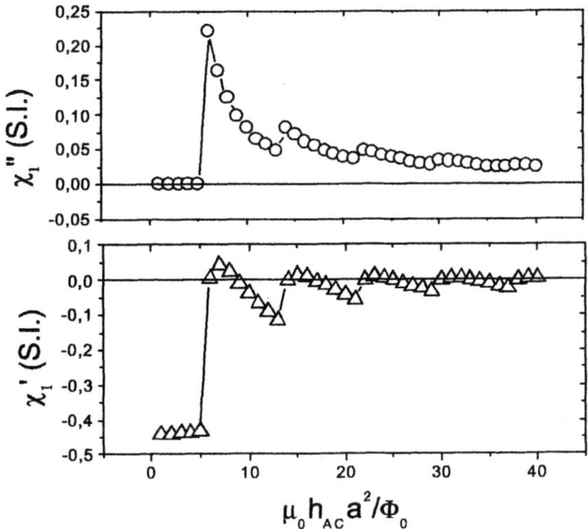

Figure 17 - Numerical result for $\chi(h_{AC})$ for T = 4.2 K, using the four-junctions model. As can be observed, there is a sharp increase in both χ' and χ'', around h_{AC}=50 mOe.

An analysis of $\chi''(T)$ confirms this picture: the losses are negligible in the Meissner-like regime and increase significantly in the reentrant regime.

Surprisingly, numerical simulations of the simple model based on a four junctions loop described before, account very satisfactorily for our experimental results, suggesting that this reentrance is dynamic in origin. However, we cannot make a completely quantitative comparison between our model and the measured array. The response of the array results from an average of the response from many loops. The flux distribution in the array is in general non-uniform, giving rise to different values of h_{AC} in different loops. As a consequence, the measured response of the arrays presents no sharp transitions. The profile of the field penetration in the whole array has been analyzed by other authors [29,30-35], but is not included in our model.

We can only make a quantitative comparison with our model in the Meissner-like state. In this case, the screening current in each loop is equivalent to a screening current flowing through a very large loop of junctions (about the dimension of the diameter of the coil).

In the reentrant state, it is instructive to compare the simulation with the measured response after subtracting the contribution of the superconducting islands. From our data, we can subtract the contribution of the niobium from the measured χ_{AC} of the array. In Figure 18 we subtract the measured niobium response (multiplied by the factor 11/46, which accounts for the fraction of volume the niobium occupies in the sample) from the data corresponding to the curve of χ_{AC} vs. T at h_{AC} = 96 mOe which we showed in Figure 12. The resulting curve shows paramagnetic response both at low and high temperature, analogous to the simulations in Figure 15. Both, the non-uniform flux distribution in the array and the contribution of the niobium islands, are responsible for the fact that our total measured response is always diamagnetic. The signature of the paramagnetic contribution from some multi-junction loops is the reentrance, but the paramagnetic contribution is never sufficiently strong to change the sign of the (total) measured $\chi'_{I}(T)$.

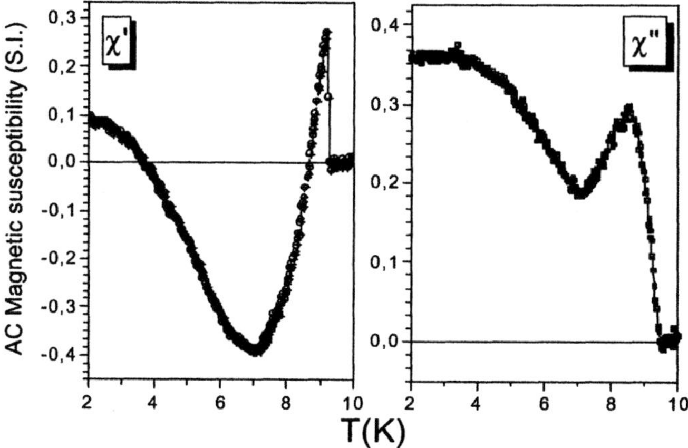

Figure 18 - Measured response of the JJA after subtracting the contribution of the superconducting Nb islands.

We measured PME from a network of conventional (non π) Josephson junctions. In our experiments, PME occurs in the form of a low-temperature reentrance in the AC magnetic susceptibility. This reentrance appears for values of h_{AC} higher than about 50 mOe, in excellent agreement with our estimated value of $LI_C\mu_0a^2$=3.7 A/m ≈ 47 mOe. This value of h_{AC} is a threshold for the screening of the sample, i.e. for the Meissner-like state. Above this value, magnetic flux penetrates the sample and is pinned, because of the high value of I_C and β_L. (High values of β_L correspond to strong pinning, i.e. strong hysteresis in the Φ_{TOT} vs. Φ_{EXT} curve.) Through numerical simulations of a simple model, we showed that the multi-junction loops are paramagnetic in the reentrant region. Moreover, by subtracting the contribution of the niobium islands from the measured reentrant susceptibility, we find a similar paramagnetic response from the multi-junction loops in our samples. We note that these results confirm measurements of PME in a niobium disk performed by Kostic and collaborators[36].

Our results can also be directly related to the PME measured in high-T_C granular superconductors. Returning to the summary of high-T_C experimental data in the introduction of this work:

(1) PME occurs in low field FC experiments because flux quanta get trapped in the voids between the grains. In our experiment, this happens for $h_{AC} >$ 50 mOe, in the reentrant regime, corresponding to states in upper overlapping branches of Figure 16. These branches are paramagnetic at small values of the field, and become diamagnetic at higher values, explaining the observed crossover.

(2) PME occurs only if there are weak links between the grains. This follows naturally from our multi-junction loop model.

(3) PME appears for strongly coupled grains because high values of β_L are required to get hysteretic Φ_{TOT} vs. Φ_{EXT} curves (see Fig. 16).

The diamagnetic response measured from these materials at small values of magnetic field in ZFC experiments can also be explained within the same scenario. In these experimental conditions, most of the loops will be in states corresponding to the diamagnetic branch crossing $\Phi_{TOT} = 0$ (see Fig. 16). In our measurements, this corresponds to the Meissner-like regime.

Perhaps the most striking discrepancy between our results and the measurements reported for granular superconductors is that granular samples are either paramagnetic or diamagnetic when measured with DC methods, with no reentrance measured in χ_{AC}. By contrast, our samples and simulations show a reentrance in χ_{AC} as temperature is lowered. There are two reasons for this:

(1) a granular system has a distribution of critical currents and loop sizes, and thus a distribution of β_L's. As can be verified in the literature, typical values of β_L in granular high-T_C superconductors are in the range 5-300[37];

(2) although the value for β_C is not known for typical grains, it is probably less than one.

In our simulations, only loops with large β_L and β_C display reentrant AC susceptibility. By contrast, to have a paramagnetic χ_{DC}, only a multi-branch solution is needed (see Fig. 16).

VI. SUMMARY

Phenomena causing the reentrance we observed in Josephson junction arrays should also exist in granular superconductors. This has been recently experimentally confirmed reported by Passos *et al.*[38] where HTS samples with controlled granularity have shown reentrance. The reentrant behavior, associated to PME, appears as an anomalous increase of dissipation at low temperature, in the case of AC susceptibility measurements. It is interesting to mention that Nielsen *et al.*[38] have confirmed the occurrence of PME in unshunted JJA samples by using the scanning SQUID microscopy technique[39]. Numerical simulations of two-dimensional Josephson junction networks with a distribution of characteristic parameters β_L and β_C would be very useful for further theoretical investigations of these phenomena.

REFERENCES

[1] A. M. Saxena *et al.*, Sol. State Comm. **14**, (1974) 799.

[2] M. L. Yu and A. M. Saxena, IEEE Trans. Mag. **11**, (1975) 674.

[3] T. D. Clark, Phys. Lett. 27A, (1968) 585.

[4] D. J. Resnick *et al.*, Phys. Rev. Lett. **47**, (1981) 1542.

[5] F. M. Araujo-Moreira, O. F. de Lima, W. A. Ortiz, Physica C, **240**, (1994) 3205.

[6] F. M. Araujo-Moreira O. F. de Lima, W. A. Ortiz., J. Appl. Phys. **80**, 6, (1996) 3390.

[7] F. M. Araujo-Moreira O. F. de Lima, and W. A. Ortiz, Physica C **311**, (1999) 98.

[8] W. A. C. Passos, P. N. Lisboa-Filho, R. Caparroz, C. C. de Faria, P. C. Venturini, F. M. Araujo-Moreira, S. Sergeenkov and W. A. Ortiz; Physica C **354**, (2000) 189.

[9] R. S. Newrock, C. J. Lobb, U. Geigenmüller, and M. Octavio, Solid State Physics **54**, (2000) 263.

[10] P. Martinoli and C. Leeman, J. Low temp. Physics **118**, (2000) 699.

[11] R. B. Goldfarb M. Lelental, and C. A. Thomson, in *Magnetic Susceptibility of Superconductors and Other Spin Systems*, p. 49, edited by R. A. Hein, T. L. Francavilla and D. H. Liebenberg, Plenum Press, New York (1992).

[12] D. Wohlleben *et al.*, Phys. Rev. Lett. **66**, (1991) 3191.

[13] Braunich *et al.*, Phys. Rev. Lett. **68**, (1992) 1908.

[14] C. Kostic *et al.*, Phys. Rev. B **53**, (1996) 791.

[15] A. K. Geim *et al.*, Nature **396**, (1998) 144.

[16] F. M. Araújo-Moreira, P. Barbara, A. B. Cawthorne, and C. J. Lobb Phys. Rev. Lett., **78**, (1997) 4625.

[17] P. Barbara, F. M. Araujo-Moreira, A. B. Cawthorne, and C. J. Lobb, Phys. Rev. B **60**, (1999) 7489.

[18] J. L. Jeanneret, G. A. Gavilano, A. Racine, Ch. Leemann, and P. Martinoli, Appl. Phys. Lett. **55**, (1989) 2336.

[19] T. Wolf and A. Majhofer, Phys. Rev. B **47**, (1993) 5383.

[20] W. Braunisch, N. Knauf, S. Neuhausen, A. Grutz, A. Koch, B. Roden, D. Khomskii, and D. Wohlleben, Phys. Rev. Lett. **68**, (1992) 1908.

[21] W. Braunisch, N. Knauf, G. Bauer, A. Koch, A. Becker, B. Freitag, A. Grutz, V. Kataev, S. Neuhausen, B. Roden, D. Khomskii, and D. Wohlleben, Phys. Rev. B **48**, (1993) 4030.

[22] L. N. Bulaevskii, V. V. Kuzii, and A. A. Sobyanim, JETP Lett. **25**, (1977) 290.

[23] F. V. Kusmartsev, Phys. Rev. Lett. **69**, (1992) 2268.

[24] H. Kauamura and M. S. Li, Phys. Rev. B **54**, (1996) 619.

[25] D. Dominguez, E. A. Jagla, and C. A. Balseiro, Phys. Rev. Lett. **72**, (1994) 2773.

[26] C. Auletta, P. Caputo, G. Costabile, R. de Luca, S. Pace, and A. Saggese, Physica C **235-240**, (1994) 3315.

[27] C. Auletta, G. Raiconi, R. De Luca, S. Pace, Phys. Rev. B. **51**, (1995) 12844.

[28] A. Barone and G. Paternò in *Physics and Applications of the Josephson Effect*, John Wiley & Sons Publ., New York (1982).

[29] D. X. Chen, and A. Sanchez, J. Appl. Phys. **70**, (1991) 5463.

[30] R. De Luca, S. Pace, G. Raiconi, Phys. Lett. A **172**, (1993) 391.

[31] D. X. Chen, J. J. Moreno, and A. Hernando, Phys. Rev. B **53**, (1996) 6579.

[32] M. Chandran, and P. Chaddah, Physica **267** C, (1996) 59.

[33] D. Reinel, W Dieterich, A. Majhofer, and T. Wolf, Physica **245** C, (1995) 193.

[34] D. X. Chen, A. Sanchez, and A. Hernando, Physica C **250**, (1995) 107.

[35] D. X. Chen, A. Sanchez, and A. Hernando, Physica C **244** (1995) 123.

[36] P. Kostic *et al.*, Phys. Rev. B **53**, 791 (1996).

[37] R. Marcon, R. Fastampa, and M. Giura, Phys. Rev. B **39** (1989) 2796.

[38] W. A. C. Passos, P. N. Lisboa-Filho, and W. A. Ortiz; Physica C **341** (2000) 2723.

[39] A. P. Nielsen, J. Holzer, A. B. Cawthorne, C. J. Lobb, R. S. Newrock, and J. Markus, Physica B **280**, (2000) 444; A. P. Nielsen *et al.*, Phys. Rev. B **62** (2000) 14380.

ACKNOWLEDGMENTS

We thank M. G. Forrester, A. W. Smith and C. B. Whan for their technical help in the experiments. We also thank A. Sanchez and D. X. Chen for useful discussions. We gratefully acknowledge financial support from U. S. Air Force Office of Scientific Research, through grant no. F496209810072 and from the National Science Foundation through grant no. 9510464. F.M.A.M. also gratefully acknowledges financial support from Brazilian Agencies CNPq and FAPESP, under grants 96/7704-6 and 98/12809-7.

LOW-NOISE QUASI-OPTICAL SUPERCONDUCTING MIXER BASED ON NIOBIUM NITRIDE FOR SUBMILLIMITER WAVELENGTHS

Yoshinori Uzawa and Zhen Wang
Kansai Advanced Research Center, CRL
588-2 Iwaoka, Iwaoka-cho, Nishi-ku, Kobe 651-2492, Japan

1. INTRODUCTION

The term submillimeter-wave applies to electromagnetic waves with wavelengths ranging from 100 μm to 1 mm, or frequencies from 300 GHz to 3 THz. This submillimeter-wave band is a relatively unexploited region of the electromagnetic spectrum because it is very difficult to produce and detect submillimeter-waves. Recently, however, submillimeter-wave technologies are becoming important in many practical and basic scientific applications such as remote sensing, radio astronomy, plasma diagnostics, imaging, and communications. The depletion of stratospheric ozone, for example, is caused by trace gases such as ClO, BrO, and HO_2, many of which have intense spectral bands in the submillimeter-wave range. Emission spectroscopy in the submillimeter-wave range is thus a most promising method for remotely measuring these trace gases as well as ozone [1]. The submillimeter-wave band is also a critical one for astronomy. The signals it contains provide spectral and spatial information on the cosmic background, on very distant newly formed galaxies, and on the early stages of star formation within gas clouds in our own galaxy [2]. These applications require the use of low-noise heterodyne receivers, because the submillimeter signals are generally very weak and their frequencies are too high for the signals to be amplified directly. Heterodyne receivers use frequency mixers, which are usually nonlinear devices, to down-convert the RF signal into one with a much lower intermediate frequency (IF) efficiently. This mixer is one of the important components in a low-noise receiver. As well known, the lower the noise of the mixer is, the better to detect the weak RF signal (i.e. the higher signal-to-noise (S/N) ratio can be obtained, which improve the sensitivity). Thus, the mixer in the low-noise receiver should have low noise characteristics and furthermore high conversion efficiency for the

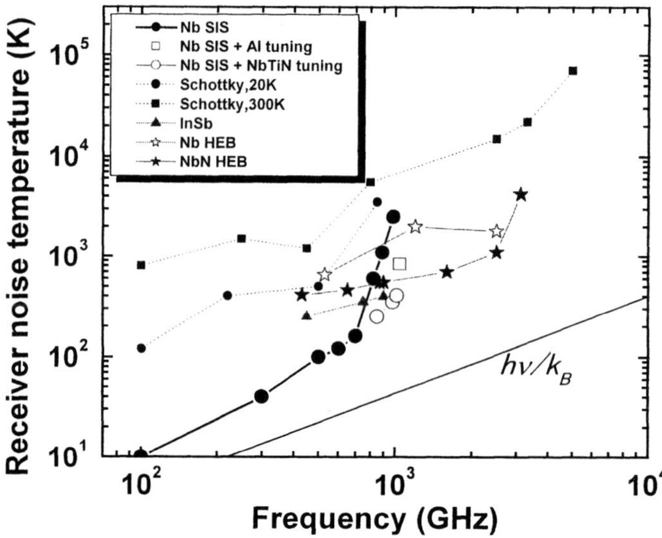

Fig. 1. State of art of heterodyne receivers at submillimeter wavelengths.

submillimeter-wave receivers.

Three types of heterodyne mixer elements have been used at millimeter and submillimeter wavelengths: Schottky-barrier diodes, superconductor-insulator-superconductor (SIS) junctions, and the recently developed superconducting hot-electron bolometric (HEB) mixers [3]-[10]. Their typical double sideband (DSB) noise performance is summarized in Fig. 1, where noise is shown as a function of LO frequency. In this figure, the solid straight line represents the quantum limit noise temperature of $h\nu/k_B$ for single-sideband (SSB), where h is Planck's constant, ν is frequency, and k_B is Boltzmann's constant. From this figure, it can be seen that SIS mixers are the most sensitive detectors and their noise temperatures are approaching the quantum limit at the millimeter- and submillimeter-wave frequencies, which usually employ Nb/AlOx/Nb tunnel junction and Nb microstripline to tune out the junction geometric capacitance. As shown in the figure, however, the noise performance degrades rapidly at around the Nb gap frequency of 700 GHz due to the onset of pair braking, and there is no excellent SIS mixers approaching the quantum limit above this frequency. To obtain higher operating frequencies, the SIS mixer must be implemented using a superconducting material that has a higher gap frequency than Nb, or alternative devices whose performance are not limited by the gap frequency, such as HEB mixers, must be developed. In this chapter, we present our recent results on SIS mixers and HEB mixers based on NbN at submillimeter-wave frequencies.

2. SUBMILLIMETER-WAVE SIS MIXERS BASED ON NBN

The low superconducting energy gap of Nb essentially limits its application as an ultralow noise SIS mixer in the terahertz frequency band because the onset of pair breaking above the gap frequency results in a rapid increase in the RF loss of superconducting electrodes and tuning circuits [11] as mentioned above. The only way to solve this problem is by using high energy gap (high T_c) superconducting materials instead of Nb as the junction electrodes and the tuning circuits. Although a high energy gap has advantages for high-frequency responses, it has disadvantages for fabricating tunnel junctions and tuning circuits. In mixer design, for example, the $\omega C_J R_N$ product, where ω is the RF angular signal frequency, C_J is the junction capacitance and R_N is the junction resistance, must be kept approximately constant (usually below 10) as ω increased for optimum response at high frequency. This requires that the junctions have a high current density J_c and a small $C_J R_N$ product. Since the junction $I_c R_N$ product scales with the superconducting energy gap, a higher energy gap junction needs to have higher current density to maintain a small $\omega C_J R_N$ product. As a result, a high current density of several tens kA/cm^2 is required for NbN tunnel junctions applied as terahertz SIS mixers. It is well known that the electrons only tunnel though a thin barrier with the thickness less than superconducting coherence length ξ so that it is very difficult to fabricate high-current-density SIS tunnel junctions with high T_c superconducting materials due to their short superconducting coherence length ξ (ξ_{NbN}~4-7 nm).

Up to now, NbN tunnel junctions have been the best choice for terahertz SIS mixers because of their high superconducting energy gap, corresponding with a gap frequency of about 1.4 THz, and ease of fabrication. NbN tunnel junctions with MgO tunnel barriers were widely studied ten years ago [12], [13], and SIS mixing experiments had been demonstrated in the millimeter wave regions [14], [15]. However, two problems limited the operating frequency and noise performances of the mixer using NbN/MgO/NbN tunnel junctions. First, the MgO has a high barrier potential and pinholes in thin tunnel barriers, which lead to difficulty in fabrication of tunnel junctions with high current density over 10 kA/cm^2 [16]. Second, NbN films fabricated on the conventional low dielectric substrates, such as quartz or Si, have a polycrystalline structure with a long penetration depth and a large surface resistance [17], [18]. These result in a large inductance and RF loss for tuning circuits, which greatly degrade the mixer noise performance. Thus, even though the NbN has a high gap frequency, there have been few reports on their use in submillimeter-wave SIS mixers because it is very difficult to fabricate high-quality and high-current density junctions and high-quality NbN film for tuning circuits as mentioned above.

Thus, it is necessary to develop a process for the deposition of single-crystal NbN thin

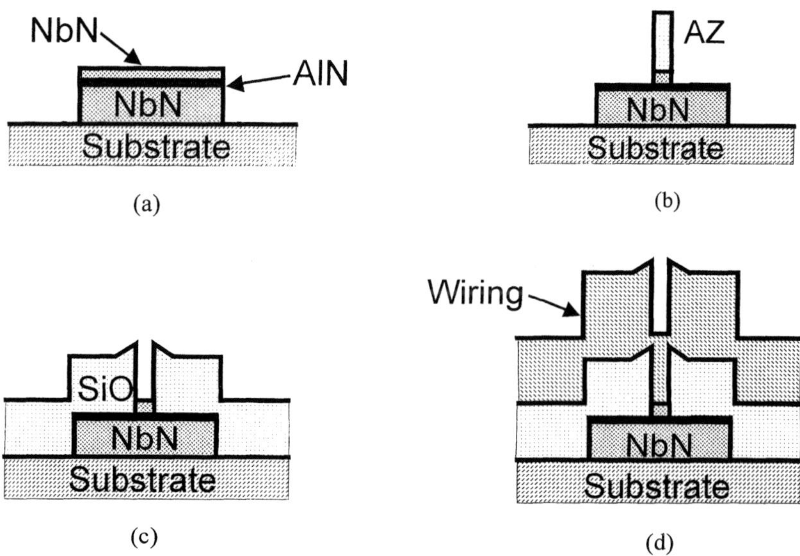

Fig. 2. Fabrication processed for NbN/AlN/NbN tunnel junctions. (a) Deposition of NbN/AlN/NbN trilayers, and (b) patterning of the base electrodes. (c) Deposition of SiO or other material and liftoff from the contact wiring window. (d) Deposition of NbN films and patterning in the contact wiring layer.

films and the fabrication of high-current-density NbN tunnel junctions with a tunnel barrier having a lower barrier potential than that of the MgO. In the following section, we first present the development of high current density NbN/AlN/NbN tunnel junctions [19]-[22]. Using these junctions, we next design quasi-optical SIS mixers at submillimeter-wave frequencies that does not have a conventional waveguide structure, and lastly investigate their experimental mixing properties at submillimeter-wave frequencies [23], [24].

2.1 High current density NbN/AlN/NbN tunnel junctions

A. Junction fabrication

The NbN/AlN/NbN trilayers were deposited *in situ* by rf magnetron sputtering in a load-locked sputtering system, and the junctions were prepared by photolithography and reactive ion etching (RIE) techniques. Figure 2 schematically shows the processes for fabricating the NbN/AlN/NbN tunnel junctions. The NbN/AlN/NbN trilayers were reactively sputtered in situ, from a 8-inch diameter Nb and Al target, in an Ar + N_2 mixture under a total pressure of 2 mTorr [Fig. 2(a)]. The deposition parameters are listed in Table

Table 1. Sputtering parameters for NbN and AlN films.

Base Pressure	< 3 x 10^{-7} Torr
Target	Nb: ϕ8 in. (99.99%)
	Al: ϕ8 in. (99.999%)
Sputtering gas	NbN: 87.5% Ar + 12.5% N_2
	AlN: 100% N_2
Total pressure	2 mTorr
RF power	NbN: 6.2 W/cm^2
	AlN: 4.6 W/cm^2
Deposition rate	NbN: 18 nm/min
	AlN: 1nm/min

1. Although the substrates were not heated intentionally, it can be estimated that the temperature rise due to the self-heating during sputtering was less than 100 °C because we were able to deposit the trilayers on the photoresist without its apparent degradation.

The junction area was patterned by using photo lithography and reactive ion etching in CF_4 [Fig. 2(b)]. Following the RIE, typically 300-nm-thick SiO or other material films were deposited and lifted off to form an electrical isolation layer between the base electrode and the contact-wiring layer [Fig. 2(c)]. Finally a contact-wiring layer of NbN or Nb was deposited and patterned by using photolithography and RIE [Fig. 2(d)].

To obtain high-quality junction with low leakage current and small gap voltage widths, both base and counter NbN films should have a uniform superconducting energy gap along the interface of the tunnel barriers. Thus it is important to fabricate uniform NbN films for the electrodes. The NbN films deposited onto MgO underlayers and single crystal MgO substrates have shown excellent superconducting and crystalline properties [19], [25] because NbN and MgO both have NaCl cubic crystal structure and have similar lattice constants (the misfit is about 4%). This work uses single-crystal MgO substrates with (100) surfaces as epitaxial substrates for base NbN films. The NbN films fabricated on the MgO substrates had best superconducting properties: a maximum critical temperature of 16 K, a minimum normal state resistivity of 62 μΩcm measured at 20 K, and a maximum residual resistivity ratio of 1.2 [19].

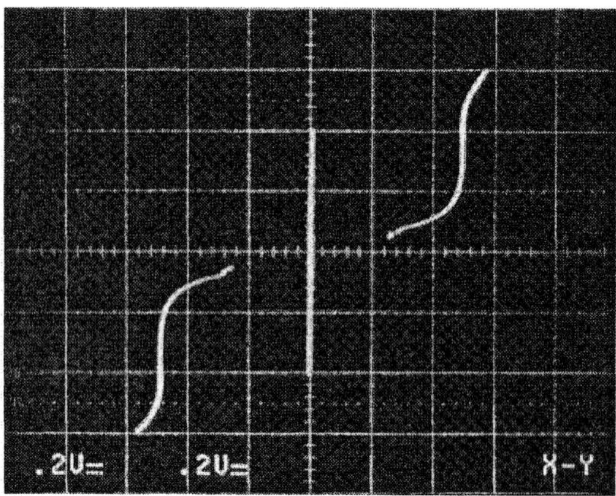

Fig. 3. Typical I-V characteristic measured at 4.2 K for a high current density NbN/AlN/NbN tunnel junction. The horizontal scale is 2 mV per division, and the vertical scale is 0.2 mA per division. The current density $J_c = I_c/A$ measured from the I-V curve and the mask size is 54 kA/cm^2.

B. I-V curve and tunneling properties

Typical I-V characteristics measured at 4.2 K for a high current density junction are shown in Fig. 3. The critical supercurrent I_c measured from the I-V curve is 0.42 mA, and the mask size of the junctions is 1 μm in diameter. This gives a current density $J_c = I_c/A$ of 54 kA/cm^2, where the junction size A is taken as the mask size. Since we used the contacted exposure photo-lithography to pattern the junction area, it is conceivable that the junction size was smaller than the mask size due to the overexposure. The current density J_c may therefore be higher than that calculated by the mask size.

Electrical parameters for some typical high J_c junctions are listed in Table 2. Figure 4 explains the definition of parameters for a SIS tunnel junction which describe the qualities of the junction. Gap voltage V_g is defined to be the voltage corresponding to the inflection point of quasiparticle current rise, and gap voltage width ΔV_g is given by the difference in the voltage for the quasiparticle current rising from 30% to 70% of the values at gap voltage. The theoretical $I_cR_N^{th}$ product was calculated from the Ambegaokar-Baratoff relation ($I_cR_N^{th} = \pi V_g/4$) in the weak coupling limit [26]. Since a strong coupling parameter $\alpha = 4.16$ was given at our work [19], the measured value of I_cR_N products were about 75% of the $I_cR_N^{th}$, which is the same factor found for strong coupling materials like Pb [27] and NbN/MgO/NbN tunnel junctions [12]. The subgap leakage factor V_m is defined as the product of I_c and subgap resistance R_{sg} measured at 4 mV, and R_{sg}/R_N is the ratio of the subgap resistance R_{sg} to the normal resistance R_N. Even though the current density of the

Table 2. Electrical parameters for typical high Jc NbN/AlN/NbN tunnel junctions measured at 4.2 K

Junction size (μm)	J_C (kA/cm^2)	V_g (mV)	ΔV_g (mV)	V_m (mV)	R_{sg}/R_N	$I_C R_N$ (mV)	$I_C R_N^{th}$ (mV)
2.5 x 2.5	5.2	5.0	0.16	25	9.3	2.7	3.9
5 x 5	8	5.0	0.18	24	8.5	2.8	3.9
ϕ1	15	5.2	0.12	26	8.8	3.0	4.1
ϕ1	20	5.1	0.12	24	7.9	3.1	4.0
ϕ2	28	5.1	0.14	15	5.2	3.0	4.0
ϕ2	35	5.1	0.12	15	5.2	3.0	4.0
ϕ1	54	5.1	0.10	14	4.5	3.1	3.9

junctions varies in the wide range of 5-54 kA/cm^2, as shown in Table 2, there is no evident degradation in the junction quality. I_cR_N products were typical around at 3.0 mV with no evident J_c dependence. The ratio of the I_cR_N to the $I_cR_N^{th}$ were about 0.8, which is the same value as reported in high-quality NbN/MgO/NbN tunnel junctions [16]. These results suggest that high current density above 54 kA/cm^2 can be achieved in the NbN/AlN/NbN tunnel junctions. We are improving the depositing process of the AlN to control the barriers thickness, and are trying to fabricate ultra-high J_c junctions above 100 kA/cm^2.

To investigate the Josephson tunneling properties of the critical supercurrents, the magnetic field and temperature dependence of the I_c have been measured. Over the J_c rang

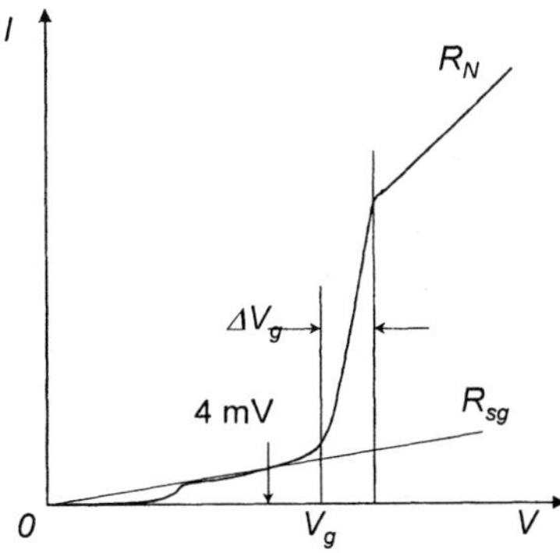

Fig. 4. The definition of SIS tunnel junction parameters. V_g and ΔV_g respectively are the gap voltage and the gap voltage width. R_N and R_{sg} are the normal-state resistance and the subgap resistance at 4 mV. These parameters describe the qualities of the junction.

of this work, the I_c of junctions exhibited BCS-like temperature dependence and an ideal Fraunhofer pattern, indicating that the high-J_c junctions are well-behaved Josephson junctions [22]. As J_c increases above 10 kA/cm^2, the junction quality, as reflected in the subgap leakage (V_m and R_{sg}/R_N value listed in Table 2), is well above any reported previously for NbN tunnel junctions and even beyond the Nb tunnel junctions [28], [29].

Photon-assisted tunneling (PAT) properties were investigated by monitoring the PAT steps induced on the *I-V* curve by a solid-state 300 GHz source and an optically pumped far-infrared (FIR) laser [23]. An array of two NbN/AlN/NbN junctions at the center of a 90-degree bow-tie antenna was fabricated for the measurements. The current density of the junctions was 6 kA/cm^2. Figure 5 (a) shows the *I-V* characteristics obtained with and without irradiation at 303 GHz. The array evidences a strong nonlinearity in the *I-V* curve without irradiation. Sharp photon-assisted tunneling steps, having a spacing of ($2h\nu/e$) of 2.5 mV in the dc voltage, below and above the gap voltage were observed on the *I-V* curve with irradiation. The *I-V* curves that were obtained when the junctions were irradiated at 584 GHz, 693 GHz, and 762 GHz are shown in Figs. 5(b), 5(c) and 5(d). Photon-assisted tunneling steps below the gap voltage, where $n = 1$, and above the gap voltage, where $n = -1$, clearly appeared on each pumped *I-V* characteristic. However, with an increase in

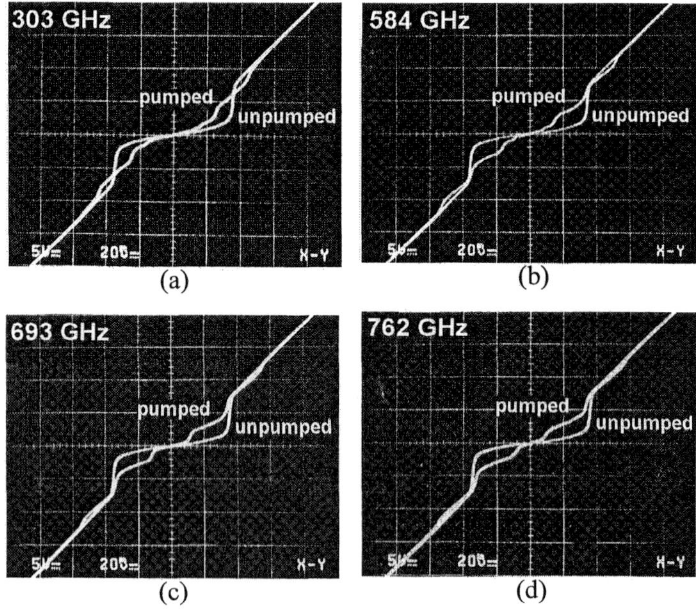

Fig. 5. Pumped and unpumped *I-V* curves for the two NbN junction array: (a) at 300 GHz; (b) at 583 GHz; (c) at 693 GHz; (d) at 762 GHz. The horizontal scale is 5 mV per division, and the vertical scale is 20 μA per division.

frequency, a slight decrease in the energy gap of the superconductor was observed. This is because of the radiation power was so large that a temperature rise occurred in the junctions. The junctions have large geometrical capacitances, which bring about a large coupling loss between the irradiated submillimeter wave and the SIS junctions. Therefore, tuning circuits (i.e. impedance matching and resonating out the capacitances) are necessary to make submillimeter-wave mixers for improving the coupling efficiency.

C. Junction Capacitance

The specific capacitance of the tunnel junctions is an important parameter in designing the tuning elements of the SIS mixers. Two-junction dc-SQUIDs have been fabricated to measure the specific capacitance in our NbN/AlN/NbN tunnel junctions. The specific capacitance was estimated by determining the resonant voltage steps resulting from the interaction between the Josephson effect and the LC resonant circuit in the dc SQUID at the voltage

$$V_r = \Phi_0 / 2\pi\sqrt{LC} \tag{1}$$

where L is the loop inductance of the SQUID and C is the junction capacitance.

Figure 6 shows typical I-V curves for a dc-SQUID with a current density of 14 kA/cm^2. Resonant steps were clearly observed on the I-V curves, and the voltage value of the first step V_{r1} and second step V_{r2} was measured at 150 mV and 300 mV, respectively. The loop inductance L was calculated from the relation of the minimum modulation I_c^{min} and the loop inductance [30]. The I_c^{min} was measured by the decrease in I_c in the I-V curve when

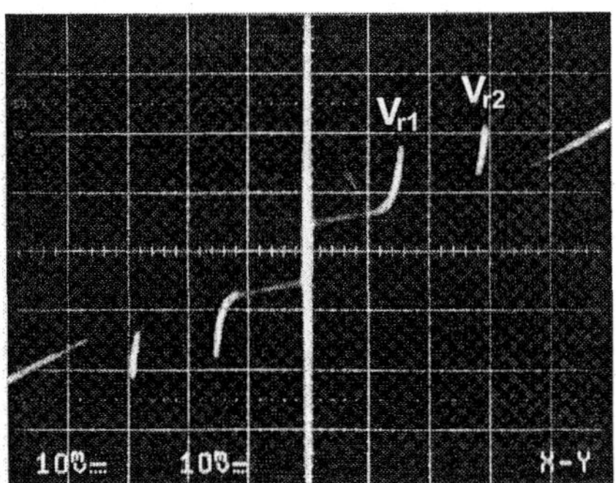

Fig. 6. Typical I-V curves for a dc-SQUID showing two resonant steps at V_{r1} = 150 μV and V_{r2} = 300 μV. The horizontal scale is 100 μV per division, and the vertical scale is 1 mA per division. The picture was taken by multiple exposure.

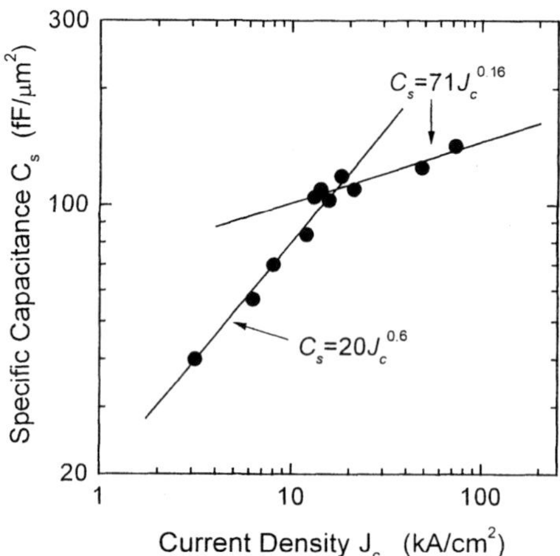

Fig. 7. The specific capacitance C_S as a function of the critical current density J_c. The lines are calculated from apparent linear fit.

half quantum of flux was put into the dc-SQUID loop. Figure 7 plots the specific capacitance C_s as a function of the critical current density J_c. As the J_c increases from 1-100 kA/cm^2, the specific capacitance C_s varies in the range of 20-200 fF/μm^2, which is the same as the results of Nb/AlOx/Nb tunnel junctions [31], [32]. It is very interesting that the C_s experimentally shows two distinct dependencies with the J_c. For the region of $J_c <$ 15 kA/cm^2 the C_s follows as

$$C_s = 20J_c^{0.6}, \tag{2}$$

while for the high-J_c region of $J_c >$ 15 kA/cm^2

$$C_s = 71J_c^{0.16}. \tag{3}$$

Exactly why the dependencies are different is not clear and more experimental date is needed to explain this phenomenon.

Note that the I_cR_N products for the junctions with $J_c >$ 15 kA/cm^2, as listed in Table 2, are larger than those for the junctions with $J_c <$ 10 kA/cm^2. Also note that the base electrodes of our NbN/AlN/NbN junctions have a single-crystal structure [19], [21]. It is conceivable, therefore, that the crystal growth of the counter electrode may be improved in the NbN/AlN/NbN trilayer system as the AlN barrier becomes thinner. This brings about a low effective barrier height for the tunnel junctions resulting in a small junction capacitance. Further investigation of the junction interface and the C_s-J_c relation is certainly needed, however. The small junction capacitance in the high-J_c region is favorable for high-frequency and high-speed device applications.

D. Theoretical mixing properties

It is important to theoretically predict the mixing properties of SIS mixers using NbN/AlN/NbN tunnel junctions. We simulated SSB receiver noise temperatures and SSB conversion gains based on Tucker's quantum theory of mixing at frequencies up to around twice the gap frequency of NbN.

In heterodyne mixing, the received signal and the output from a local oscillator (LO) are mixed in a non-linear circuit element to generate an intermediate-frequency (IF) signal. Many higher harmonics are produced when this occurs, which causes higher-order noise to be down-converted. But in SIS junctions, the high-frequency component is short-circuited due to the relatively high junction capacitance. This led us to use a quasi 5-port model enabling us to handle up to second-order frequencies, so we could use computer simulation derive the noise temperature characteristics of SIS superconductor receivers implemented with NbN/AlN/NbN junctions [7]. The *I-V* curve of a NbN/AlN/NbN tunnel junction used in our simulation of noise performance was an experimental *I-V* curve shown in Fig. 8. The current density was about 30 kA/cm^2. A small subgap leakage current and large gap voltage were observed. The gap voltage was about 5.4 mV, corresponding to a gap frequency of 1.31 THz. The mixer noise consists of thermal noise and shot noise. The IF load impedance viewed from the junction was set at 50 Ω, and the RF source admittance was set at $1/R_N\ \Omega^{-1}$. Here we assume that the junction capacitance is tuned by an external circuit at each simulation frequency. We also assume that the noise temperature of the IF amplifier at the next stage of the mixer is 7 K and the IF frequency was 1.5 GHz. The

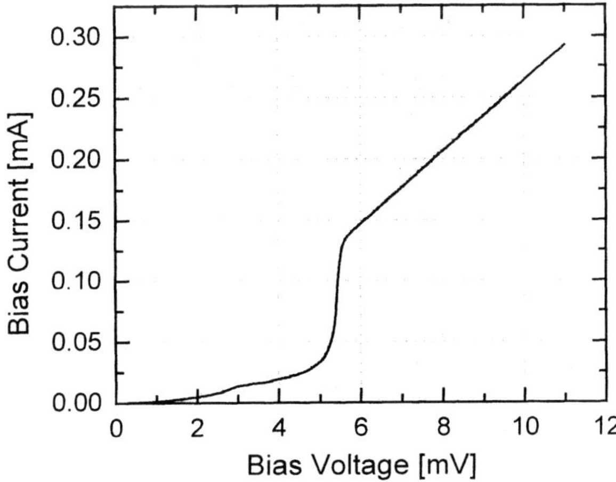

Fig. 8. *I-V* curve of a practical NbN/AlN/NbN tunnel junction for the performance simulation of the SIS receiver.

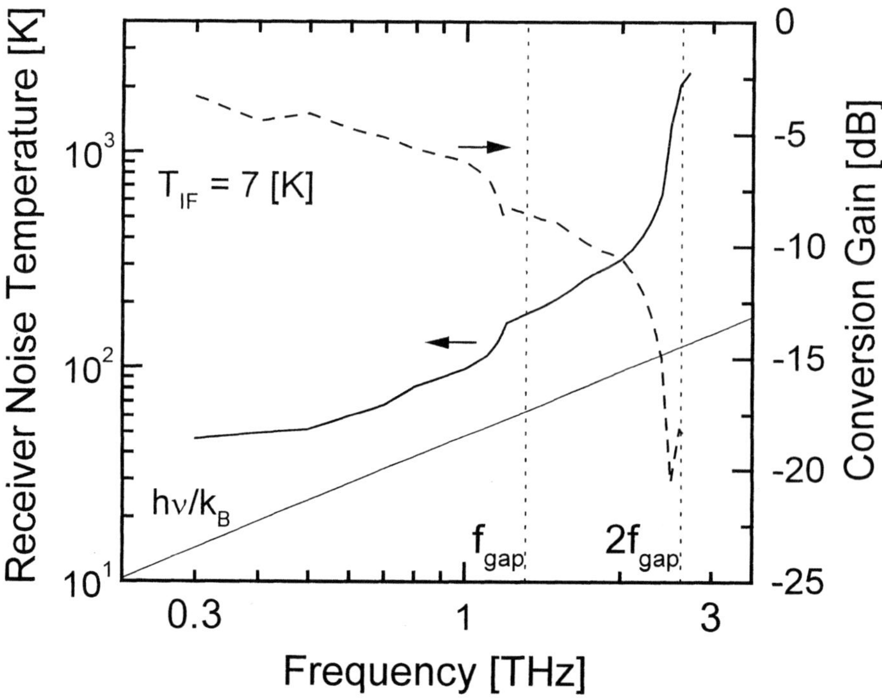

Fig. 9. Theoretical sensitivity of the NbN SIS receiver based on the practical I-V curve.

resulting optimum receiver noise temperatures at the physical temperature of 4.2 K are shown in Fig. 9. The strength of the LO power and the bias voltage were optimized at each frequency to yield the smallest receiver noise temperature. As can be seen in the figure, the simulation predicted that the sensitive SIS receiver noise temperatures approaching $h\nu/k_B$ and high conversion gains can be achieved by using NbN/AlN/NbN junctions, at least for frequencies up to the gap freuqncy.

These excellent temperature noise characteristics for NbN junctions demonstrate their ability to reach terahertz frequencies exceeding the limits of conventional Nb-based junctions. We then proceeded to design and fabricate quasi-optical mixers integrating a tuning circuit with the NbN junctions, and evaluate the noise characteristics of the device when used as a receiver.

2.2 Mixer design

A. Quasi-optics

Although mixers are often implemented using waveguides, it is extremely difficult to fabricate waveguides, mechanical tuners, horn antennas, and other components due to the extremely short wavelength at terahertz frequencies. Another drawback is that waveguide losses increase with increasing frequency. These considerations led us to adopt the quasi-optical approach for implementing our mixer [36]. In a quasi-optical mixer, a lens or some other optical method is used to focus an incident beam onto a thin-film microantenna that is fabricated on the same substrate as the SIS junctions. This approach can not only be implemented at relatively low cost, but the bandwidth can also be readily increased by choosing an appropriate thin-film antenna.

Figure 10 shows a schematic of the quasi-optical RF coupling structure. Parallel beams collimated by an off-set parabolic mirror are further focused by an MgO hyperhemispherical lens without aberration onto a mixer chip that is mounted on the back of the lens. Because the dielectric constant of MgO is relatively large 9.6 (n = 3.1), the reflection loss is about 26% (-5.8 dB) or the transmission is 74% when there is no matching layer. A $\lambda/4$ antireflection cap made of Kapton-JP polyimide film is put on the lens to reduce the reflection loss at the surface of the lens [37]. This material has not only a dielectric constant of 3.46, necessary for the $\lambda/4$ matching layer, but also excellent

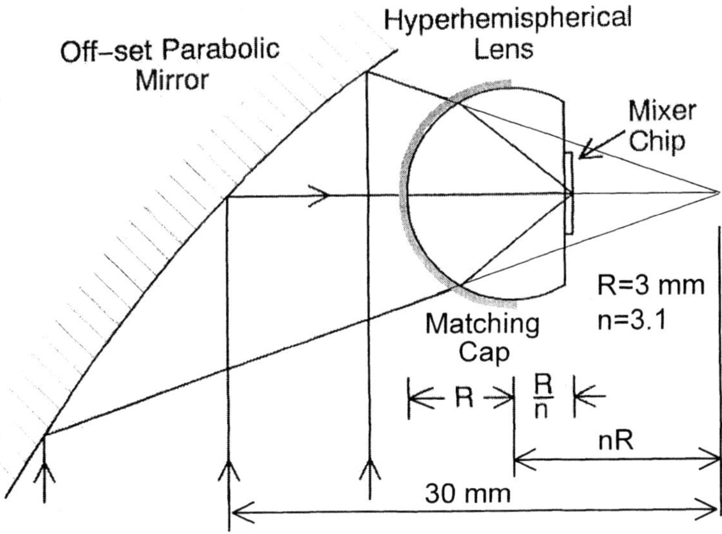

Fig. 10. Schematic diagram of RF optics. A planar antenna on a MgO substrate is placed on the back of a MgO hyperhemispherical lens.

physical, electrical, and mechanical properties at cryogenic temperatures. By choosing the film thickness, a center frequency can be determined. At this frequency, the coupling efficiency will be improved by a factor of 1.35 (= 1/0.74).

A focused beam is received by a planar thin-film microantenna which is fabricated on the mixer chip using optical lithography. The mixer chip substrate and the hyperhemispherical lens are implemented out of the same material. By thus eliminating the optical discontinuity between the two elements, the substrate can essentially be regarded as part of lens. Generally, as the wavelength in the mixer chip substrate becomes smaller, the thickness of the substrate becomes problematic, and losses increase due to surface waves propagating in the substrate. However, since this substrate-lens structure is essentially the same as placing an antenna on a half-space of dielectric, there is no problem on the surface wave. In addition, a strong beam pattern on the lens side is obtained, so losses are minimized. For the antenna, we employed a self-complementary log-periodic antenna. It is common knowledge that the impedance Z_0 of a self-complementary antenna in a free-space observes the Mushiake's relationship and operates over a wide bandwidth since the impedance is a constant value of 60π Ω that is not dependent on frequency. In this work we used MgO for the substrate and lens which have a dielectric constant of ε_r = 9.6. Impedance Z_m then becomes Z_0/n_m = 82 Ω since it is given for the case where

$$n_m = [(\varepsilon_r + 1)/2]^{1/2}. \tag{4}$$

B. Superconductive microstripline

One of the advantages of using superconducting transmission lines for tuning circuits is the extremely lowloss characteristic. The lower the losses in the transmission line used in their tuning circuit, the more efficiently the line works. The geometry of a superconducting microstrip line is shown in Fig. 11. The characteristics of the microstrip line are derived from incremental inductance considerations [38], [39]. Assuming that the loss of the dielectric is negligible, the series impedance and shunt admittance, per unit length, can be written as [40]

$$Z = j\omega\mu_0 g_1 + g_2(Z_{s1} + Z_{s2}) \tag{5}$$

$$Y = j\omega\varepsilon_0\varepsilon_p / g_1 \tag{6}$$

where g_1 and g_2 are geometrical factors, ε_p is the effective dielectric constant of the corresponding perfectly-conducting microstrip line, and Z_{s1} and Z_{s2} are the surface impedances of the two materials. The characteristic impedance and propagation constant of this superconducting transmission line are:

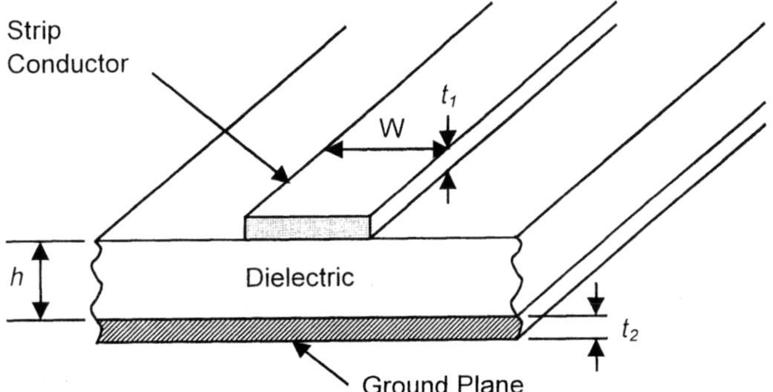

Fig. 11. Geometry of a microstrip line.

$$Z_{line} = \sqrt{\frac{Z}{Y}} = \frac{g_1\eta_0}{\sqrt{\varepsilon_0}}\sqrt{1 - j\frac{g_2(Z_{s1}+Z_{s2})}{\omega\varepsilon_0 g_1 \eta_0{}^2}} \tag{7}$$

$$\gamma = \sqrt{ZY} = jk_0\sqrt{\varepsilon_p}\sqrt{1 - j\frac{g_2(Z_{s1}+Z_{s2})}{\omega\mu_0 g_1}}, \tag{8}$$

where η_0 is the free space impedance. For the corresponding perfectly conducting line, Z_{s1} = 0 and Z_{s2} = 0. Its characteristic impedance is

$$Z_p = \frac{g_1\eta_0}{\sqrt{\varepsilon_0}} \tag{9}$$

Applying the (9) to (7) and (8), we obtain

$$Z_{line} = Z_p\sqrt{1 - j\frac{g_2\eta_0(Z_{s1}+Z_{s2})}{\omega\mu_0 Z_p\sqrt{\varepsilon_p}}} \tag{10}$$

$$\gamma = jk_0\sqrt{\varepsilon_p}\sqrt{1 - j\frac{g_2\eta_0(Z_{s1}+Z_{s2})}{\omega\mu_0 Z_p\sqrt{\varepsilon_p}}}. \tag{11}$$

Closed form analytical expression for g_2 is given in [38], which is

$$g_2 = \frac{\left[\left(w' + \frac{w'}{\pi(w'/2+0.94)}\right)\left(1 + \frac{1}{w'} + \frac{\ln(2h/t_1) - t_1/h}{\pi w'}\right)\right]}{\left[h\left(w' + \frac{2}{\pi}\ln\left(2\pi e\left(\frac{w'}{2}+0.94\right)\right)\right)^2\right]} \tag{12}$$

where

$$w' = \frac{w + \frac{t_1}{\pi}\ln\left(\frac{2h}{t_1} + 1\right)}{h} \tag{13}$$

and e is the Naperian base.

At frequencies below the gap, the loss of the line can be ignore, so we can write Z_s using the two fluid model. At frequencies near the gap, the loss and the effective penetration depth of the line should be considered. In this case, surface impedance must be calculated by the Mattis-Bardeen theory. We used the two fluid model in the 300-GHz mixer design, and the Matthis-Bardeen theory in the terahertz mixer design.

C. 300-GHz mixer

To investigate the mixing properties of NbN/AlN/NbN tunnel junctions in detail, we first designed an SIS mixer for the lower submillimeter-wave frequency band around 300 GHz, which make it possible to utilize conventional circuit design and reliable materials for tuning circuits, i.e. SIS junctions is considered as lumped elements and Nb material can be used for the wiring layer [41].

An optical micrograph and a cross-section view of the mixer chip designed at the center frequency of 300 GHz are shown in Fig. 12. On a 0.3-mm-thick single-crystal MgO substrate, two NbN/AlN/NbN junctions in series were integrated with a single-crystal NbN planer self-complementary log-periodic antenna and Nb tuning circuits. The mixer chip is placed on the back of a 3-mm-radius MgO hyperhemispherical lens with a 125-μm-thick Kapton JP AR cap (the quarter-wave length at 323 GHz), which is approximately equivalent to the mixer chip onto a half-space of MgO with the dielectric constant of about 9.6. Thus, according to the equation (4), the antenna has a frequency-independent impedance Z_{ant} of about 82 Ω over several octaves. We used a mirror-symmetric design which will halve the driving impedance as seen by the circuit, i.e., the effective driving impedance of the antenna seen by each half of the circuit is $Z_{ant}/2$ = 41 Ω. The lower driving impedance of the antenna makes a broadband tuning design easily possible, because the impedance of microstripline for tuning circuits we can realize are usually below 30 Ω (limited by fabrication processes, for example, a minimum strip width of 2 μm and a maximum insulator thickness of 300 nm).

Figure 13 shows the equivalent nonsymmetric circuit of Fig. 12. We assumed that NbN/AlN/NbN tunnel junctions had the size of 1 μm in diameter for the design. Because the $J_C R_N A$ product is about 350 kVμm^2/cm^2 (=3.5 mV) for such a recent-batch NbN SIS junction, the normal state resistance is 22 Ω. Here J_C is the current density in kA/cm^2, R_N is

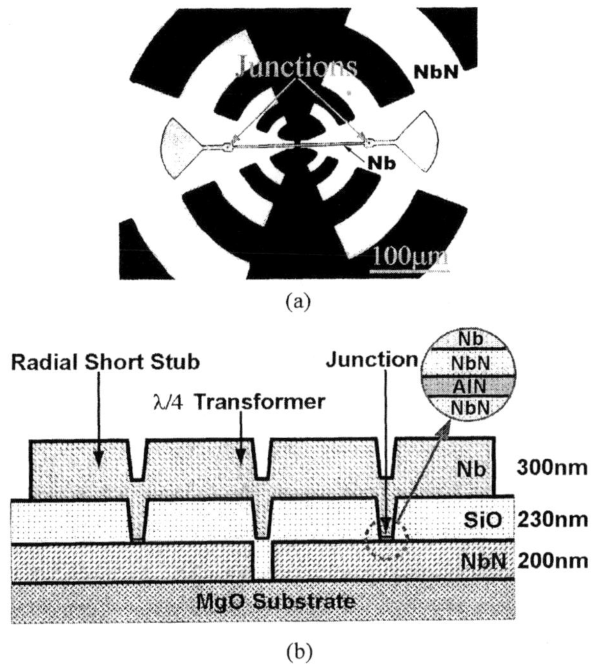

(a)

(b)

Fig. 12. (a) Optical micrograph and (b) cross section view of the NbN/AlN/NbN mixer. Each junction is approximately 1 μm in diameter.

the normal state resistance in Ω, and A is the area of the junction in μm^2. The capacitance of the junction was 90 fF, which was obtained from a specific capacitance value of 115 fF/μm^2 estimated by measuring a dc SQUID resonant voltage step for the junctions with a current density of 20 kA/cm^2 according to the equation (3). The resulting $\omega C_J R_N$ product was 3.7 at 300 GHz. The optimum RF admittance of SIS junctions biased on their first photon step below the gap is expressed as follows [42].

$$G_{opt} = G_N \left(\frac{1}{2} + \frac{1}{4(f / f_g)} \right) \tag{14}$$

where G_N is the normal state admittance of the junction, f is the operating frequency, and f_g is the gap frequency of the junctions. This gives the RF admittance of 13.9 Ω for $f_g = 1.3$ THz (typical gap frequency of NbN junctions) for the operating frequency of 300 GHz. The tuning circuit incorporates a radial short stub tuner. A microstrip inductance having the characteristic impedance of 9 Ω and the electrical length of 33 degree was placed in parallel with the junction for resonating out the junction capacitance at the center

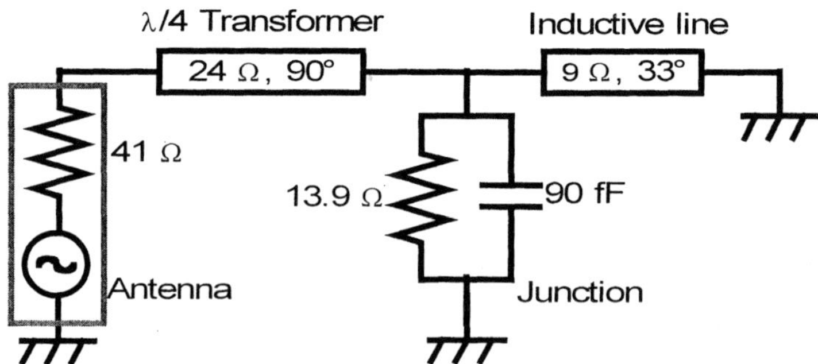

Fig. 13. The equivalent nonsymmetric tuning circuit. The electrical lengths of the λ/4 transformer (90 degree) and the inductive line (33 degree) are values at the center frequency of 300 GHz.

frequency by using the radial stub as a RF short circuit. The impedance of this circuit is thus only the real part of the junction of 13.9 Ω. To match the junction resistance to the antenna impedance, a $\lambda/4$ impedance transformer was used. The characteristic impedance of the transformer Z_{trans} was determined by the following simple transmission line's formula.

$$Z_{trans} = \sqrt{41 \cdot 13.9} \approx 24 . \qquad (15)$$

Figure 14 shows (a) the circuit's calculated return loss from the antenna to the tuning circuit with (b) the Smith chart plot of the RF impedance, which predicts a good match, better than –10 dB, from 255 to 350 GHz.

These tuning structures utilize superconducting microstriplines that use the arms of the antenna as a ground plane. Because we had not measured the magnetic penetration depth of NbN thin films fabricated on SiO underlayers and there are large RF losses in the NbN film, we used reliable material of Nb to make microstrip for the tuning structures. Generally, sputtered Nb films on SiO show good superconducting characteristics, which are not dependent on the underlayer. While ground plane is the single-crystal NbN film on the MgO substrate. Therefore, Nb/SiO/NbN microstriplines should show extremely low-loss characteristics as predicted theoretically. Since the operating frequency is very low, far from the gap frequenceis of Nb and NbN, the two-fluid model was used to calculate the size of the Nb(300 nm)/SiO(250 nm)/NbN(200 nm) microstripline with a London penetration depth λ_{Nb} = 84 nm for Nb [43], λ_{NbN} = 180 nm for NbN [19], [25] and the dielectric constant of 5.5 for SiO. Using the two fluid model, the surface impedance of the Nb and NbN thin films are:

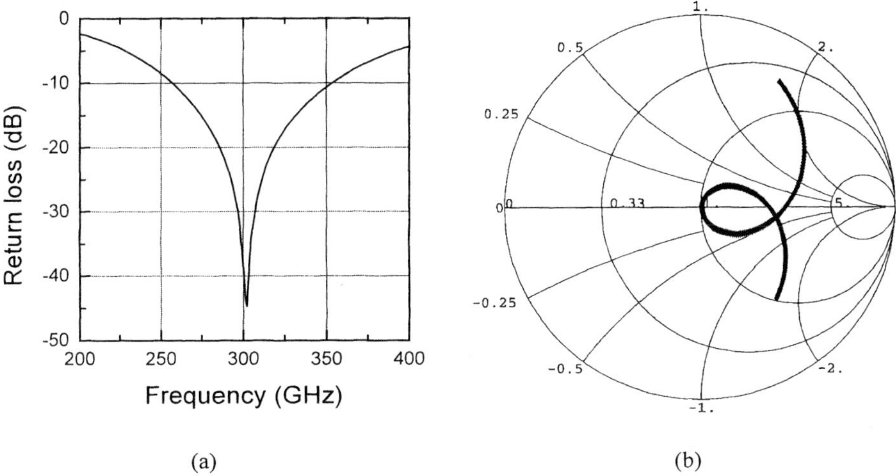

Fig. 14. (a) The circuit's calculated return loss from the antenna to the tuning circuit. (b) The Smith chart plot of the RF impedance, which is normalized to 41 Ω, i.e. the driving impedance of the circuit.

$$Z_{s,Nb} = j\omega\mu\lambda_{Nb}\coth\frac{t_{Nb}}{\lambda_{Nb}} \qquad \text{for Nb} \tag{16}$$

$$Z_{s,NbN} = j\omega\mu\lambda_{NbN}\coth\frac{t_{NbN}}{\lambda_{NbN}}, \qquad \text{for NbN} \tag{17}$$

where t_{Nb} and t_{NbN} are the thickness of the Nb and NbN films. The characteristic impedance and the propagation constant can be calculated by using expressions (10) and (11) as follows:

$$Z_{line} = Z_p\sqrt{1 - j\frac{g_2\eta_0(Z_{s,Nb} + Z_{s,NbN})}{\omega\mu_0 Z_p\sqrt{\varepsilon_p}}} \tag{18}$$

$$\gamma = jk_0\sqrt{\varepsilon_p}\sqrt{1 - j\frac{g_2\eta_0(Z_{s,Nb} + Z_{s,NbN})}{\omega\mu_0 Z_p\sqrt{\varepsilon_p}}}\,. \tag{19}$$

The length and the width of the microstrip line for the tuner were found to be 27 μm and 6 μm by numerical calculations. For the transformer, the size was found to be 76-μm long and 2-μm wide.

D. Terahertz mixer

In the terahertz region, there is following problems for making SIS mixers. One is to fabricate low-loss NbN tuning circuits. Usually, the polycrystalline structure of NbN films is a problem because this structure has a large surface resistance on conventional insulators, such as SiO. This results in a large RF loss, which greatly degrades the mixer noise performance [44]. To solve this problem, we recently developed NbN/MgO/NbN microstriplines epitaxially grown on a single-crystal MgO substrate that can be used in extremely low-loss tuning circuits [45].

Another problem is that the relatively long penetration depth of NbN (even if NbN is a single crystal of, let's say 200 nm) makes the circuit design difficult because the large slow-wave factor shortens the circuits at terahertz frequencies. It was found that a conventional integrated tuning circuit did not work as effectively as it was designed to work because the spreading inductances introduced by the embedding of small junctions in the wide microstrip lengthened the tuning section [46]. Thus, a novel tuning circuit design is needed for the terahertz SIS mixer. We design an SIS mixer with a self-compensated or self-tuned NbN/AlN/NbN tunnel junction [47], [48]. This mixer dose not have a conventional tuning structure consisting of a microstrip inductor and a SIS junction as a lumped element [41], such as the former mixer design at 300 GHz.

An optical micrograph of our mixer chip is shown in Fig. 15. A rectangular NbN/AlN/NbN junction and an NbN wiring on an MgO insulator were integrated with an NbN log-periodic antenna on a 0.3-mm-thick single-crystal MgO substrate. We designed the tuning circuit so that it had a center frequency of 1 THz. In this frequency region, NbN-based μm-sized SIS junctions must be considered as distributed elements, not lumped elements, because the slow-wave factor in these junctions is much larger than that in Nb junctions [49], [50]. This means that an NbN junction must be treated as a microstrip

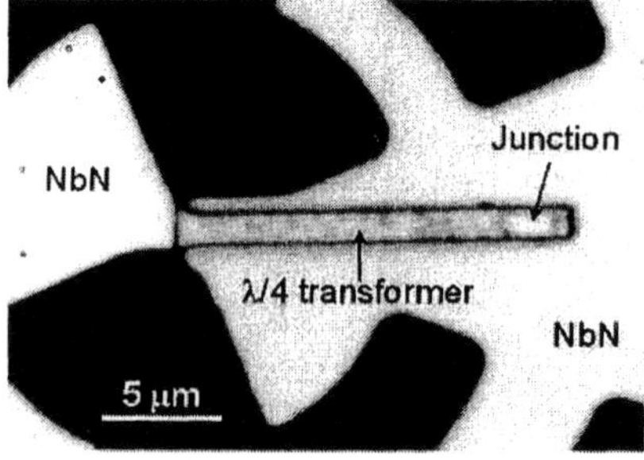

Fig. 15. Optical micrograph of the mixer. A rectangular NbN/AlN/NbN tunnel junction, as a distributed element, and a λ/4 NbN/MgO/NbN microstrip-impedance transformer were integrated with a self-complementary NbN log-periodic antenna.

transmission line with a quasi-particle tunneling loss. It is well known that an imaginary part of the impedance in transmission lines becomes zero at a length of n × λ/4, where n is an integer. For distributed SIS junctions, this means that the junction capacitance is tuned out, and the input impedance becomes the only real component. The impedance must be the largest at the λ/2 length for an open-ended junction.

We calculate the size of the junction, including the NbN/MgO/NbN microstrip overhanging layer, as shown in Fig. 16. For the calculation, since the design frequency is near the gap frequency of NbN, the Matthis-Bardeen theory of the anomalous skin effect must be used to obtain the surface impedance of the NbN films. Matthis and Bardeen derived the following results on the complex conductivity $\sigma = \sigma_1 + \sigma_2$ (scaled to the normal state conductance σ_n) [51].

$$\frac{\sigma_1}{\sigma_n} = \frac{2}{\hbar\omega}\int_{\Delta}^{\infty}[f(E) - f(E+\hbar\omega)]g(E)dE + \frac{1}{\hbar\omega}\int_{\Delta-\hbar\omega}^{-\Delta}[1-2f(E+\hbar\omega)]g(E)dE \tag{20}$$

$$\frac{\sigma_2}{\sigma_n} = \frac{1}{\hbar\omega}\int_{\Delta-\hbar\omega,-\Delta}^{\Delta}\frac{[1-2f(E+\hbar\omega)][E^2+\Delta^2+\hbar\omega E]}{[\Delta^2-E^2]^{1/2}[(E+\hbar\omega)^2-\Delta^2]^{1/2}}dE, \tag{21}$$

where $f(E)$ is the equilibrium Fermi-Dirac distribution at the ambient temperature T that is

$$f(\eta) = \frac{1}{1+e^{\eta/k_B T}} \tag{22}$$

and

$$g(E) = \frac{E^2+\Delta^2+\hbar\omega E}{(E^2-\Delta^2)^{1/2}[(E+\hbar\omega)^2-\Delta^2]^{1/2}}. \tag{23}$$

The first term on the right side of (20) represents conduction by thermally excited quasiparticles, and the second term describes the scattering of quasiparticles created by Cooper-pair breaking and therefore is zero unless the photon energy $\hbar\omega > 2\Delta$, in which case the lower limit of the integral in equation (21) is $-\Delta$, instead of $\Delta - \hbar\omega$. As is the case in the two-fluid model, equation (21) describes the 'kinetic inductance' of the surface caused by the response of the Cooper pairs. Using this complex conductivity, the specific surface impedance Z_S per square of a superconducting film with thickness d is expressed as follows:

$$Z_s(\omega) = \sqrt{\frac{j\omega\mu_0}{\sigma}}\coth\left(d\sqrt{j\omega\mu_0\sigma}\right). \tag{24}$$

Because the NbN/MgO/NbN microstripline and the high-current-density NbN/AlN/NbN tunnel junction were grown epitaxially [52], we assumed in the mixer design process that the upper and lower NbN films had the same gap frequency of 1.27 THz and a normal state conductivity of 1.5 x $10^6\ \Omega^{-1}m^{-1}$. These values were obtained from our measurements.

To determine the characteristics of the SIS-junction transmission line, the series impedance per unit length, Z, and the distributed shunt admittance per unit length, Y, must be known. For a given distributed-SIS-junction geometry, Z and Y are defined as

$$\frac{1}{Z} = 1\Big/\left[j\omega\mu_0 \frac{s}{W_J} + \frac{1}{W_J}(Z_{SU} + Z_{SL}) \right] + 1\Big/\left[j\omega\mu_0 \frac{d}{W_S - W_J} + \frac{1}{W_S - W_J}(Z_{SU} + Z_{SL}) \right], \quad (25)$$

$$Y = \frac{W_J}{R_{rf}} + j\omega C_S W_J + j\omega\varepsilon_r\varepsilon_0 \frac{W_S - W_J}{d}, \quad (26)$$

where Z_{SU} and Z_{SL} are, respectively, the surface impedance of the upper and lower electrodes, W_J and W_S are the widths of the NbN/AlN/NbN junction and the NbN/MgO/NbN microstripline, respectively, and R_{rf} represents the specific resistance of the quasi-particle loss due to the tunneling current through the insulator. The specific junction capacitance, C_S (per unit area), was calculated based on the equation (3) in Sec. 2.1.C. The tunneling barrier thickness, s, was assumed to be 1 nm based on the observation of the high-current-density junction by using a transmission electron microscope (TEM) [52]. We used the relative dielectric constant, ε_r, of 9.6 for MgO, and the thickness was expressed as d.

Table 3. Parameters used in the 1 THz mixer design.

NbN upper electrode thickness:	200 nm
NbN lower electrode thickness:	200 nm
MgO insulator thickness, d:	200 nm
NbN/AlN/NbN junction width, W_J:	0.7 μm
length:	2.25 μm
current density, J_C:	40 kA/cm^2
$J_C R_N A$ products	350 kVμm^2/cm^2
NbN/MgO/NbN microstripline width, W_S:	1.7 μm

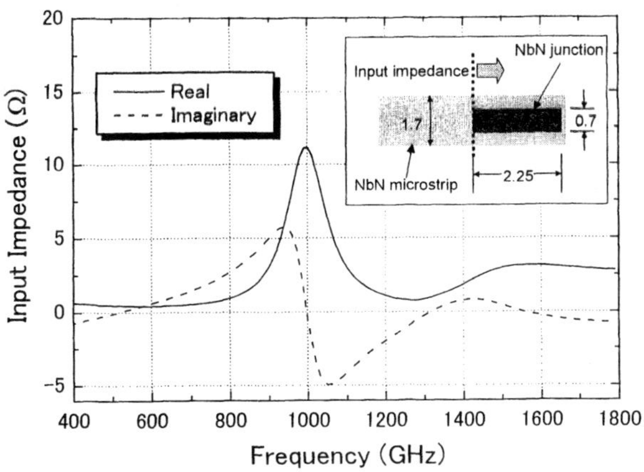

Fig. 16. Input impedance of the self-compensated junction tuning circuit. The junction capacitance was tuned out at the λ/2 wavelength frequency of 1 THz

The input impedance of this lossy transmission line was calculated using the parameters described in Table 3. We decided that the width of the junction should be small to prevent transverse mode propagation. Figure 16 shows the real and imaginary parts of the input impedance. It can be seen that the junction capacitance is self-compensated at around 1 THz, i.e., the imaginary part becomes zero, while the real part becomes the maximum

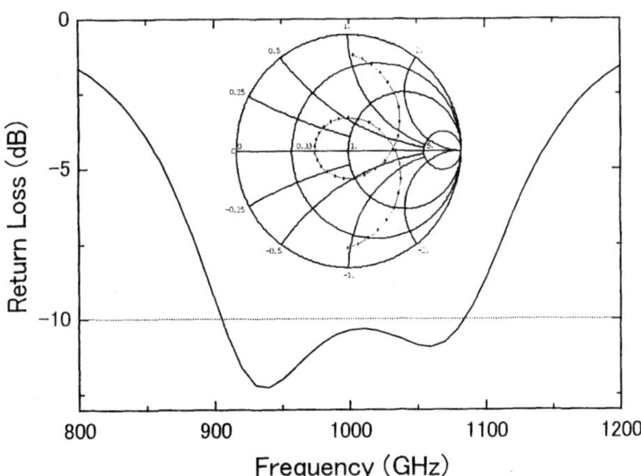

Fig. 17. Return loss characteristics from the antenna to the tuning circuit. Impedance chart was normalized to antenna source impedance of 80 Ω.

value of about 11 Ω. This relatively large resistance can be easily matched to the source impedance of the log-periodic antenna on the MgO lens (about 80 Ω) by using a λ/4 impedance transformer. The 13.5-μm length of the NbN/MgO/NbN microstripline enables good matching of below −10 dB at frequencies ranging from 900 to 1080 GHz, as shown in Fig. 17. Theoretically, a relative bandwidth of 18% can be achieved at terahertz frequencies.

2.3 Mixer performance

A. Measurement set-up

A schematic layout of the noise measurement system is shown in Fig. 18. In the heterodyne receiver measurements at the 300-GHz band, a Schottky varactor tripler pumped by a mechanically tunable Gunn oscillator was used as a local oscillator (LO). For the terahertz band, an optically pumped far-infrared laser and a backward-wave oscillator were used as the LO. The LO was introduced into the signal path with a Mylar film which acts as a partial reflector. In case of LO polarization with an electric field perpendicular to plane of incidence, the reflectivity R is calculated by the following formula.

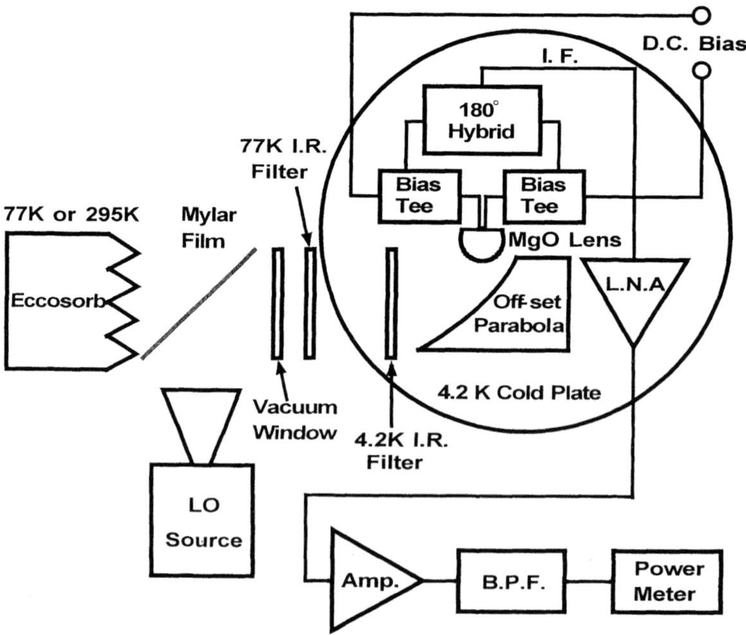

Fig. 18. Schematic layout of the measurement set-up. The anti-reflection cap fromed from polyimide film is put on the MgO hyperhemispherical lens.

$$R = \frac{4r^2 \sin^2(\Delta/2)}{\left(1-r^2\right)^2 + 4r^2 \sin^2(\Delta/2)} \tag{27}$$

where

$$r = \frac{\sin(i_1 - i_2)}{\sin(i_1 + i_2)} \tag{28}$$

$$\Delta = \frac{4\pi n d \cos i_2}{\lambda_0} \tag{29}$$

n represents the refractive index of the reflector, which is 1.69 for the Mylar film, and d represents the thickness of the reflector. i_1 and i_2 represent the incident and the refractive angles. In case of $i_1 = \pi/4$ rad (45 °), i_2 becomes 0.43169 rad (24.734 °), according to Snell's law. For example, using 25-μm thick Mylar film, the reflection power from the LO source is about 4% at 300 GHz. While the transmissivity of the signal power is 96%. Our signal sources were microwave absorbers at room temperature (295 K) and dipped in liquid nitrogen (77 K) to measure receiver noise temperature with a standard hot/cold load Y-factor technique.

The incoming signal and LO entered the dewar through a 0.5-mm-thick Teflon vacuum window and Teflon or Zitex infrared filters cooled to 77 K and 4.2 K, and received by the quasi-optical SIS mixer. The IF signal, down-converted by the mixer, was brought out in the balanced method and transformed into an unbalanced signal for the IF amplifier by using a 180-degree hybrid coupler [53], [54] and coupled to a 1.25-1.75 GHz HEMT IF amplifier. The IF signal was amplified by another room temperature IF amplifier and detected by a power meter or a calibrated square law detector through a 500-MHz band-pass filter centered at 1.5 GHz. DC bias to the mixer was applied through two bias tee. A magnetic field was applied perpendicular to the junction to suppress the unwanted noise caused by the Josephson effect by using a superconducting coil. No corrections were made for losses in front of the reciever.

B. Experimental results on 300-GHz mixer

The mixer chip was fabricated as described in Sec.2.1.A. First, an NbN/AlN/NbN trilayer was deposited *in situ* onto an ambient temperature MgO substrate having dimensions of 15 mm × 15 mm × 0.3 mm by rf magnetron sputtering. The log-periodic antenna was formed by conventional photolithography and RIE. Then, the two junction areas were patterned by photolithography and RIE. After a 250 nm-thick SiO film was deposited to electrically isolate the base electrodes and contact-wiring layer (microstripline), a 300 nm-thick Nb film was deposited and patterned in the contact-wiring

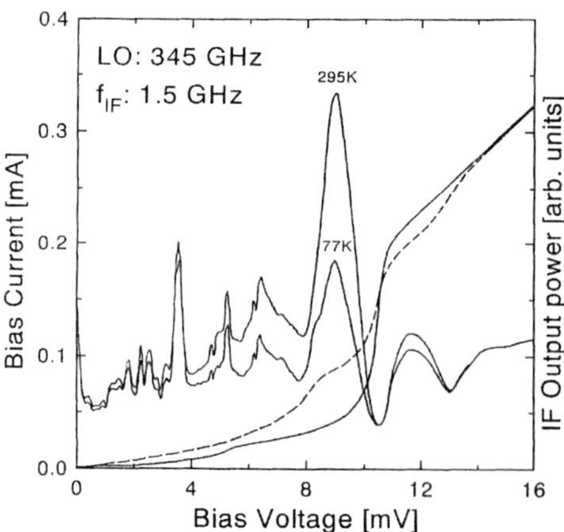

Fig. 19. Heterodyne response of the receiver at 345 GHz. Shown are the I-V characteristics for the array of two NbN/AlN/NbN tunnel junctions with (solid line) and without (dashed line) LO power. Also shown is the IF power as a function of bias voltage for hot (295 K) and cold (77 K) loads. The IF bandwidth is 500 MHz.

layer to form the tuning circuits and to contact the junctions. The substrate is cut into the mixer-chip dimensions of 4 mm × 4 mm × 0.3 mm by a dicing saw.

The Josephson critical current of the prepared junctions was 140 μA, and the junction size was about 1 μm in diameter for the fabricated 300-GHz-mixer chip. This gives a current density of about 20 kA/cm^2 for these junctions. Figure 19 shows *I-V* characteristics for the receiver at 345 GHz with and without LO power. The receiver IF output in response to hot and cold loads is also shown in Fig. 19 as a function of bias voltage. Even though the junctions have a very high current density, the *I-V* characteristics without LO show excellent tunneling characteristics with a large gap voltage, a small gap voltage width, and a low subgap leakage current. From the normal-state resistance R_N = 22 Ω for each junction. Photon-assisted tunneling steps were clearly observed with LO applied. The distinct IF responses to hot and cold loads show a maximum *Y*-factor of about 1.81, corresponding to a double sideband (DSB) receiver noise temperature of 192 K.

Figure 20 shows the DSB noise temperature of the receiver as a function of LO frequency. The average receiver noise temperature measured in the frequency band from 297 to 356 GHz was about 200 K, and the minimum receiver noise temperature of 167 K was obtained around 303 GHz. The poorer performance around 330 GHz is due to the 0.48-mm thick Teflon vacuum window and cold filters which are $3\lambda/4$ thick at that

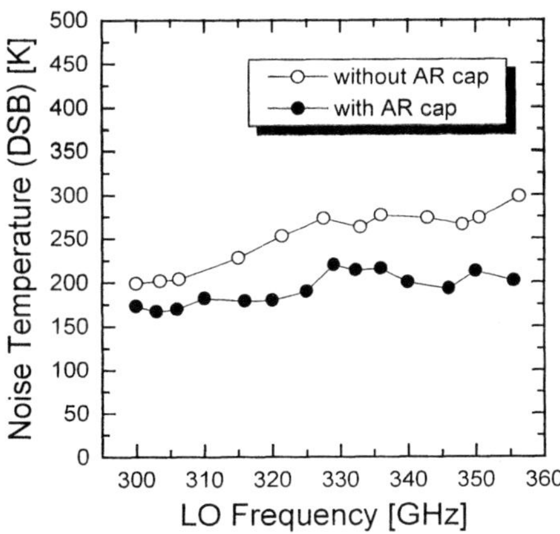

Fig. 20. DSB noise temperature of the receiver with and without the AR cap as a function of LO frequency. (No data was obtained below 297 GHz and above 356 GHz because no LO source was available.

frequency. For comparison, the result without the AR cap is also shown in the figure. It is interesting to note that a difference in the receiver noise temperature between the two RF optic configurations was observed. The improvement is by a factor of about 1.3, which is almost consistent with the noise performance predicted in the design. Thus, putting a matching layer on the lens is important for reducing the entire receiver noise temperatures.

To investigate the noise contribution to the receiver, standard measurement techniques were used that separate receiver noise T_{RX} into three major components: noise due to input losses T_{inp}, mixer noise T_{mix}, and multiplied IF system noise LT_{IF}, where L is the conversion loss of the mixer. The shot-noise produced by the linear resistance of the junction I-V curve above the gap voltage was used to calibrate the IF system [55]. From this calibration, the IF system noise of T_{IF} = 6.5 K was obtained. This allows the mixer noise and conversion loss including input losses to be estimated from the measured receiver IF responses to hot and cold loads at each frequency. T_{inp} can be determined by the intersecting lines technique described in [56] and [57]. This involves making receiver IF output power measurements for hot/cold loads as a function of LO power level. We can plot receiver noise temperature against relative conversion loss, which is inversely proportional to the difference between the receiver outputs in the hot/cold load measurements [58]. Figure 21 shows the result at 345 GHz as an example. The experimental points fall into a straight line. The noise temperature corresponding to a point at which the straight line intersects the vertical axis is

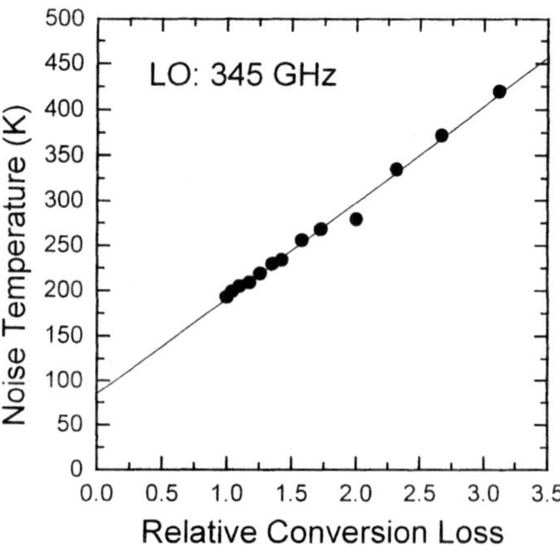

Fig. 21. Receiver noise temperature at different LO power levels plotted against relative conversion loss. This gives an input noise of 84 K.

the noise due to input losses T_{inp}. From the figure, the losses in front of the mixer element result in an input noise contribution of about 84 K.

Figure 22 shows the noise contribution to the receiver separated into its three components using these procedures mentioned above. In our receiver, the multiplied IF

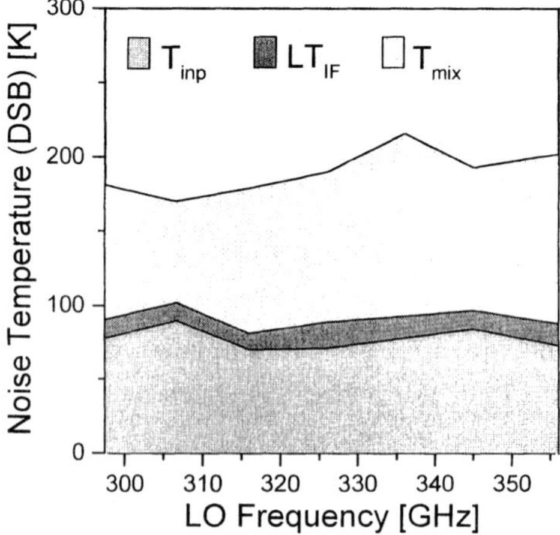

Fig. 22. A breakdown of the receiver noise into input noise T_{inp}; multiplied IF noise LT_{IF}; and mixer noise T_{mix}.

system noise is quite small, which is about 15 K. This result indicates that a good IF matching, and high mixer conversion efficiency were achieved. However, the large input noise contribution of around 80 K can be seen. Thus, there is considerable room for improvement in our receiver, because waveguide receivers have shown input noise of about 30 K for a similar frequency band [59]. The most likely reason for the high input noise is that the lens-coupled log-periodic antenna has a poorer beam pattern due to a high level of the side-lobe than that of waveguide horns. To reduce the input noise, antennas with the beam pattern as good as waveguide horns should be developed. The mixer itself, T_{mix}, shows noise of around 100 K as shown in Fig. 22, which is a bit larger than the theoretical prediction. This may result from the leakage current relating to the subgap structure in the *I-V* curve. The performance would be improved if reducing this gap structure. More investigation must be done to interpret this noise performance.

Finally, to reduce the input reflection loss (0.48 mm-thick Teflon was thick for the 300 GHz band as described in the former section), Teflon infrared filters cooled at 77 K and 4.2 K were changed to Zitex filters, which is a porous form of Teflon (very thin film). Figure 23 shows the measured receiver noise temperatures from 300 GHz to 350 GHz. The minimum receiver noise temperature of 136 K was obtained at 303 GHz, which is better than that of the former experimental set-up by about 30 K. In the band, also, about 30 K was reduced on the average by changing the cooled filters. Therefore, it is expected that the input loss will be reduced by changing the Teflon vacuum window to thin film having a low dielectric constant.

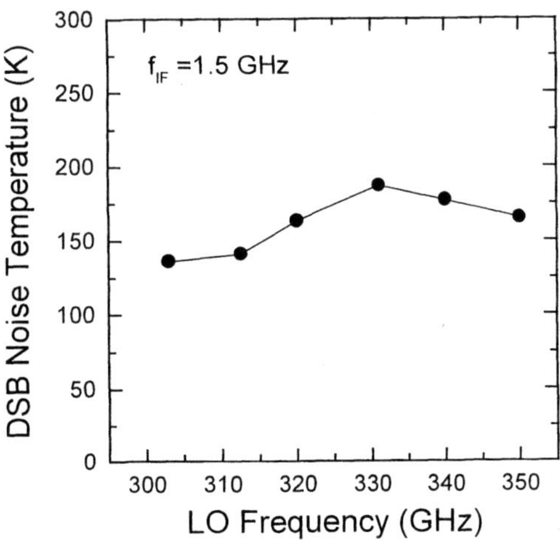

Fig. 23. DSB receiver noise temperature of the receiver using cooled Zitex filters instead of Teflon filters as a function of LO frequency.

C. Experimental results on terahertz mixer

The mixer chips were prepared by the fabrication process described in Sec. 2.1.A. First, NbN/AlN/NbN trilayers were deposited in situ on an ambient-temperature MgO substrate by rf-magnetron sputtering in a load-locked sputtering system. One side of the antennas was formed by conventional photolithography and reactive ion etching (RIE). Then, rectangular junctions were defined using photolithography and RIE. After the MgO films were deposited by rf- and dc-magnetron sputtering to electrically insulate the base electrodes and the contact-wiring layer [45], an NbN film was deposited and patterned on the other side of the antennas and the contact-wiring layer to form a tuning circuit and to make contact with the junction. The actual thicknesses of the lower NbN, the lower MgO, and upper NbN were 200, 180, and 350 nm, respectively. The measured width was about 1.4 μm. The Josephson critical current of the junction was 950 μA at 4.2 K, and the junction was 0.8 μm wide and 2.4 μm long. This resulted in a current density of about 50 kA/cm^2. In the liquid helium measurement, we observed a slightly lower gap voltage of about 5 mV with backbending characteristics because of the large current through the junction. The normal state resistance was about 4 Ω.

To roughly define the tuning frequency, we also observed resonance-induced Josephson steps, as shown in Fig. 24. A dc current step could be seen at a voltage of about 1.9 mV, which corresponds to 920 GHz according to the Josephson frequency relationship, $f_J = 2eV_0/h$. This result indicates that the tuning frequency was slightly lower than the design frequency of 1 THz, because the fabricated junction was a little bit longer and the overhanging layer was a little bit narrower than the design junction and layer.

The receiver set-up was basically the same as the one described in Sec. 2.3.A, except that an isolator was used instead of the 180-degree hybrid coupler. LO power was introduced

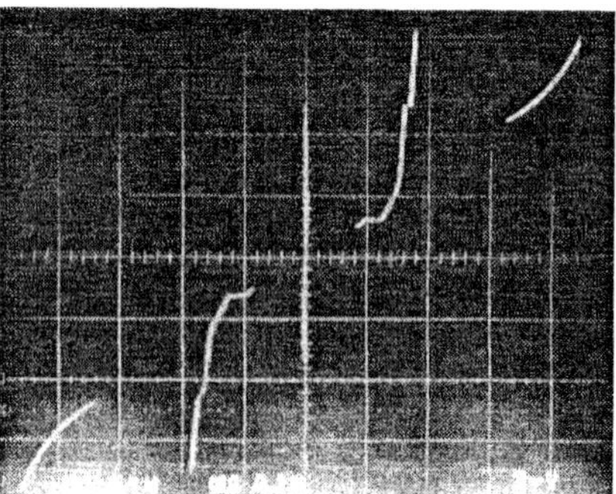

Fig. 24. Observed resonance-induced Josephson steps in the *I-V* curve. The horizontal scale is 1 mV per division, and the vertical scale is 0.1 mA per division.

into the signal path through 9-μm and 16-μm-thick Mylar beam splitters. Figure 25 shows the *I-V* characteristics of the receiver at 909 GHz with and without the LO power. The LO source was the BWO, and the 9-μm-thick Mylar beam splitter was used. As can be seen there was a suppression of the gap voltage in the pumped *I-V* curve. In addition, the width of the voltage measured at the PAT step, which was 3.25 mV, was narrower than the estimated width of 3.76 mV. These results show that the SIS junction suffered from the heating effects of the applied DC and RF power, because NbN has a low thermal conductivity. This local overheating increased the sub-gap leakage current of the junction and the loss in the NbN electrodes, which would degrade the noise performance. To reduce this heating effect, a normal metal, such as Al, should be deposited on the top electrode [60]. Figure 25 also shows the receiver IF output in response to the hot and cold loads as a function of the bias voltage. At voltages above the half-gap voltage of about 2.7 mV, the maximum *Y*-factor was about 1.25, which corresponded to the DSB receiver noise temperature of 800 K. Below the half-gap voltage, the *Y*-factor was 1.35 (550 K) at 2.2 mV. However, the IF outputs around the first-Shapiro-step voltage of 1.88 mV were not stable because of the high dynamic resistance of the pumped *I-V* curve, as mentioned above. To obtain a better performance, a stronger magnetic field should be applied to the junction to suppress unwanted noise resulting from the Josephson effect.

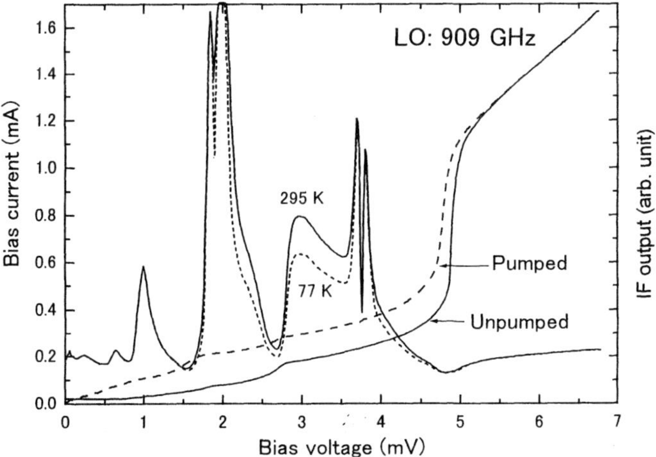

Fig. 25. Heterodyne response of the receiver at 909 GHz. Shown are the I-V characteristics for the self-compensated NbN/AlN/NbN tunnel junctions with and without the LO power. Also shown is the IF power as a function of the bias voltage for hot (295 K) and cold (77 K) loads.

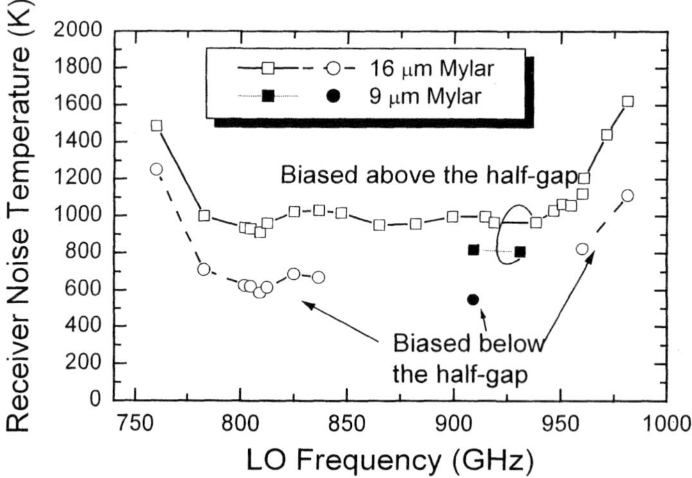

Fig. 26. DSB-receiver noise temperatures as a function of the LO frequency.

The frequency dependence of the receiver noise temperature was investigated at frequencies ranging from 760 to 980 GHz. We used two BWOs for the 800-980 GHz measurements and an optically pumped CH_2F_2 far-infrared laser for the 760- and 783-GHz measurements. Figure 26 shows the receiver noise temperature as a function of the LO frequency. The 16-μm-thick Mylar beam splitter was used at all the measured frequencies, and a bias voltage was applied to the junction above the half gap to obtain stable operation. Wideband and flat-noise characteristics were observed at frequencies ranging from 780 to 960 GHz. Although the measured center frequency of about 870 GHz was lower than the design frequency, the characteristics of the tuning bandwidth were almost consistent with the design ones, which means that the NbN films fabricated for the tuning circuits had no significant losses at terahertz frequencies. The fact that the center frequency was slightly lower than the observed resonance frequency of 920 GHz may be attributed to the increase in the effective penetration depth of the NbN due to the increase in the effective temperature in the NbN films by the DC and LO power.

At the higher and lower ends of the tuning frequencies, we could apply a bias voltage below the half-gap easier because the dynamic resistance of the pumped *I-V* curve became slightly smaller than that around the center frequency. Figure 27 shows the heterodyne response at 809 GHz when the 16-μm thick Mylar beam splitter was used. The maximum *Y*-factors below and above the half-gap voltage were, respectively, 1.33 (584 K) and 1.22

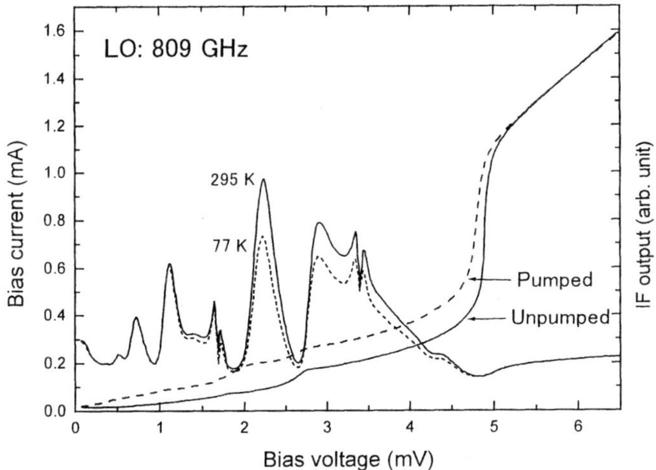

Fig. 27. Heterodyne response of the receiver at 809 GHz. Shown are the I-V characteristics for the self-compensated NbN/AlN/NbN tunnel junctions with and without the LO power. Also shown is the IF power as a function of the bias voltage for hot (295 K) and cold (77 K) loads.

(914 K), including the Mylar-beam-splitter loss and other optical losses. Using the intersecting lines technique [56], [57], we estimated the input loss around this frequency from the IF responses biased below the half-gap voltage. A large input-noise contribution of about 350 K was derived that included the loss from the 16-μm-thick Mylar beam splitter. The most likely reason for the high input noise is the following. In the fabricated log-periodic antenna structure, at terahertz frequencies, the size of the teeth in the active region on the base (ground) side of the antenna was approximately 15% smaller than that on the other (wiring) side, due to shrinking by overetching. This unsymmetrical antenna structure results in an impedance mismatch and a poor beam pattern.

Using a high-efficiency input-coupling configuration, such a quasi-optical twin-slot antenna, can improve the performance. We believe that all of the improvements described here can reduce the noise temperatures. We also believe that self-compensated NbN/AlN/NbN junctions and NbN/MgO/NbN microstriplines are more capable of low-noise and wideband operation at terahertz frequencies of up to 1.4 THz than conventional mixers that use lumped elements.

3. TERAHERTZ HEB MIXER BASED ON NBN

Even if NbN-based SIS mixers are developed, the operating frequency will be limited by a gap frequency of up to 1.4 THz [47], [48]. Superconducting HEB mixers are more promising as low-noise, wide-band heterodyne sensors operating at terahertz frequencies because their performance is not limited by the gap frequency of the superconducting material.

Two types of HEB mixers are under development: a diffusion-cooled type [61] and a phonon-cooled type [62]. In the former type, the hot electrons are cooled by out-diffusion into normal metal contact pads attached to the HEB strip. If a typical Nb material, for example, is used for the strip, the length of the strip must have a critical dimension of less than 0.1 μm to obtain an instantaneous bandwidth >10 GHz [63]. In the latter type, the hot electrons are cooled by their interaction with the phonons, which then escape into the substrate. The upper limit for the instantaneous bandwidth is determined by the inverse electron-phonon interaction time, which is about 10 GHz for NbN films [64]. The film must be thin enough, less than 5 nm thick, to enable quick escape of the nonequilibrium phonons into the substrate to obtain a large instantaneous bandwidth. The performance achieved of both types of mixers has largely been similar at frequencies from 0.6 to 2.5 THz. Recent results demonstrated noise performance of 650 K at 0.53 THz [6] and 1800 K at 2.5 THz [7] for diffusion-cooled (Nb-based) HEB mixers and 440-480 K at 0.65 THz [8], 550 K at 0.87 THz [9], 700 K at 1.6 THz, and 1100 K at 2.5 THz [10] for phonon-cooled (NbN-based) HEB mixers as shown in Fig. 1. Which type of HEB mixer is better is still open question.

We investigated the performance of a quasi-optical phonon-cooled HEB mixer based on NbN. The instantaneous bandwidth was measured using two phase-locked signal sources in the 100-GHz band. The noise temperature was measured from 0.9 to 2.5 THz by using a quasi-optical receiver configuration, which made it possible to evaluate the broadband characteristics [65].

3.1 Mixer chip and characteristics of HEB element

The HEB mixer chip we investigated was an ultra-thin superconducting NbN strip located at the feed point of a thick normal-conducting Au log-spiral antenna on a high-resistivity Si substrate. An optical microphotograph and a cross-sectional view of the mixer chip are shown in Fig. 28. The 3-nm-thick NbN film was deposited on a 0.35-mm-thick Si substrate by DC magnetron sputtering with intentional heating; the fabrication was done at the Moscow State Pedagogical University. The fabrication was fully described elsewhere [66]. To reduce the local-oscillator (LO) power requirements for measurements at high frequencies, the device volume had to be reduced. Electron beam lithography was used to form a lift-off mask for contact pads consisting of 5-nm-thick Ti (to obtain good adhesion between the NbN and Au) and 80-nm-thick Au, so that the

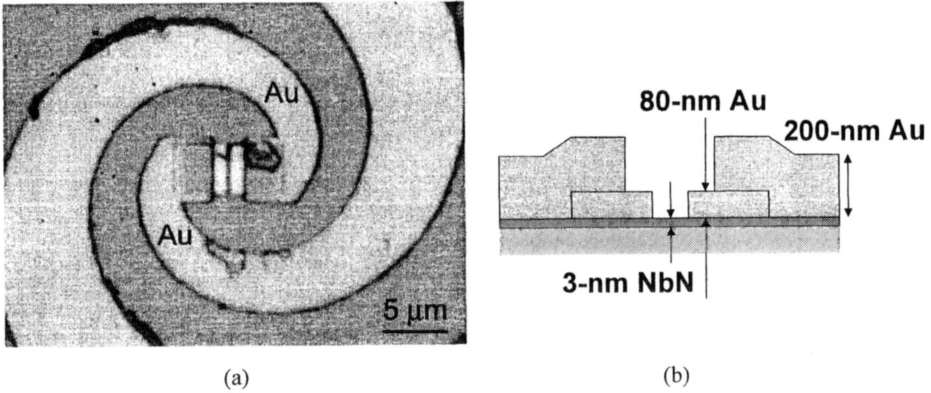

Fig. 28. (a) Optical microphotograph and (b) cross-sectional view of HEB mixer based on NbN.

spacing between the Ti and Au was at the submicron level. After fabricating a log-spiral antenna consisting of 5-nm-thick Ti and 200-nm-thick Au, an NbN microbridge was defined using electron beam lithography and Ar ion beam etching. The active area of the fabricated HEB was 3 nm thick, 0.4 μm long, and 4 μm wide.

The resistance-temperature (RT) curve of the HEB mixer is shown in Fig. 29. The critical temperature, T_C, was about 8.5 K with a transition width of about 2 K. The resistance below T_C was mainly from the two-terminal measurement, not from the contact resistance between the Au + Ti and the NbN. The mixer's room-temperature resistance was 130 Ω, and the normal-state resistance, R_N, near T_C was 175 Ω. The critical current, I_C, was

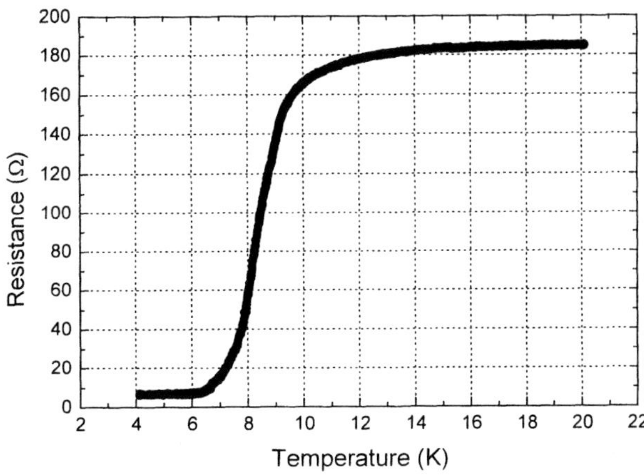

Fig. 29. Resistance-Temperature curve of HEB mixer device around critical temperature.

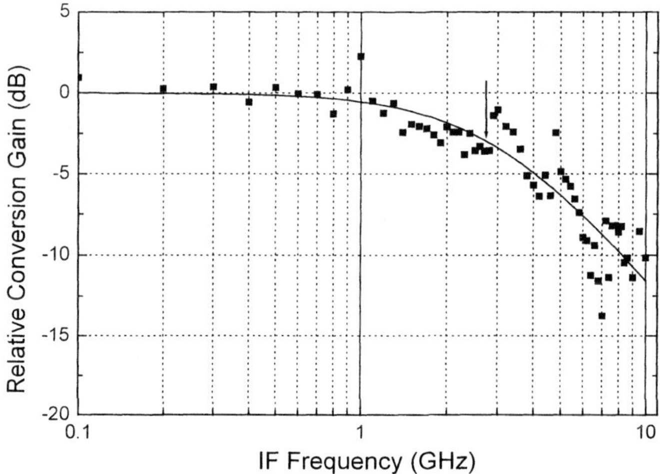

Fig. 30. Measured IF gain bandwidth of HEB mixer operating at around 100 GHz. The 3-dB rollover frequency is marked with arrow.

85 μA at 4.2 K, and the critical current density was about 0.7 MA/cm^2.

The IF gain bandwidth is an important parameter in characterizing HEB mixers. To determine the IF bandwidth of our NbN HEB mixer, we performed heterodyne mixing experiments with two phase-locked 100-GHz sources, one used as a signal source, and the other one as an LO source. The signal power and frequency were kept constant as the IF was increased from 0.1 to 10 GHz by changing the LO source frequency and regulating the LO power to maintain a fixed operating point. Figure 30 shows the results of measuring the gain response. The solid line is the best fit to the function $\eta(f) = (1+(f/f_C)^2)^{-1}$, where parameter f_C is usually called the 3-dB rolloff frequency. This gives a 3-dB IF bandwidth of about 2.75 GHz, which is enough for the IF band of our receiver noise measurements and many practical applications. However, a larger IF bandwidth, up to the theoretical limit of about 10 GHz, would be better.

3.2 Noise performance

The receiver set-up was basically the same as the one described in Sec. 2.1.A, except that an isolator was used instead of the 180-degree hybrid coupler. To investigate the frequency dependence of the noise temperature from 0.9 to 2.5 THz, we used a BWO and a far infrared laser pumped by a CO_2 laser as the LO source. A CH_2F_2 laser was used to measure at 1.26 and 1.4 THz, and a CH_3OH laser was used at 2.52 THz. The other

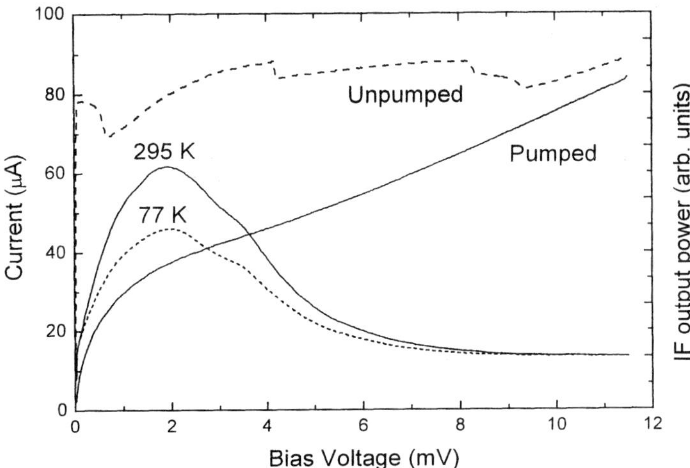

Fig. 31. Current-Voltage characteristics of HEB mixer with and without LO power at 917 GHz. Also shown is receiver IF power as a function of bias voltage in response to hot (295 K) and cold (77 K) loads. A 9-μm-thick Mylar beam splitter was used.

measurements of 0.9 to 1.1 THz were made using the BWO. The LO power was adjusted by rotating a wire grid in front of the LO source. A DC voltage bias source was also used.

Figure 31 shows an unpumped *I-V* curve of the receiver and a pumped *I-V* curve with a corresponding IF output at 917 GHz. A 9-μm-thick Mylar beam splitter was used. The LO power was adjusted to obtain the minimum noise temperature. From these *I-V* curves, we estimated the LO power absorbed in the mixer by using the isothermal technique used by Kroug and colleagues [67]. The estimation showed that about 360 nW of the LO power were needed to pump the NbN HEB strip. This low LO power requirement would make it possible to use solid-state sources.

The IF responses to the hot and cold loads showed a maximum *Y*-factor of 1.35 at around 2 mV, which corresponds to the DSB receiver noise temperature of 550 K, including the Mylar-beam-splitter loss and other optical losses. We also estimated the conversion gain using the method described by Kawamura and colleagues [68], which uses the calibrated IF output when the device is in superconducting state (zero bias voltage without LO power). The estimated conversion gain was –14 dB at around 2 mV. We did not see any changes in the pumped *I-V* curves for hot and cold loads, so it seems that the receiver is not affected by direct-detection effects.

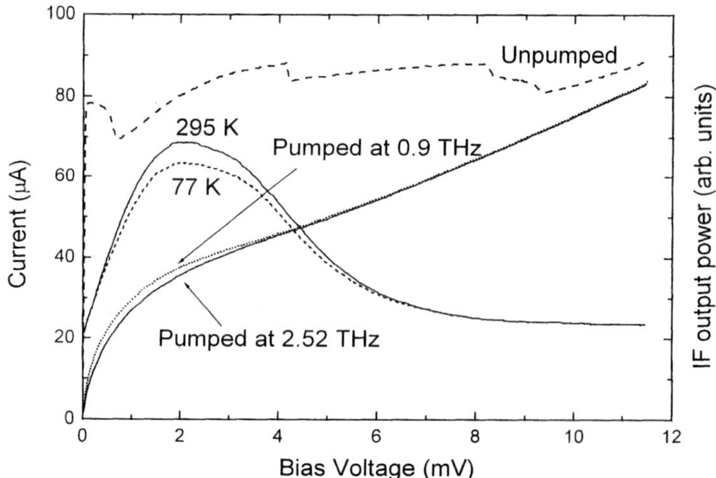

Fig. 32. Current-Voltage characteristics of HEB mixer with and without LO power at 2.52 THz. Also shown is receiver IF power as a function of bias voltage in response to hot (295 K) and cold (77 K) loads. A 6-μm-thick Mylar beam splitter was used.

We investigated the frequency dependence of the receiver noise temperature at frequencies ranging from 0.9 to 2.52 THz using the same receiver configuration. The only difference was the thickness of the Kapton-JP matching cap used above and below 1.1 THz. The 9-μm-thick Mylar beam splitter was used at all measured frequencies, except for 2.52 THz. Because this Mylar is a bit thick for the beam splitter at 2.52 THz (the calculated signal transmission was about 75% with signal polarization using an electrical field perpendicular to a plane with 45-degree incidence), a 6-μm-thick Mylar beam splitter was used instead (the transmission was about 86%). The noise temperature was corrected using Callen & Welton's law [69] at 2.52 THz because the difference between the radiated power temperature of a black body and its physical temperature is large above 2 THz.

Figure 32 shows the heterodyne response of the receiver at 2.52 THz. The DSB receiver noise temperature was about 2300 K, and the estimated conversion gain was about –24 dB. One reason for this degradation in conversion gain can be attributed to the optic losses, which increased with the frequency. We are planning to measure these losses, such as the MgO and Kapton-JP ones at 4.2 K, by using a Fourier transform infrared spectrometer.

However, the conversion gain still seems too low if the optical losses are considered. To find another possible reason for the degraded performance at higher frequencies, we compared the pumped *I-V* curves between 0.9 and 2.5 THz, as shown in Fig. 32. Although

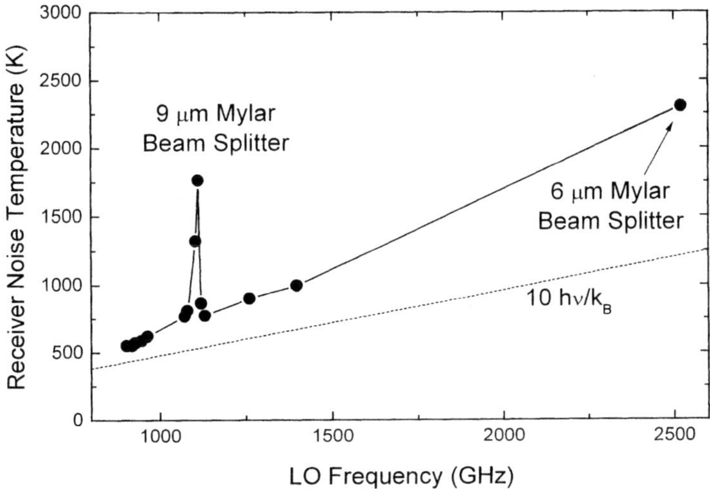

Fig. 33. DSB receiver noise temperatures as a function of LO frequency.

the pumping levels were almost the same in the higher bias voltage region, the *I-V* characteristics at 2.52 THz were suppressed more by irradiation than at 0.9 THz in the lower bias region. Thus, the absorption mechanism of the RF power in the HEB device was possibly changed by the frequency. We should thus take into account the frequency-dependence parameters to be used in HEB theory, and more investigations are needed.

Figure 33 shows the DSB receiver noise temperature as a function of the LO frequency. Below 1 THz, the noise temperature was close to 10 $h\nu/k_B$; it gradually rose to 2300 K at 2.52 THz. The jump in noise temperature at around 1.1 THz was presumably due to the absorption in the atmosphere, because the receiver set-up was in air. These noise temperatures below 1 K/GHz over the measured frequency range demonstrated the low-noise and wide-band characteristics of our quasi-optical NbN-based HEB mixer.

4. CONCLUSION

In this chapter, our latest development of quasi-optical superconducting mixers have been described with the aim of achieving low-noise characteristics at submillimeter-wave frequencies beyond the gap frequency of Nb (700 GHz). First, we presented SIS mixers based on a high-gap material of NbN instead of conventional Nb. The high-current density NbN/AlN/NbN tunnel junctions with good *I-V* characteristics were developed on a single-crystal MgO substrate. Investigation of the junction characteristics, such as the submillimeter-wave response, the specific capacitance, and theoretical mixing properties, indicated that our developed NbN/AlN/NbN tunnel junctions can be applied as the submillimeter-wave mixers.

We designed, fabricated and tested quasi-optical SIS mixers using the NbN junctions at frequencies below and above the gap frequency of Nb. The SIS mixer designed at 300 GHz showed that the NbN SIS junctions had good mixing properties of the minimum DSB receiver noise temperature of 136 K at around the design frequency, which was comparable to the results of conventional Nb SIS receivers at the same frequency range. These evidences suggest that the design parameters such as the capacitance of NbN/AlN/NbN junctions are well characterized and the fabrication processes are well controlled.

The terahertz SIS mixer employing no conventional tuning structure of the self-compensated SIS junction successfully demonstrated the low noise and wideband characteristics beyond the gap frequency of Nb. We obtained the minimum receiver noise temperature of 550 K including the estimated input loss of about 300 K and flat noise characteristics from 780 to 960 GHz as predicted in the design process. This mixer had a simple structure and was fabricated by using conventional photolithography. Our fabricated NbN devices remained stable throughout more than ten repetitions of the thermal cycle. We believe that the all-NbN SIS mixers with these excellent characteristics can be effectively used in many practical applications such as global environmental observation, radio astronomy, and so on. One unsettled problem is that the mixer noise was somewhat higher than the theoretical prediction. This may result from the leakage current relating to the subgap structure in the *I-V* curve of NbN/AlN/NbN tunnel junctions, which have never been observed in conventional Nb/AlOx/Nb tunnel junctions. More investigation must be done to interpret this noise performance and this gap structure must be reduced to build high-performance terahertz SIS mixers.

On the other hand, to explore the possibility of alternative devices that can be used in the submillimeter-wave range, the performance of a phonon-cooled HEB mixer based on NbN was investigated. A relatively large 3-dB IF bandwidth of about 2.75 GHz was obtained, which was sufficient for many practical applications. Receiver noise measurements from 0.9 to 2.5 THz by using a quasi-optical receiver configuration showed low-noise and wide-band characteristics. The lowest receiver noise of 550 K was measured at 917 GHz, which is almost the same as that of NbN-based SIS mixers in the same frequency range. These results suggest that HEB mixers are also usable as the front-end of low-noise terahertz receivers. Although the HEB mixers seems to be one of the good alternatives as seen in this experiment and other results on HEB mixers, their behavior has not been fully understood yet. Much work must be done to improve the mixer performance at frequencies up to the end of the submillimeter-wave frequency and more.

ACKNOWLEDGEMENTS

We thank Dr. A. Kawakami of our group for his helpful assistance in fabricating NbN SIS mixers, Drs. C. E. Tong and R. Blundell of the Harvard-Smithsonian Center for Astrophysics for giving us their knowledge of SIS mixers, and Prof. T. Noguchi and Dr. S.-C. Shi of the Nobeyama Radio Observatory for valuable discussions on SIS mixers. We also thank Drs. M. Kroug, P. Yagoubov of the Chalmers University of Technology, and Prof. G. Gol'tsman of Moscow State Pedagogical University for providing NbN HEB devices. This work was supported in part by the ALMA Joint Research Fund of the National Astronomical Observatory of Japan.

REFERENCES

[1] J. Mees, S. Crewell, H. Nett, G. de Lange, H. van de Stadt, J. J. Kuipers, and R. A. Panhuyezen, *IEEE Trans. Microwave Theory and Tech.*, **43** (1995) 2543.

[2] T. G. Phillips and J. Keene, *Proc. IEEE*, **80** (1992) 1662.

[3] J. Carlstrom and J. Zmuidzinas, *Review of Radio Science 1993-1996*, edited by W. Ross Stone (The Oxford University Press, Oxford, 1996).

[4] J. Kawamura, J. Chen, D. Miller, J. Kooi, J. Zmuidzinas, B. Bumble, H. G. LeDuc, and J. A. Stern, *Appl. Phys. Lett.*, **75** (1999) 4013.

[5] B. D. Jackson, A. M. Baryshev, G. de Lange, J. R. Gao, S. V. Shitov, N. N. losad and T. M. Klapwijk, *Appl. Phys. Lett.*, **79** (2001) 436.

[6] A. Skalare, W. R. McGrath, B. Bumble, H. G. LeDuc, P. J. Burke, A. A. Verheijen, R. J. Schoelkopf and D. E. Prober, *Appl. Phys. Lett.*, **68** (1996) 1558.

[7] R. A. Wyss, B. S. Karasik, W. R. McGrath, B. Bumble and H. G. LeDuc, *Proceedings of the 10th International Symposium on Space Terahertz Technology*, Charlottesville, VA, USA (1999) 215.

[8] P. Yagoubov, M. Kroug, H. Merkel, E. Kollberg, J. Schubert, H. W. Huebers, S. Svechnikov, B. Voronov, G. Gol'tsman and Z. Wang, *Proceedings of Europ. Conf. on Appl. Superconductivity*, Barcelona, Spain, **2** (1999) 687.

[9] C. E. Tong, J. Kawamura, T. R. Hunter, D. C. Papa, R. Blundell, M. Smith, F. Patt, G. Gol'tsman and E. Gershenzon, *Proceedings of the 11th International Symposium on Space Terahertz Technology*, Ann Arbor, MI, USA, (2000) 49.

[10] M. Kroug, S. Cherednichenko, H. Merkel, E. Kollberg, B. Voronov, G. Gol'tsman, H. W. Huevers and H. Richter, *IEEE Trans. Appl. Supercond.*, **11** (2001) 962.

[11] J. Zmuidzinas, N. G. Ugras, D. Miller, M. Gaidis, H. G. LeDuc, and J. A. Stern, *IEEE Trans.*

Appl. Supercond., **5** (1995) 3053.

[12] A. Shoji, M. Aoyagi, S. Kosaka, F. Shinoki, and H. Hayakawa, *Appl. Phys. Lett.*, **46** (1985) 1098.

[13] D. Hunt, H. G. LeDuc, S. R. Cypher, and J. A. Stern, *Appl. Phys. Lett.*, **55** (1989) 81.

[14] A. Stern, B. D. Hunt, H. G. LeDuc, A. Judas, W. R. McGrath, S. R. Cypher, and S. K. Khanna, *IEEE Trans. Magn.*, **25** (1989) 1054.

[15] R. McGrath, J. A. Stern, H. H. S. Javadi, S. R. Cypher, B. D. Hunt, and H. G. LeDuc, *IEEE Trans Magn.*, **27** (1991) 2650.

[16] A. Shoji, M. Aoyagi, S. Kosaka, and F. Shinoki, *IEEE Trans. Magn.*, **23** (1987) 1464.

[17] S. Kubo, M. Asahi, M. Hikita, and M. Igarashi, *Appl. Phys. Lett.*, **44** (1984) 258.

[18] M. Pham Tu, K. Mbaye, and L. Wartski, *J. Appl. Phys.*, **63** (1988) 4586.

[19] Z. Wang, A. Kawakami, Y. Uzawa, and B. Komiyama, *J. Appl. Phys.*, **79** (1996) 7837.

[20] Z. Wang, A. Kawakami, Y. Uzawa, and B. Komiyama, *Appl. Phys. Lett.*, **64** (1994) 2034.

[21] Z. Wang, A. Kawakami, Y. Uzawa, and B. Komiyama, *IEEE Trans. Appl. Supercond.*, **5** (1995) 2322.

[22] Z. Wang, A. Kawakami, and Y. Uzawa, *Appl. Phys. Lett.*, **70** (1997) 114.

[23] Y. Uzawa, Z. Wang, A. Kawakami, and B. Komiyama, *Appl. Phys. Lett.*, **66** (1995) 1992.

[24] Y. Uzawa, Z. Wang, and A. Kawakami, *Appl. Phys. Lett.* **69** (1996) 2435.

[25] B. Komiyama, Z. Wang, and M. Tonouchi, *Appl. Phys. Lett.*, **68** (1996) 562.

[26] V. Ambegaokar and A. Baratoff, *Phys. Rev. Lett.*, **10** (1963) 486.

[27] T. A. Folton and D. E. McCumber, *Phys. Rev.*, **175** (1968) 585.

[28] E. Miller, W. H. Mallison, A. W. Kleinsasser, K. A. Delin, and E. M. Macedo, *Appl. Phys. Lett.*, **63** (1993) 1423.

[29] W. Kleinsasser, W. H. Mallison, and R. E. Miller, *IEEE Trans. Appl. Supercond.*, **5** (1995) 2318.

[30] A. Barone and G. Paterno, *Physics and applications of the Josephson effects*, Wiley, New York, USA (1982) 380.

[31] P. A. A. Booi, Ph.D. thesis, University of Twente, December 1995.

[32] H. S. J. van der Zant, T. P. Orland, and A. W. Kleinsasser, *IEEE Trans. Appl. Supercond.*, **5** (1995) 3333.

[33] J. R. Tucker, *IEEE J. Quantum Electron.*, **QE-15** (1979) 1234.

[34] S. C. Shi, J. Inatani, T. Noguchi, and K. Sunada, *Int. J. Infrared Millimeter Wave*, **14** (1993) 1273.

[35] Y. Uzawa, Z. Wang, A. Kawakami, S. C. Shi, and T. Noguchi, *Physica C*, **282-287** (1997) 2549.

[36] Y. Uzawa, A. Kawakami, Z. Wang, and T. Noguchi, *IEICE Trans. Electron.*, **E-79** (1996) 1237.

[37] Y. Uzawa, Ph.D. thesis, Tokyo Institute of Technology, April 2000.

[38] R. A. Pucel, D. J. Masse, and C. P. Hartwig, *IEEE Trans. Microwave Theory Tech.*, **MTT-16** (1968) 342: see also correction in **MTT-16** (1968) 1064.

[39] C. E. Tong, R. Blundell, S. Paine, D. C. Papa, J. Kawamura, X. Zhang, J. A. Stern, and H. G. LeDuc, *IEEE Trans. Microwave Theory Tech.*, **MTT-42** (1994) 698.

[40] K. C. Gupta, R. Garg, and I. J. Bahl, *Microstrip Lines and Slotlines*, Artech House: Dedham, MA, USA (1979).

[41] Y. Uzawa, Z. Wang, and A. Kawakami, *IEEE Trans. Appl. Supercond.*, 7 (1997) 2574.

[42] Q. Ke and M. J. Feldman, *IEEE Trans. Microwave Theory Tech.*, **41** (1993) 600.

[43] R. L. Kautz, *J. Appl. Phys.*, **49** (1978) 308.

[44] Y. Uzawa, Z. Wang, and A. Kawakami, *Appl. Supercond.*, **6** (1998) 465.

[45] A. Kawakami, Z. Wang, and S. Miki, *IEEE Trans. Appl. Supercond.*, **11** (2001) 80.

[46] B. D. Jackson, G. de Lange, W. Laauwen, J. R. Gao, N. N. Iosad, and T. M. Klapwijk, *Proceeding of the 11th International-Symposium on Space Terahertz Technology*, University of Michigan, Ann Arbor, MI, USA, (2000).

[47] Y. Uzawa, A. Kawakami, S. Miki, and Z. Wang, *IEEE Trans. Appl. Supercond.*, **11** (2001) 183.

[48] Y. Uzawa, Z. Wang, A. Kawakami, and S. Miki, To appear in *Proceedings of the 12th International Symposium on Space Terahertz Technology*, Humphrey's Half Moon Inn, San Diego, CA, USA (2001).

[49] V. Yu. Belitsky and E. L. Kollberg, *J. Appl. Phys.*, **80** (1996) 4741.

[50] V. Yu. Belitsky and E. L. Kollberg, *Proceeding of the 7th International-Symposium on Space Terahertz Technology*, University of Virginia, Charlottesville, VA,USA (1996).

[51] D. C. Matthis, J. Bardeen, *Phys. Rev.*, **111** (1958) 412.

[52] Z. Wang, H. Terai, A. Kawakami, and Y. Uzawa, *Appl. Phys. Lett.*, **75** (1999) 701.

[53] A. I. Harris, K. –F. Schuster, and L. Tacconi, *Int. J. Infrared & MM Waves*, **15** (1993) 715.

[54] Y. Uzawa, T. Noguchi, A. Kawakami, and Z. Wang, *Supercond. Sci. Technol.*, **9** (1996) A140.

[55] D.P. Woody, R.E. Miller, and M.J. Wengler, *IEEE Trans. Microwave Theory Tech.*, **33** (1985) 90.

[56] R. Blundell, R.E. Miller, and K.H. Gundlach, *Int. J. IR & MM Waves*, **13** (1992) 3.

[57] Q. Ke and M.J. Feldman, *IEEE Trans. Microwave Theory Tech.*, **42** (1994) 752.

[58] C.-Y. E. Tong and R. Blundell, Harvard-Smithsonian Center for Astrophysics, Cambridge, MA, USA, personal communication (1994).

[59] C.-Y. E. Tong, R. Blundell, S. Paine, D. C. Papa, J. Kawamura, X, Zhang, J. A. Stern, and H. G. LeDuc, *IEEE Trans. Microwave Theory Tech.*, **44** (1996) 1548.

[60] P. Dieleman, T. M. Klapwijk, S. Kovtonyk, and H. van de Stadt, *Appl. Phys. Lett.*, **69** (1996) 418.

[61] D. E. Prober, *Appl. Phys. Lett.*, **62** (1993) 2119.

[62] E. M. Gershenzon, G. N. Gol'tsman, I. G. Gogidze, Y. P. Gusev, A. I. Elantev, B. S. Karasik, and A.D. Semenov, *Sov. Phys. Superconductivity*, **3** (1990) 1582.

[63] P. J. Burke, R. J. Schoelkopf, D. E. Prober, A. Skalare, W. R. McGrath, B. Bumble, and H. G. LeDuc, *Appl. Phys. Lett.*, **68** (1996) 3344.

[64] G. N. Gol'tsman, A. D. Semenov, Y. P. Gusev, M. A. Zorin, I G. Gogidze, E. M. Gershenzon, P. T. Lang, W. J. Knott, and K. F. Renk, *Superconduct. Sci. Technol.*, **4** (1991) 453.

[65] Y. Uzawa, S. Miki, Z. Wang, A. Kawakami, M. Kroug, P. Yagoubov, and E. Kollberg, To appear in *Superconduct. Sci. Technol.*

[66] P. Yagoubov, M. Kroug, H. Merkel, E. Kollberg, G. Gol'tsman, A. Lipatov, S. Svechnikov, and E. Gershenzon, *Proceedings of the 9th International Symposium on Space Terahertz Technology*, Pasadena, CA, USA (1998) 131.

[67] M. Kroug, P. Yagoubov, G. Gol'tsman, and E. Kollberg, *Inst. Phys. Conf. Ser.*, **158** (1997) 405.

[68] J. Kawamura, R. Blundel, C. E. Tong, G. Gol'tsman, E. Gershenzon, B. Voronov, and S. Cherednichenko, *Appl. Phys. Lett.*, **70** (1997) 1619.

[69] A. R. Kerr, M. J. Feldman, and S. K. Pan, *Proceedings of the 8th International Symposium on Space Terahertz Technology*, Cambridge, MA, USA (1997) 101.

SUBJECT INDEX
VOLUME-43